ANIONIC
SURFACTANTS

SURFACTANT SCIENCE SERIES

CONSULTING EDITORS

MARTIN J. SCHICK

Diamond Shamrock Chemical Company
Process Chemicals Division
Morristown, New Jersey

FREDERICK M. FOWKES

Chairman of the Department of Chemistry
Lehigh University
Bethlehem, Pennsylvania

OTHER VOLUMES IN PREPARATION

ANIONIC SURFACTANTS

(IN TWO PARTS)

PART I

edited by **Warner M. Linfield**

Eastern Regional Research Center
U.S. Department of Agriculture
Philadelphia, Pennsylvania

MARCEL DEKKER, INC.   New York and Basel

MARCEL DEKKER, INC.
270 Madison Avenue, New York, New York 10016

LIBRARY OF CONGRESS CATALOG CARD NUMBER: 75-22777
ISBN: 0-8247-6158-8
Current printing (last digit):
10 9 8 7 6 5 4 3 2 1

PRINTED IN THE UNITED STATES OF AMERICA

PREFACE

This two-part volume deals primarily with the organic chemistry of anionic surfactants. Among the various types of surface-active agents, this class of surfactants is probably the one of greatest importance from both an economic and a scientific viewpoint. The vast literature on the subject is covered here in such a manner as to present each major type of anionic surfactant in a separate chapter. The lipid and petrochemical antecedents of these surfactants are each treated in a separate chapter. There is also a chapter on the mechanisms of sulfonation and sulfation reactions, since so many synthetic routes towards anionic surfactants include either of these reactions. An introductory chapter on soap and lime-soap dispersing agents is included because this subject has never been adequately discussed before, and, as the reader will see, such combinations tend to behave like the surfactants covered in some of the subsequent chapters.

The difficulties encountered in the compilation of this volume were numerous, such as the relocation of some authors as well as what is referred to as an "act of God." The untimely deaths of three contributors created a hiatus which fortunately was bridged by the unselfish devotion and cooperation of professional colleagues of the deceased. Since the individual chapters were not all completed at the same time, the cutoff dates of the chapters are not uniform. While all chapters cover the literature through 1974, some also cover most or all of 1975.

At this point I wish to take the opportunity to thank my friends and professional colleagues for their splended cooperation in contributing individual chapters to this work. They gave generously of their spare time and technical expertise, and my meetings and scientific discussions with them have been a stimulating and enriching personal experience. Much of this work was created during a period of economic stress on the chemical profession, and I am thus all the more indebted to the individual authors, their spouses, and others who lent a hand, as well as to their employers who for the most part furnished clerical help, library support, and others forms of assistance and encouragement.

I would like to pay special tribute to three authors, George E. Hinds, Samuel Shore, and Alexander J. Stirton who unfortunately were not privileged to live long enough to see these volumes appear in print. Special thanks are due to Dr. Martin J. Schick and Mr. Marcel Dekker who entrusted me with the editing of these volumes, and whose patience and encouragement have been much appreciated. Finally I owe a special debt of gratitude to my wife Shirley R. Linfield who has been bearing with me patiently through periods of agonizing over these volumes, and who assisted me with many chores connected with this work. Without her cheerful encouragement I might not have undertaken this task.

Warner M. Linfield

CONTENTS OF PART I

v

Cumulative Indexes appear in Part II.

CONTRIBUTORS TO PART I

DANIEL R. BERGER, The Richardson Company, Melrose Park, Illinois

BERNARD A. DOMBROW, Nopco Chemical Division, Diamond Shamrock Chemical Company, Morristown, New Jersey

BEN E. EDWARDS, Department of Chemistry, University of North Carolina at Greensboro, Greensboro, North Carolina

GEORGE C. FEIGHNER, Petrochemical Department, Continental Oil Company, Saddle Brook, New Jersey

GEORGE E. HINDS, Continental Oil Company, Ponca City, Oklahoma

WARNER M. LINFIELD, Eastern Regional Research Center, U. S. Department of Agriculture, Philadelphia, Pennsylvania

FRANK SCHOLNICK, Eastern Regional Research Center, U.S. Department of Agriculture, Philadelphia, Pennsylvania

SAMUEL SHORE, Mazer Chemicals, Inc., Gurnee, Illinois

ALEXANDER J. STIRTON, Eastern Regional Research Center, U.S. Department of Agriculture, Philadelphia, Pennsylvania

JAMES K. WEIL, Eastern Regional Research Center, U.S. Department of Agriculture, Philadelphia, Pennsylvania

CONTRIBUTORS TO PART I

DANIEL R. BERGER, The Hilderbrison Company, Melrose Park, Illinois

BERNARD A. DOMBROW, Nopco Chemical Division, Diamond Shamrock Chemical Company, Morristown, New Jersey

BEN PLEISSNIKS, Department of Chemistry, University of North Carolina at Greensboro, Greensboro, North Carolina

GEORGE C. FEIGHNER, Petrochemical Department, Clabberland Oil Company, Bound Brook, New Jersey

GEORGE E. HURST, Continental Oil Company, Ponca City, Oklahoma

WALTER M. LINFIELD, Eastern Regional Research Center, U. S. Department of Agriculture, Philadelphia, Pennsylvania

FRANK SCHOLNICK, Eastern Regional Research Center, U.S. Department of Agriculture, Philadelphia, Pennsylvania

SAMUEL SICKEL, Ninol Chemicals, Inc., Chicago, Illinois

ALEXANDER J. STIRTON, Eastern Regional Research Center, U.S. Department of Agriculture, Philadelphia, Pennsylvania

JASPER K. WEIL, Eastern Regional Research Center, U.S. Department of Agriculture, Philadelphia, Pennsylvania

CONTENTS OF PART II

CHAPTER 1

SOAP AND LIME-SOAP DISPERSING AGENTS

Warner M. Linfield

Eastern Regional Research Center
U.S. Department of Agriculture
Philadelphia, Pennsylvania

While soaps, i.e., the water-soluble salts of fatty acids, are properly classified as anionic surfactants, it has become customary to think of soap as
belonging to a different category of surface-active materials. The reason for
this is probably that one thinks of anionic and other types of surfactants as
being synthetic materials rather than simple derivatives of natural products.
The preponderance of the modern literature on soaps, in fact, deals with the
fields of soapmaking technology and physical chemistry, neither of which are
the subject matter of this book.

In recent years much emphasis in the field of surfactant and detergent
chemistry has been placed on the investigation of phosphate-free heavy-duty
laundry detergents. The controversy about the possibility of detergent phosphates contributing to the eutrophication of various bodies of water is beyond
the scope of this book. However, because of the importance of the development of phosphate-free detergents research workers at the Eastern Regional
Research Center (ERRC) of the U.S. Department of Agriculture are investi-
gating the suitability of soap in heavy-duty detergent formulations. Under
proper formulation conditions soap can be made to behave very much like
"synthetic" anionic surfactants, and thus this chapter provides a fitting introduction into the field of anionic surfactants.

As is generally known, the purpose of the Regional Research Centers of
the U.S. Department of Agriculture is to stimulate the industrial utilization of
certain agricultural raw materials such as tallow. In light of the need for
good phosphate-free detergents it appeared logical to investigate the suitability
of tallow soap in heavy-duty detergent formulations. Except for the occasional
anomalous period of extremely high prices and even scarcity of agricultural
commodities, tallow for several decades has been the least expensive natural
fat and, in fact, has been one of the cheapest chemical raw materials available.

The use of soap as a cleaning agent goes back to the dawn of civilization and has had a long record of safety and efficacy. When used as a detergent, soap has only two important drawbacks, its poor solubility in cold water and the insolubility in water of the calcium and magnesium soaps. In an effort to remove these drawbacks the detergent industry decided after the end of World War II to shift away from soap and adopt various synthetic surfactants as the active ingredients of its detergent formulations.

More than twenty years ago Borghetty and Bergman [1] discovered that the precipitation of lime soap could be prevented through the addition of lime-soap dispersing agents (LSDA) to the soap. Their formulations always contained sodium pyrophosphate, thus their investigations cannot be applied to the alleviation of the present detergent phosphate problem which, of course, was unrecognized at that time. The principle of a combination of soap and LSDA could also be applied to industrial laundry detergents without addition of phosphates [2].

The fatty acid amides of N-methyltaurine and the fatty acid esters of isethionic acid were described by Mayhew and Burnette [3] as especially effective LSDA which appeared particularly suitable in bar-soap or heavy-duty detergent formulations. However, in the latter formulations these workers still used sodium pyrophosphate.

The molecular structure of an effective LSDA is characterized by a bulky hydrophilic polar end which is much larger than that of simple, only feeble LSDA surfactants. The mode of action has been depicted graphically by Stirton and co-workers [4] as a micellar phenomenon shown in Fig. 1. The soap molecules which are oriented on a surface or interface (a) form a typical soap micelle (b) in soft water. In the presence of calcium ions an inversion of the micelle takes place (c) which leads to the familiar precipitation of the lime soap. In the presence of an LSDA this inversion is prevented because the mixed micelle, consisting of soap and LSDA, retains the proper curvature (c) due to the larger bulk of the LSDA. Thus no lime-soap precipitation occurs.

It would thus appear obvious that a high lime-soap dispersing power of an LSDA is a prerequisite for high detergency of a soap-based detergent. This does not mean, however, that an extremely effective LSDA automatically leads to high detergency. Apparently detergency and lime-soap dispersing ability are independent of each other.

Since the measurement of the lime-soap dispersing ability of the LSDA, to be subsequently discussed, is of special importance, it is proper here to mention the methodology developed by Borghetty and Bergman [1]. The LSDR (lime-soap dispersant requirement) is the minimum number of grams of the test LSDA which will just prevent a solution of 100 g of sodium oleate in 320

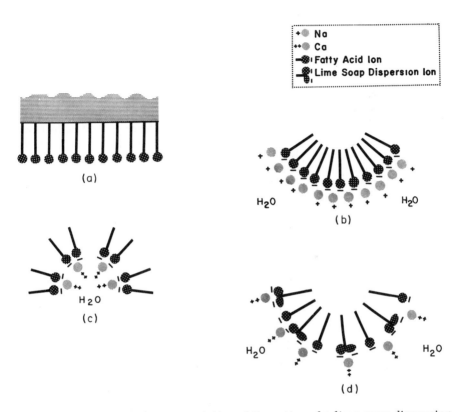

FIG. 1. Schematic representation of the action of a lime-soap dispersing agent according to Stirton et al. [4].

ppm hard water from precipitating. Accordingly, a low LSDR represents a high dispersing ability and vice-versa.

In the investigations discussed below it was decided on an empirical basis that an LSDR of 10 or less was adequate to suggest a more detailed study of the test compound. A comparison of the LSDR of various surfactants can be seen in Table 1. Incidentally, the LSDR of 40 for linear alkylbenzenesulfonate (LAS), the active ingredient of many current laundry detergents, indicates clearly the inadequacy of LAS as an LSDA.

Bistline and co-workers [5] carried out a number of wash tests which showed that certain LSDA had the ability to potentiate the detergency of soap. Addition of suitable builders, such as sodium silicates, sodium tripolyphosphate or sodium nitrilotriacetate (NTA), could increase the detergency of a soap-LSDA blend even further. A systematic investigation of three-component systems soap-LSDA-builder showed that a detergency maximum could be attained which corresponded to a certain fixed ratio of components as is shown graphically in Fig. 2. This figure summarizes the detergency tests carried out with the system tallow soap, sodium metasilicate, and sodium methyl α-sulfotallowate (TMS). The latter is a commercially available LSDA. Detergency is shown in terms of ΔR values (i.e., difference in light reflectance before and after washing). It should be noted that the triangular graph represents only the lower right-hand portion of a three-component diagram as shown in the small triangular graph. Thus the concentration ranges covered

TABLE 1

Lime-Soap Dispersant Requirements (LSDR) of Various Types of
Lime-Soap Dispersing Agent (LSDA) Surfactants

LSDA, Formula	LSDR
$RCOOCH_2CH_2SO_3Na$	10
$RCH(SO_3Na)COOCH_3$, TMS	8
$RCH(SO_3Na)COOCH_2CH_2SO_3Na$	5
$RCON(CH_3)CH_2CH_2SO_3Na$, IgT	6
$RO(CH_2CH_2O)_3SO_3Na$	4
$RCONHCH_2CH(CH_3)OSO_3Na$, TAM	4
$R'C_6H_4SO_2NHCH_2CH_2OSO_3Na$	6
$R'C_6H_4COCH(SO_3Na)CH_2COOCH_3$	8
$C_9H_{19}C_6H_4(OCH_2CH_2)_{9.5}OH$	5
$RCONH(CH_2CH_2O)_{15}H$	3
$RN^+(CH_3)_2CH_2CH_2CH_2SO_3^-$	3
$RCONHCH_2CH_2CH_2N^+(CH_3)_2CH_2CH_2CH_2SO_3^-$	3
$R'C_6H_4SO_3Na$*	40

R represents an alkyl group derived from tallow and R' represents an alkyl group in the C_{11}-C_{13} range corresponding to commercial detergent alkylates).
*LAS.

by the large diagram are 50-100% soap, 0-50% sodium metasilicate, and
0-50% of the dispersant TMS. The curves connect points of equal detergency
(ΔR) on EMPA-101 artificially soiled cotton (EMPA = Eidgenössische Mate-
rialprüfungsanstalt). The plateau of maximum detergency thus corresponded
to an approximate composition of 75% soap, 10% TMS, and 15% metasilicate.
Similar results were obtained with several other detergent builders. The
tests were carried out at 120°F and at 300 ppm water hardness which is well
above that of U.S. urban water supplies. The use of other artificially soiled
test cloths such as the cotton cloth of U.S. Testing Co., or Testfabrics, Inc.,
cotton-polyester blends led to triangular graphs whose detergency maximum
had slightly shifted. However, the principle of detergency potentiation of the
soap by the LSDA and the builder was always evident.

Noble and co-workers [6] prepared successfully a series of soap-based
detergent formulations whose detergency was equal to or surpassed leading
commercial household detergents containing almost 50% tripolyphosphte.
In these formulations sodium methyl α-sulfotallowate (TMS), the sodium
salt of the tallowamide of N-methyltaurine (IgT) or the sodium salt of

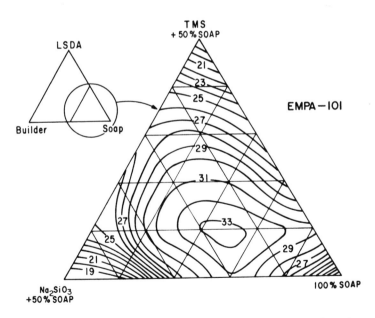

FIG. 2. Detergency (ΔR) behavior of the three-component system: Soap,
sodium metasilicate, sodium methyl α-sulfotallowate (0.2% total detergent
concentration, 120°F, 300 ppm water hardness, EMPA-101 soiled cotton cloth.

the sulfated isopropanolamide of tallow fatty acid (TAM) were used as the LSDA. The most effective formulations of that study had the following composition:

Tallow soap	64%
LSDA	19%
Sodium silicate (1 Na_2O:1. 6 SiO_2)	14%
Sodium carboxymethyl cellulose	1%
Impurities, brighteners	2%

The superior detergency of the above formulation was determined by single- and multiwash Tergotometer tests with artificially soiled cloths and confirmed by several actual laundry-bundle tests several of which were conducted by large detergent manufacturers. It was also found that this type of formulation could be processed by spray-drying, drum-drying, or by various soap-drying processes.

In an effort to enhance the understanding of the function of the LSDA, a variety of other fat-derived LSDA were synthesized through the reaction of propanesultone with fatty acids, fatty alcohols, or fatty amides. These compounds were highly effective LSDA and mixtures containing these LSDA (particularly the tallow derivatives), soap, and a glassy silicate according to the previously mentioned general formulation possessed high detergency [7].

Salts of alkylated or acylated iminodiacetic acids of the structures $RN(CH_2COONa)_2$ or $RCON(CH_2COONa)_2$, where R is a long-chain alkyl group, were found to be poor lime-soap dispersants in spite of their structural relationship to salts of ethylene diamine tetraacetic acid (EDTA) [8]. However, upon reaction of these compounds with propanesultone, disulfonates were obtained whose LSDR values of 4 to 8 for the C_{12} to C_{18} derivatives were satisfactory, and again soap-based detergents made from the LSDA according to the general formulation listed above were found to possess high detergency.

It should be emphasized here that the field of useful LSDA is by no means confined to fat derivatives. Sulfonamide derivatives of alkylbenzenes such as the commercial detergent alkylates were synthesized as shown in Eq. (1). The alkylbenzenesulfonic acid [1] or the alkylbenzene [2] were treated with chlorosulfonic acid to yield the corresponding sulfonyl chloride [3]:

$$R-C_6H_4-SO_3H + HSO_3Cl$$
[1]

$$R-C_6H_5 + 2HSO_3Cl$$
[2]

$$R-C_6H_4-SO_2Cl$$ (1)
[3]

Compound [3] could then be reacted with various aminoalkylsulfonic acids, including N-methyltaurine or aminoalkyl esters of sulfuric acid such as 2-aminoethyl sodium sulfate, according to Eq. (2):

$$R-C_6H_4-SO_2Cl + H_2NCH_2CH_2OSO_3Na \longrightarrow$$

[3]
$$R-C_6H_4-SO_2NHCH_2CH_2OSO_3Na \qquad (2)$$

Incidentally, compounds of this type have been recommended as active ingredients of phosphate-free detergents in a study carried out at the IIT Research Institute [9] under the sponsorship of the U.S. Environmental Protection Agency [10, 11]. The alkylbenzenesulfonamide derivatives were found to be efficient lime-soap dispersants [12]. The dispersing power of this series of compounds depended primarily upon the length of the alkyl side chain, whereas the nature of the substituents on the sulfonamide nitrogen atom had but a slight effect upon the surface-active properties of the compounds. The derivatives of commercial dodecyl- or tridecylbenzene were found to be the most effective LSDA of this class of compounds with LSDR values of about 7. A soap-based detergent formulation containing, for example, the tridecylbenzenesulfonamide derived from 2-aminoethyl hydrogen sulfate (from monoethanolamine and sulfuric acid) according to the formulation listed above proved to be about equal in detergency to a leading commercial phosphate-containing detergent which was customarily used as a control in all wash tests.

Another class of anionic surfactants studied by the IIT Research Institute [9, 10] are the alkylbenzoylsulfopropionates of the structure $R-C_6H_4-COCH-(SO_3Na)CH_2COOCH_3$, where R was preferably a C_{10}-to-C_{13} straight-chain group; they proved to be satisfactory LSDA's [13]. These compounds could again be derivatives of commercial alkylbenzenes, and soap-based detergents containing these LSDA according to the above general formula were found to be excellent detergents on a par with the commercial control.

It thus appears that a great variety of different types of anionic surfactants possessing the required hydrophilic bulk can function as effective LSDA and give rise to the potentiation of the detergency of tallow soap. In contrast to the anionics, the nonionic surfactants show an entirely different behavior. Typical ethylene oxide adducts with alkylphenols or higher-molecular alcohols showed LSDR values of 5 to 8. It seemed thus appropriate to investigate the possible potentiation of the detergency of soap by nonionics. Weil and co-workers [14] reported on the behavior of various ethylene oxide adducts to mono- and diethanolamides of various fatty acids. These, surprisingly, in spite of low LSDR values did not potentiate the detergency of soap and exhibited possibly some antagonism.

On the other hand, amphoteric surfactants proved to be especially suitable LSDA. Such compounds could be synthesized by the reaction between primary, secondary, or tertiary fatty amines and propanesultone [15], according to Eq. (3):

$$RNR'R'' + \underset{\underset{\displaystyle O}{\rule{2cm}{0.4pt}}}{CH_2CH_2CH_2SO_2} \longrightarrow RN^+(R'R'')CH_2CH_2CH_2SO_3^- \qquad (3)$$

R = long-chain alkyl group

R', R" = H or CH_3

Another type of amphoteric surfactant investigated was a fatty amide derivative which was prepared by the amidation of a methyl ester [4] of a fatty acid with 1-(N,N-dimethylamino)-3-aminopropane. The resulting amide [5] was converted to an amphoteric compound [6] by reaction with propanesultone according to Eqs. (4) and (5):

$$RCOOCH_3 + H_2NCH_2CH_2CH_2N(CH_3)_2 \xrightarrow{Na}$$

[4]

$$RCONHCH_2CH_2CH_2N(CH_3)_2 \quad (4)$$

[5]

$$RCONHCH_2CH_2CH_2N(CH_3)_2 + \overset{\displaystyle CH_2CH_2CH_2SO_2}{\underset{\displaystyle O}{\rule{0pt}{0pt}}} \longrightarrow$$

[5]

$$RCONHCH_2CH_2CH_2N^+(CH_3)_2CH_2CH_2CH_2SO_3^- \quad (5)$$

[6]

The LSDR values for all of the amphoteric surfactants (with alkyl side chains from C_{12} to C_{18}) were in the range of 2 to 4. These represent the lowest values obtained so far in the investigation of lime-soap dispersants. When standard detergent formulations, according to the general formulation shown above, were prepared with the amphoteric LSDA, the resulting products exhibited generally high detergency about on a par with that of the commercial control [16].

In contrast to those soap-based detergents prepared from anionic LSDA, those made with amphoteric compounds exhibited a much greater latitude with respect to detergent composition, i.e., the relative ratios of components could be varied over a substantial range without discernible change of detergency performance. Amphoteric types of formulations also tolerated the addition of sodium sulfate without great loss of detergency in contrast to anionic LSDA-soap formulations upon which the addition of sodium sulfate generally has an adverse effect.

If the LSDA used in soap-based detergent formulations is biodegradable, the resulting detergent is also degradable. The biodegradability of three spray-dried soap-based detergent formulations where TMS, TAM, and IgT, respectively, were the dispersants used, was examined under aerobic and microaerophilic conditions [17]. All three were found to degrade readily under both conditions. The three formulations were also found to be nontoxic to mammals and fish, nonirritating to skin or eyes, and were nonsensitizing.

Swatches washed 25 consecutive times with various soap-LSDA formulations were examined under a scanning electron microscope for buildup of foreign matter and were compared with those washed with phosphate- and carbonate-built commercial detergents, as well as with plain soap. The carbonate-built detergent and soap showed comparatively large accumulations of deposits whereas the phosphate-built detergent and the soap-LSDA type of detergent showed essentially no buildup of foreign matter [18]. Thus the soap-LSDA type of detergent was also found to have no adverse effects upon flameproofing finishes, unlike soap by itself or carbonate-built detergents [19].

This chapter represents a review of the research carried out through 1974 at ERRC on soap-based detergents. While not all of the LSDA mentioned here are currently available commercially, it is appropriate to point out that many of them could be readily manufactured at relatively low cost if a demand for such materials should arise. The price of the five or six LSDA which are presently manufactured as relatively expensive specialty chemicals would undoubtedly come down if there should be a large-volume demand.

This research demonstrated that it is feasible to produce soap-based detergents which do not contribute to eutrophication or other forms of water pollution, and which possess excellent detergency characteristics to those of the leading phosphate-built products. In addition, these formulations have the added advantage of preparation from readily replenishable agricultural products rather than from natural resources, such as phosphate rock or petroleum, whose depletion poses an increasing threat to our economy.

REFERENCES

1. H. C. Borghetty and C. A. Bergman, J. Am. Oil Chemists' Soc., 27, 88-90 (1950).
2. W. M. Linfield, Soap Chem. Specialties, 35 (3), 51-52, 110-111 (1959).
3. R. L. Mayhew and L. W. Burnette, Soap Chem. Specialties, 38 (8), 55-57 (1962).
4. A. J. Stirton, F. D. Smith, and J. K. Weil, J. Am. Oil Chemists' Soc., 42, 114-115 (1965).
5. R. G. Bistline, Jr., W. R. Noble, J. K. Weil, and W. M. Linfield, J. Am. Oil Chemists' Soc., 49, 63-69 (1972).
6. W. R. Noble, R. G. Bistline, Jr., and W. M. Linfield, Soap, Cosmet. Chem. Specialties, 48 (7), 38-42, 62 (1972).
7. N. Parris, J. K. Weil, and W. M. Linfield, J. Am. Oil Chemists' Soc., 49, 649-651 (1972).
8. T. J. Micich, M. K. Sucharski, J. K. Weil, and W. M. Linfield, J. Am. Oil Chemists' Soc., 49, 652-655 (1972).
9. W. M. Linfield, K. A. Roseman, H. G. Reilich, P. C. Adlaf, and C. C. Harlin, Jr., J. Am. Oil Chemists' Soc., 49, 254-258 (1972).
10. K. A. Roseman and W. M. Linfield, Water Pollution Control Series, U.S. Government Printing Office, 16080 DVF 12/70.
11. H. G. Reilich, Water Pollution Control Series, U.S. Government Printing Office, 16080 DVF 02/72.

12. R. G. Bistline, Jr., W. R. Noble, and W. M. Linfield, J. Am. Oil
    Chemists' Soc., 51, 126 (1974).
13. W. N. Marmer, D. E. Van Horn, and W. M. Linfield, J. Am. Oil
    Chemists' Soc., 51, 174 (1974).
14. J. K. Weil, F. D. Smith, and W. M. Linfield, J. Am. Oil Chemists'
    Soc., 49, 383 (1972).
15. N. Parris, J. K. Weil, and W. M. Linfield, J. Am. Oil Chemists' Soc.,
    50, 509 (1973).
16. W. M. Linfield, W. R. Noble, and N. Parris, Proceedings of the 59th
    Mid-Year Meeting, Chemical Specialties Manuf. Assoc., Washington,
    D.C., May 1973, p. 85.
17. E. W. Maurer, T. C. Cordon, J. K. Weil, and W. M. Linfield, J. Am.
    Oil Chemists' Soc., 51, 287 (1974).
18. W. R. Noble, J. K. Weil, R. G. Bistline, Jr., S. B. Jones, and W. M.
    Linfield, J. Am. Oil Chemists' Soc., 52, 1 (1975).
19. R. J. Brysson, A. M. Walker, and W. M. Linfield, Am. Dyestuff
    Reptr., 64, 19 (1975).

CHAPTER 2

PETROLEUM-BASED RAW MATERIALS
FOR ANIONIC SURFACTANTS

George E. Hinds[*]

Continental Oil Company
Ponca City, Oklahoma

*Deceased.

11

I. INTRODUCTION

The class of anionic surfactants alone, as reported on the 100% weight basis for the year 1973, constitutes about 68% of the total U.S. production of surfactants [1]. The other three classes, totaling approximately the remaining 32%, are nonionic, cationic, and amphoteric.

The following anionic surfactant raw materials are covered in this chapter: alkylated aromatics, higher acyclic hydrocarbons, chlorinated higher acyclic hydrocarbons, higher acyclic alcohols, and higher acyclic monocarboxylic acids.

The more basic starting materials from which the anionic detergent raw materials, listed in the preceding paragraph, are derived are not covered in this chapter, except to show where they are required in various methods of preparation or production. These basic starting materials are all derived entirely or predominantly from petroleum; for example, benzene, phenol, toluene, xylene, naphthalene, ethylene, propylene, butylenes, and n-paraffins.

II. ALKYLATED AROMATICS: C_{10} AND HIGHER ALKYLBENZENES

The principal aromatic chemicals employed as starting materials in the syntheses of anionic surfactants are benzene and phenol. Benzene is generally considered the most important because it is converted to alkylbenzenes which are the intermediates converted by sulfonation to alkylbenzenesulfonates. Furthermore, it serves as the starting material for synthetic phenol produced from cumene or by other synthetic routes.

Phenol is converted by alkylation to alkylphenol, followed by oxyalkylations, often with ethylene oxide, to form polyoxyethylene alkylphenols. The 1-to-6-mol adducts are the principal raw materials for the conversion to sulfates. These are the most important anionic surfactants based on phenol.

United States production statistics for benzene and phenol, shown in Table 1 for 1967 [2], 1970 [3], and 1974 [4], illustrate the increasing preponderance of benzene and phenol produced from petroleum sources. (All synthetic phenol is produced from benzene or benzene-derived cumene.)

Monoalkylbenzenes, in which the side chain contains about 10 to 15 carbon atoms, are commonly referred to in industry as "detergent alkylates."

TABLE 1

U.S. Production of Benzene and Phenol in 1968, 1970, and 1974

	Unit quantity	1967	1970	1974
Total benzene	1000 gal	969.3	1133.5	1490.6
Petroleum derived	1000 gal	878.7	1040.0	--
Total phenol	million lb	1356	1755	2426
Synthetic	million lb	1297	1708	2399

The number of carbon atoms per chain in different grades may vary from an average of about 11.5 to 13.

Two types of detergent alkylates have reached industrial importance as raw materials for production of anionic surfactants by further processing to alkylarylsulfonates. The older, branched-chain type, which had such a rapid growth for over ten years during and after World War II, is now commonly referred to as "hard detergent alkylate." Its use by domestic producers of synthetic detergents was voluntarily discontinued in the summer of 1965 because of voluminous and persistent foam caused by the lack of biodegradation of its alkylbenzenesulfonate derivative (known as ABS) in sewage disposal systems, rivers, and streams. The use of ABS has also been discontinued or regulated in most West European countries and Japan. Many other countries, especially in South America, still use this stable foam-producing alkylbenzenesulfonate.

Since the use of ABS was discontinued in many countries, alkylbenzenesulfonates produced from the newer biodegradable "linear detergent alkylates" (or soft detergent alkylates) have largely taken its place wherever regulations require a more rapid and complete degradation. These "linear alkylbenzenesulfonates" have become known as LAS.

A. Branched-Chain-Type Detergent Alkylates

1. Alkylation with Partially Chlorinated Kerosene

Branched-chain detergent alkylates were first produced by alkylation of benzene with a chlorinated hydrocarbon made by partial chlorination of the suitable kerosene or white-oil fraction of petroleum or a Fischer Tropsch reaction product [5-8]. These were commonly called "kerylbenzene." Some of these kerylbenzenes might also be classified as somewhat linear-type detergent alkylates because a kerosene or mineral oil derived from a highly paraffinic feedstock could contain a substantial proportion of linear constituents.

However, an appreciable percentage of nonlinear isomers were generally present in the kerosene cuts used before the advent of molecular-sieve extraction and urea-adduction processes by which most of these nonlinear molecules could be effectively removed. Although the side chain of kerylbenzene is of branched configuration to some extent, it is not nearly as highly branched as the side chain of the propylene-tetramer-type detergent alkylate from which ABS is produced.

The rate at which kerylbenzenesulfonate is degraded by bacteria is between that of the very slowly degradable ABS and that of the more rapidly and more completely degradable LAS.

Equations (1) and (2) illustrate the preparation of kerylbenzene.

$$C_nH_{2n+2} + Cl_2 \longrightarrow C_nH_{2n+1}Cl + HCl \tag{1}$$

Kerosene fraction   Kerylchloride
(n = 10 - 15)

$$C_nH_{2n+1}Cl + C_6H_6 \xrightarrow[\text{Catalyst}]{AlCl_3} C_nH_{2n+1}C_6H_5 + HCl \tag{2}$$

Kerylchloride   Benzene   Kerylbenzene

Feighner [9] points out the importance of avoiding overchlorination of kerosene to minimize the formation of undesirably large amounts of by-products, such as diphenylalkanes, during the alkylation reaction.

2. Alkylation with Propylene Polymers

With the advent of catalytic cracking shortly before World War II as a method of making high-octane gasoline, the problem of profitably utilizing large amounts of low-molecular-weight olefins arose. A helpful solution to that problem was found by the development of catalytic polymerization of propylene contained in streams from the catalytic cracking process which were composed of about 30-50% propylene mixed with about 50-70% propane and other light hydrocarbons. Thus lower-molecular-weight olefins, such as propylene, were polymerized to higher-molecular-weight olefins suitable for use in gasoline blends. Certain of these polymers, especially the tetramer and pentamer were found to be of special value in the manufacture of detergent alkylates [9]. Further information on the derivation of these propylene polymers is presented in Sec. IV, B of this chapter.

Equations (3) and (4) illustrate the production of highly branched detergent alkylate from propylene and benzene.

$$C_3H_6 \xrightarrow[\text{Catalyst}]{\text{Polymerize}} C_{12}H_{24} \quad \text{and} \quad C_{15}H_{30} \tag{3}$$

Propylene   Dodecene   Pentadecene
                            (Propylene   (Propylene
                             tetramer)   pentamer)

$$\text{C}_{12}\text{H}_{24} \text{ or } \text{C}_{15}\text{H}_{30} + \text{C}_6\text{H}_6 \xrightarrow{\text{AlCl}_3 \text{ or HF}} \text{C}_{12}\text{H}_{25}\text{C}_6\text{H}_5 \text{ or } \text{C}_{15}\text{H}_{31}\text{C}_6\text{H}_5 \quad (4)$$

| | Tetramer-type | Pentamer-type |
| Benzene | dodecylbenzene | pentadecylbenzene |

In actual practice, two detergent alkylates are the principal products used for production of ABS. One is the tetramer-type dodecylbenzene shown in Eq. (4). The side chain on that product has an average of about 12 carbon atoms. The other is a tridecylbenzene produced by alkylating benzene with a propylene tetramer-pentamer cut of such composition that the side chain averages about C_{13}.

The alkylation process represents the addition of benzene across the double bond of the olefin (tetramer or tetrapentamer). As illustrated by Eq. (4), it appears straightforward. However, competing reactions take place and proper control of alkylating conditions must be exercized to achieve satisfactory yields and quality of the detergent alkylate. Factors such as the mole ratio of benzene to olefin, reaction time, and temperature are important.

Examples of undesirable reactions which must be controlled to the extent possible during the alkylation step may be illustrated by Eq. (5), which is of the fragmentation type. Fragmentation results in formation of lower-molecular-weight monoalkylbenzenes shown in Eqs. (6) and (7), and dialkylbenzenes shown in Eq. (8).

$$(5)$$

$$(6)$$

$$(7)$$

$$(8)$$

In addition to being fragmented, the tetramer may also be isomerized, resulting in olefins of different branched-chain configurations, as well as branched-chain saturated hydrocarbons.

Most U.S. industrial detergent-alkylate processes of the type under discussion used $AlCl_3$ as the alkylation catalyst, although HF was used by one major producer. Alkylation conditions for those two catalysts are similar, except that HF is used in greater than catalytic amounts and must be recovered for recycle. Aluminum chloride is used in catalytic amounts with promoters such as H_2O or HCl and is not recycled since it becomes deactivated. Also, when $AlCl_3$ is used, the alkylation can be conducted at essentially atmospheric pressure, while with HF a slightly higher pressure must be maintained to keep the catalyst in the liquid phase [9].

Benzene is used in excess, usually from 5 to 10 moles per mole of olefin, in order to minimize unwanted side reactions, especially dialkylation. Temperatures used for the alkylation are usually in the range of 20 to 50°C. Residence time in the reactor varies widely from a few minutes to hours, depending on whether a batch or continuous process is used [9], and upon other reaction conditions.

The final detergent alkylate is separated by fractionation from excess benzene and by-products having higher and lower boiling ranges than the desired primary product.

Crude alkylated aromatic products can be purified by treatment with sulfuric acid [10]. This treatment is said to reduce corrosion in alkylate recovery equipment as well as to improve the quality of the desired detergent alkylate.

Reference [11] describes an alkylation process whereby high yields of the desired alkylaryl hydrocarbon are obtained by separating the alkylated by-products by fractionation and recycling them back into the alkylation step.

Continuous processes for production of branched-chain detergent alkylate are described in Refs. 12 and 13.

B.   Linear-Type Detergent Alkylates (LAB)

The term "linear" is used to designate the essentially straight-chain configuration of the aliphatic side chain. Actually, it is a relative term used to denote a much less highly branched structure than the side chain, for example, of the older propylene tetramer-type detergent alkylate described above.

Linear detergent alkylates of commercial importance today are derived by alkylation of benzene with either a chloro-n-paraffin mixture or linear olefins. In chloro-n-paraffins, the chlorine attachment is distributed in all possible positions on the hydrocarbon chain.

There are two principal types of linear olefins. One type is the linear α-olefin, the double bond of which is between the first and second carbon atoms of the carbon chain; the other is the linear internal olefin in which the position of the double bond is distributed down the chain.

Figure 1 illustrates alternate routes to linear detergent alkylates.

Since the development of processes for separating n-paraffins from hydrocarbons of branched chain and cyclic structures, n-paraffins have become the most widely used raw material from which alkylating agents are derived for the production of linear detergent alkylates in the United States, West Germany, and Japan.

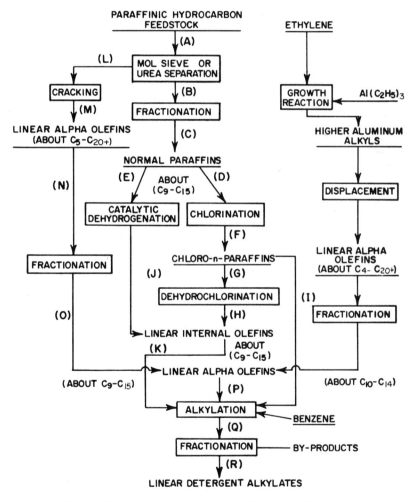

FIG. 1. Alternate routes to linear detergent alkylates.

The following companies produced linear detergent alkylates in the United States during the first half of 1975:

    Continental Oil Company
    Monsanto Company
    Union Carbide Corporation
    Witco Chemical Company

The most widely used processes for production of linear detergent alkylates based on n-paraffins utilize either chloro-n-paraffins or linear internal olefins as the alkylating agent. These alternative schemes are designated in Fig. 1 by routes, starting after (C), marked (D), (F), (I), (K), (Q), (R); and (E), (J), (K), (Q), (R); or the less widely used route (D), (F), (G), (H), (K), (Q), (R).

The basic steps by which α-olefins are produced via cracking a paraffinic wax or other suitable heavy petroleum fractions are shown in Fig. 1 by the route marked (L), (M), (N), (O). α-olefins derived by that route were used for the production of linear detergent alkylate for a short time in the United States. About 1966, that route was discontinued. It is, however, practiced by Shell in England. The quality of the linear α-olefins has been greatly improved by subjecting the heavy-paraffin feedstock to urea adduction prior to cracking.

The linear α-olefins derived from ethylene by reaction with $Al(C_2H_5)_3$ are also shown in Fig. 1. This process is technically sound but is presently economically not advantageous as a way to make olefins to be used for alkylation of benzene.

The chain length of the chloro-n-paraffins, linear internal olefins, or linear α-olefins used to alkylate benzene may include a range as broad as C_9 or C_{10} to C_{15-17}. The average chain length is usually about C_{12} or C_{13}, depending on the grade detergent alkylate desired.

Processes relating to the production of linear detergent alkylates are widely covered in the patent literature as evidenced by Refs. 14-40.

1. Alkylation with Partially Chlorinated n-Paraffins

References 14-24 describe processes for alkylating benzene with chloro-n-paraffins.

A process for alkylating benzene with chloro-n-paraffins (containing unreacted n-paraffins), which can be operated either batch-wise or continuously, is described in Ref. 14 and illustrated by Fig. 2. A strong alkylating catalyst such as $AlCl_3$ is preferred, but the effect it might have for causing isomerization of n-paraffins to isoparaffins is prevented by the presence of the high percentage of unchlorinated n-paraffins present in the alkylating mixture. Thus the n-paraffins are suitable for recycling to the chlorination reactor. The mole ratio of benzene to the chloro-n-paraffin content of the alkylating

mixture is preferred to be in the range of 5:1 to 10:1. A preferred temperature range for alkylation is about 40–50° C. Sulfonation of the linear–type
detergent alkylate to produce effective and biodegradable anionic detergent
agents is also described in Ref. 14.

Reference 15 describes a process similar to that of Ref. 14 except for
treatment of the raw n-paraffin feed with 10–30% oleum prior to the chlorination step. This removes cyclic hydrocarbons, especially those of the aromatic types, present in the feed n-paraffins in small amounts. If not removed, these alicyclic and aromatic hydrocarbons can cause quality problems
in the final detergent alkylate product.

The oleum treatment is not necessarily required if the n–paraffin feed is
substantially free of the contaminants mentioned in the preceding paragraph.

Another continuous process based on a mixture of chloro-n-paraffins
and unchlorinated hydrocarbons as the alkylation agent is covered in Ref. 16.
The specific improvement embodied in that process is treatment of the feed
n-paraffins with spent $AlCl_3$ catalyst complex or directly with an aluminum

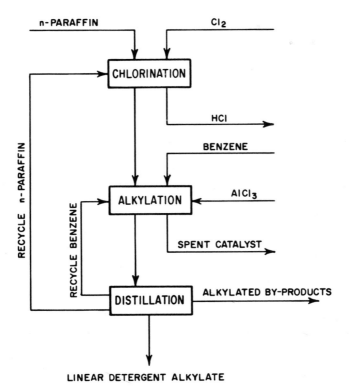

FIG. 2. Alkylation of benzene with chloroparaffins.

halide to separate nonparaffinic contaminants prior to passing it to the chlorination zone. The chloro-n-paraffins so obtained are essentially free of small amounts of contaminants which have a deleterious effect on the degree of conversion and rate of catalyst deactivation in the alkylation zone.

Yet another continuous process for the alkylation of benzene with chloro-n-paraffins is described and specifically claimed in Ref. 17. This involves essentially a two-step liquid-phase alkylation technique utilizing a special $AlCl_3$ catalyst system. An excess of preferably 5 to 15 mols of benzene per mol of chloro-n-paraffins is employed. The temperature range in the first alkylation zone is preferably 150-170° F, while the temperature in the second alkylation zone lies in the range of 100 to 185° F. Reaction time in the first zone is between 3 and 15 min, while it may be between 0.3 and 15 min in the second zone.

References 18-20 are listed as examples of other modifications of processes to alkylate benzene with chloro-n-paraffins to produce a standard linear-type detergent alkylate.

References 21-23 relate to the alkylation of benzene with chloro-n-paraffins and employing techniques which will increase the percentage of internal isomers normally present in the alkyl sidechains of the resulting linear-type detergent alkylates so produced.

Mixtures of essentially linear phenylalkanes, the alkyl groups of which are from C_{12} to C_{13} and in which the total 2-phenyl isomers constitute a maximum of only about 22%, are claimed as compositions of matter [24].

2. Alkylation with Linear Olefins

Alkylation of benzene with a linear α-olefin or with linear internal olefins gives a mixture of linear alkylbenzenes. Although the resulting dodecylbenzene is a mixture of 2, 3, 4, 5, and 6-phenyl isomers, when either linear internal olefin or α-olefin is used as the alkylating agent, the distribution of isomers may differ, depending on which type is used. References 25-41 relate to alkylation of benzene with linear olefins.

Alkylation of benzene with a linear internal olefin-paraffin mixture derived by partial halogenation of n-paraffin followed by dehydrohalogenation is described in Ref. 25. A similar process is described in Ref. 26 using HF as the alkylation catalyst.

A process for alkylating benzene with fractions of C_{10}-C_{20} α-olefins derived from the cracking of paraffin wax is described in Ref. 27.

Reference 28 covers a process for alkylation with C_9-C_{15} linear olefins in which the double bonds are located between the third and fourth and the third and fifth carbon atoms. This type linear internal olefin mixture is derived from the isomerization of the corresponding linear internal olefins wherein the double bond is in the 1, 2, 5, 6, or 7 positions or of linear α-olefins in the liquid phase with the aid of alkali metals, oxides, or hydroxides thereof or an amide thereof on a suitable support. The points of attachment of the C_9-C_{15} side chains to the benzene nucleus resulting by use of such isomerized linear olefins for alkylation are on the third and fifth carbon atoms.

Another modification of a process designed to produce a linear detergent alkylate of low 2-phenyl isomer content is described in Ref. 29. The olefin isomerization may be carried out in the vapor phase over catalysts such as alumina, chromia-alumina and magnesia-alumina. Alkylation is carried out in the presence of an appropriate Friedel-Crafts catalyst and catalyst-modifier combination.

A detergent alkylate composition containing controlled proportions of the 2- and 3-phenyl isomers is covered in Ref. 30.

Other process patents for producing controlled 2-phenyl isomer content detergent alkylates have been described in Refs. 31-34.

A process for alkylation of benzene in two steps is described in Ref. 35. In the first step, alkylation is accomplished with a mixture of primary and secondary chloro-n-paraffins. The unreacted primary chloro-n-paraffins which react more slowly and less completely are separated from the alkylation product and passed through a dehydrochlorination zone where they are converted to linear α-olefins. These olefins are then used to alkylate more benzene to produce an additional quantity of desired detergent alkylate.

Diphenylalkanes commonly produced as by-products in the production of linear-type detergent alkylates can be upgraded to more valuable products by further alkylation of the diphenylalkanes with chloro-n-paraffins or linear olefins in presence of Friedel-Crafts catalyst. The more valuable products are monophenylalkanes, dialkylbenzenes, and/or alkylated tetralins [36].

Alkylation of benzene with chloro-n-paraffins or linear internal olefins obtained by catalytic dehydrogenation of n-paraffins is described in Ref. 37.

A new route [38] to linear alkylbenzenes was developed by the Universal Oil Products Company (U.O.P.). Linear paraffins obtained by the Molex process are partially dehydrogenated to form linear internal olefins. These olefins in admixture with unconverted n-paraffins are used to alkylate benzene. This process is called the U.O.P. Pacol process.

References 39 and 40 describe other schemes for the alkylation of mononuclear aromatic compounds, particularly benzene, with linear olefins derived from the catalytic dehydrogenation of n-paraffins.

Production of linear-type detergent alkylate and other alkylated aromatics via the HF catalyzed alkylation with linear olefins (in admixture with unreacted n-paraffins) which are derived from catalytic dehydrogenation of n-paraffins is described in Ref. 41. Besides benzene, other aromatic compounds suitable for alkylation may, for example, include phenol, toluene, xylene, or naphthalene.

C. Other Alkylaryl Hydrocarbons

Other aromatic hydrocarbons typical of those from which anionic surfactants
may be produced include those derived by alkylation of benzene, toluene,
xylene, or naphthalene. These are generally classed as alkylaryl hydrocarbons. Specific examples are:

pentylbenzene   pentylnaphthalene
hexylbenzene   nonylnaphthalene
octylbenzene   dodecylnaphthalene
butylnaphthalene   dodecyltoluene

These alkylaryl hydrocarbons may be produced by processes similar to
those used to manufacture detergent alkylates. The length of the aliphatic
side chain required to impart the desired properties is controlled by selection of the proper chain-length olefins or chloroalkanes used to alkylate the
aryl hydrocarbon. In the case of the alkylation of benzene, the basic reactions are the same as illustrated in Eqs. (1) to (4). The same reactions
apply to the toluene and naphthalene derivatives except that the proper aryl
hydrocarbon is substituted for benzene.

References 6, 7, 9, 13, 28, 35, 37, and 39 previously cited are not
limited to the alkylation of benzene but also include toluene, xylene, naphthalene, and other aromatic hydrocarbons.

D. Alkylphenols

As a class, alkylphenols are important industrial chemicals. The alkyl
groups of different alkylated phenolic compounds vary over a wide range,
e.g., from C_2 to at least C_{12}. Alkylated cresols, xylenols, and other phenolic materials are included in the class.

This subsection deals only with those particular alkylphenols which are
most widely used as raw materials for surfactants. These are primarily
monoalkylated phenols, the alkyl groups of which may range from about C_5
to C_{12} in length. Certain dialkylphenols are also included.

Alkylphenols are produced by alkylation of phenol with olefins such as
pentenes, hexenes, octenes, propylene trimer, propylene tetramer, linear
α-olefins, etc. Processes utilizing alkyl halides as the alkylating agent have
been devised but are not employed as widely as processes based on alkylation
with olefins.

Although phenol reacts thermally with olefins without a catalyst, the
yield is poor. Catalysts which improve the alkylation reaction include boron

trifluoride, boron trifluoride complexes, aluminum chloride, fullers earth, activated clay, and others, and acids such as sulfuric, boric, oxalic, and toluene sulfonic; BF_3 is believed to be most widely used.

The reaction of phenol with dodecene (either propylene tetramer or 1-dodecene) will serve as a typical example, see Eq. (9).

$$C_{12}H_{24} + \underset{\text{PHENOL}}{\underset{\text{OH}}{\bigcirc}} \xrightarrow{\text{CATALYST}} \underset{\text{DODECYLPHENOL}}{\underset{\text{OH}}{\bigcirc}} C_{12}H_{25} + \underset{\text{DIDODECYLPHENOL}}{\underset{\text{OH}}{\bigcirc}} (C_{12}H_{25})_2 \qquad (9)$$

OLEFIN   PHENOL   DODECYLPHENOL   DIDODECYLPHENOL

Continuous methods for the alkylation of phenolic materials with olefins are described in a patent in Ref. 43. The most satisfactory type of catalysts employed according to this invention comprises a sulfonated polymer of a monovinyl aryl compound sufficiently crosslinked, e.g., with a polyvinyl aryl compound to provide satisfactory dimensional stability without excessively limited porosity. Typical monovinyl aryl compounds include styrene, vinyl toluenes, vinyl naphthalenes, vinyl ethyl benzenes, etc. Typical polyvinyl aryl compounds include divinyl benzenes, divinyl toluenes, divinyl xylenes, divinyl naphthalenes, etc.

Because of the relatively mild alkylating conditions used [43], the continuous process is said to produce improved yields and quality of alkyl phenols as compared with batch processes [42] previously employed.

An ortho alkylation process employing olefins as the alkylating agent is the subject of another U.S. patent [44]. According to this invention the phenol and olefin react above atmospheric pressure in the liquid phase in a preferred temperature range of 100–200° C in the presence of 0.005–0.01 mole of a promoter per mole of phenol. The promoters are aluminum compounds such as aluminum ethyl dichloride, aluminum diethyl chloride, diisobutyl aluminum chloride, etc. Preferred compounds are the aluminum dialkyl-halides.

Another process for obtaining relatively relatively high-molecular-weight alkylphenols of high purity involves separation of unreacted olefin from the alkylated phenol by solvent extraction with a selective solvent such as nitromethane or methanol [45].

Examples of patented processes utilizing chlorocarbons to alkylate phenols are described in Refs. 46 and 47. Alkylation catalysts employed are alkali metal aluminum hydrides [46] or aluminum chloride or reaction products thereof [47].

III. PROPERTIES OF ALKYLATED AROMATICS

A. C_{10} and Higher Alkylbenzenes

These alkylbenzenes are water white, oily liquids. Typical physical properties of several of the branched-chain and linear products of this class (detergent alkylates) are listed in Tables 2 and 3, respectively.

Based on feeding tests with rats, the oral toxicity is low. Inhalation of
vapors by test animals at room temperature produces no obvious ill effects
after several hours. Also, irritation produced in the eyes of rabbits was
minor [49].

By far the most important use of C_{10} and higher alkylbenzenes is in the
production of sulfonates by reaction with sulfuric acid, oleum, or sulfur trioxide followed by neutralization with a base. The sulfonates so produced
constitute the largest volume anionic surfactants derived from petroleum
sources. They are manufactured and used by virtually all companies en-
gaged in the formulation and distribution of household detergents. Well over
100 formulations containing dodecyl and tridecylbenzenesulfonates are listed
by trademarks applications and company name by J.W. McCutcheon, Inc.,
[52].

TABLE 2

Typical Physical Properties of Branched-Chain-Type
Detergent Alkylates

Property	Conventional C_{12} alkylate[a]	Conventional C_{13} alkylate[b]
Average molecular weight	245	262
Specific gravity	0.85–0.86[c]	0.85–0.86[c]
Color, Saybolt	25–28	25–28
Bromine number	0.05	0.05
Flash point ° F, COC	260	260
Distillation range, ° F	530–565	540–600
Reference	49	49

[a]Dodecylbenzene.
[b]Tridecylbenzene.
[c]20°/20° C.

TABLE 3

Typical Physical Properties of Linear–Type Detergent Alkylates

Property	Continental Oil Company			Union Carbide Corp.			Monsanto Company		
	Nalkylene 500[a]	Nalkylene 550[a]	Nalkylene 600[a]	Ucane 11[a]	Ucane 12[a]	Ucane 13[a]	A-215[a]	A-222[a]	A-230[a]
Average molecular weight	238	243	261	236	244	258	237	244	259
Specific gravity	0.859[b]	0.865[b]	0.858[b]	0.861[c]	0.861[c]	0.862[c]		0.85–0.86[b]	
Bromine number	0.03	0.003	0.03	0.03	0.03	0.03		<0.05	
Pour point, °F	<–70	––	<–70	––	––	––	––	––	––
Viscosity, SUS, at 100° F	––	––	––	40	41	44	––	––	––
Flash point, °F, COC	310	>310	340	280	285	325	––	––	––
Fire point, °F, COC	330	––	355	––	––	––	––	––	––
Distillation, °F									
IBP[d]	540	––	580	550–620	550–620	590–650	––	––	––
5%	550	––	589	––	––	––	––	––	––
50%	562	––	600	––	––	––	––	––	––
95%	588	––	610	––	––	––	––	––	––
Side–chain–homologue distribution, %									
C9 or less, %	––	––	––	2 max	0.5 max	––	––	––	––
C10	14.1	12.3	0.6	10–15	5–15	––	7	3	––

25

TABLE 3 (Continued)

Property	Continental Oil Company			Union Carbide Corp.			Monsanto Company		
	Nalkylene 500[a]	Nalkylene 550[a]	Nalkylene 600[a]	Ucane 11[a]	Ucane 12[a]	Ucane 13[a]	A-215[a]	A-222[a]	A-230[a]
Side-chain-homologue distribution, % (Continued)									
C_{11}	38.4	31.9	2.6	25-50	35 max	5 max	56	37	4
$C_{10} + C_{11}$	--	44.2	--	50 min	30-50	--	--	--	--
C_{12}	40.3	26.6	16.1	25 min	--	10-30	--	27	24
$C_{10} + C_{11} + C_{12}$	--	--	--	85 min	--	--	--	--	--
C_{13}	6.7	16.4	47.1	15 max	30 max	30 max	4	21	49
C_{14}	0.2	9.6	33.6	2 max	10 max	20 min	--	2	23
$C_{13} + C_{14}$	--	26.0	--	--	30 max	--	--	--	--
C_{15}	--	0.4	--	--	0.5 max	5 max	--	--	--
Reference	48	50	48	49	49	49	50	50	50

[a]Trademarks.
[b]60°/60°F.
[c]20°/20°C.
[d]Initial boiling point.

26

Table 4 shows the annual reported U.S. production of principally C_{10}-C_{14} linear alkylbenzenes for the years 1971 [53], 1972 [54], 1973 [55], and 1974 [4].

B. Other Alkylaryl Hydrocarbons

Numerous alkylaryl hydrocarbons other than dodecyl and tridecylbenzenes are also sulfonated to produce anionic surfactants. Physical properties of a few such alkylaryl hydrocarbons are presented in Table 5. These data are not all derived from commercial grades but are representative of relatively pure products.

Formulations in which the surface-active ingredients are sulfonates of alkylaryl hydrocarbons are listed by J.W. McCutcheon, Inc. [52].

C. Alkylphenols

Octyl-, nonyl-, and dodecylphenol are among the more important members of this class of products from the standpoint of usage as raw materials for the production of surfactants; their physical properties are listed in Table 6. Other alkylphenols, not included in Table 6, may also be employed for that purpose. Examples are butylphenol, dibutylphenol, pentylphenol, dipentyl-phenol, dinonylphenol, and didodecylphenol.

The higher alkylphenols, such as the three shown in Table 6, should be considered as moderate toxicants and skin irritants. Eye protection is con-sidered mandatory in handling these chemicals, especially those that are liquids. They cause moderately severe irritation on contact with the eye [57]. According to Ref. 58 nonylphenol is considered to be an extremely serious eye irritant [59].

TABLE 4

U.S. Production of Linear
Detergent Alkylates

Year	Million lb
1971	550.1
1972	524.0
1973	497.8
1974	534.6

TABLE 5

Physical Properties of Selected Pure Alkylaryl Hydrocarbons [56]

Property	Benzene			Naphthalene			
	n-pentyl	n-hexyl	n-octyl	1-n-butyl	1-n-pentyl	1-n-nonyl	1-n-dodecyl
Boiling point, °C at 760 mm Hg	205.4	226.1	264.5	289.34	307	372	415
Refractive index, at 25°C	1.4855	1.4842	1.4824	1.5798	1.5704	1.5456	--
Density, at 25°C, g/ml	0.8546	0.8537	0.8525	0.9732	0.9622	0.9339	--
Freezing point, °C, in air, at 1 atm	-75	-61	-48	-19.76	-22	8	27

28

TABLE 6

Typical Physical Properties of Selected Alkylphenols

Property	Octylphenol	Nonylphenol	Dodecylphenol
Formula	$C_{14}H_{22}O$	$C_{15}H_{24}O$	$C_{18}H_{30}O$
Appearance	White flakes	Clear sparkling liquids	
Color, APHA	40[a]	35	60
Specific gravity			
40°C	--	0.928	0.932
60°C	--	0.914	0.918
85°C	0.9199	--	--
Refractive index			
27°C		1.5110	1.5055
42°C		1.5060	1.4998
Distillation range, °C at 10 mm Hg	150–175[b]	--	--
Boiling range,			
°C at 23 mm Hg	--	155–202	185–217
°C at 760 mm Hg	--	293–314	314–334
Viscosity			
cP			
80°C	22.5	--	
83°C	17	--	
cSt			
100°F	--	325	560
210°F	--	6.2	8.6
Solubility			
Water	--	Insoluble	Insoluble
Alcohol	--	Soluble	Soluble
Benzene	--	Soluble	Soluble
Congealing point, °C	77.5	--	--
Pour point, °C	--	–6 to +1	+5 to +10
Flash point, °F, COC	--	295	325
Reference	57	58	59

[a]Molten.
[b]90%.

All phenols are subject to oxidation when exposed to the atmosphere. The result is development of color, the rate of which increases with increasing temperature. The materials from which equipment, lines, and storage vessels are constructed may also affect the rate of color increase. When possible, 316 or 304 stainless steel or phenolic-resin-lined mild steel should be used as material for construction wherever contacted by alkylphenols. Certain other materials are also suitable. The alkylphenols should also be blanketed with an inert gas such as nitrogen during storage and transfer [57].

Alkylphenols are versatile chemical intermediates. A large number of patents and the literature references covering reactions of alkylphenols and applications of derivatives are listed and briefly described in technical brochures [58-60].

The most important alkylphenol derivatives used as surfactants are produced by reaction of the alkylphenol with ethylene oxide. These are usually called ethylene oxide adducts or ethoxylates. Generally, those adducts which contain more than 6 moles of ethylene oxide per mole of alkylphenol constitute an important series of nonionic surfactants. Preparation, properties, and applications of such nonionic surfactants are covered by C. R. Enyeart [42].

Polyoxyethylene alkylphenols resulting from condensation with about 3 to 6 moles of ethylene oxide per mole of alkylphenol may be sulfated to produce anionic surfactants. Preparation and applications of such surfactants are discussed by L. W. Burnette [61].

Since the mid-1960s, the use of alkylphenols with linear side chains have partially replaced their branched-chain counterparts as hydrophobes for the polyoxyethylene derivatives of the types used as nonionic surfactants, as well as those that are converted to anionic surfactants by sulfation. Although surfactants performance of branched-chain and linear alkylphenol derivatives are similar, the latter are more biodegradable. This is an important consideration for many applications today. Reference 62 describes the relationship of structure to biodegradability.

IV. HIGHER ACYCLIC HYDROCARBONS

A. n-Paraffins

Since about 1964, n-paraffins (synonymous with "linear paraffins") of carbon chain lengths ranging predominantly from about C_{10} to C_{16} have become the most widely used acyclic hydrocarbon raw materials from which anionic surfactants are manufactured. They are derived from suitable petroleum fractions, such as kerosene, by separation from hydrocarbons of other structures, such as branched-chain and ring types, present in petroleum fractions.

One general type of commercial separation method widely used today utilizes the principle of selective adsorption of n-paraffins on adsorbent materials of porous structures called "molecular sieves." The other method is based on clathrate complex formation of n-paraffins with urea.

Selection of the type of separation process is primarily one of choice which is influenced by the desired molecular-weight range of the n-paraffins. The molecular-sieve method tends to yield a fraction in which the concentration of lower n-paraffins predominate. The fraction obtained from the urea adduction method contains a comparatively greater proportion of higher n-paraffins as compared to the lower homologues.

1. Molecular-Sieve Separation Method

Any suitable solid porous material may be used for sieving purposes. Such adsorbent materials are the metal alumina-silicates of the zeolite type. Those suitable for use in commercial separation processes today are dehydrated alkali and alkaline earth zeolitic silicates, the crystalline structures of which contain pores of about 5 Å in diameter. This is slightly larger than the critical diameter of the n-paraffin molecules, but somewhat smaller than the diameter of branched-chain paraffins, cycloparaffins, and aromatics. Thus, it is possible to separate n-paraffins from the mixture [63].

A review and presentation of work done from 1930 to 1950 on naturally occurring zeolites and to about 1960 on synthetic crystalline zeolites is given in Ref. 64.

There are six molecular-sieve adsorption processes now in commercial operation:

a. The Molex process developed by Universal Oil Products Company.
b. The Isosieve process developed by Linde Division of Union Carbide Corporation.
c. The Ensorb process developed by Esso Research and Engineering Company.
d. The TSF process of Texaco, Inc.
e. The process of British Petroleum Company.
f. The process of Shell Oil Company.

These separation methods require three principal processing steps for practical continuous separation. First, the feedstock is contacted with the sieves. This is followed by a step in which hydrocarbons of other than linear chain structures are separated from the sieves which contain the n-paraffins. The third step is the one in which the n-paraffins are recovered by desorbing from the sieves. The sieves may then be used again by repeating the operation. Those separation processes are widely covered in the literature and by the patents in Refs. 65-94.

The Molex process is unique in that it is a liquid-phase, isothermal process. One fixed-bed adsorber chamber may be employed as described in Refs. 65 and 66. It was originally developed for quite another purpose, but it has been placed in commercial operation for the separation of C_{11}-C_{14} n-paraffins from virgin crude fractions for use as a raw material in biodegradable detergent manufacture. A description of the process is given in Ref. 67.

Reference 68 describes a process employing a fixed-bed solid sorbent chamber containing at least four serially interconnected zones having fluid-flow connecting means between the outlet of one terminal zone and the inlet of the other terminal zone in the series to provide cyclic fluid flow in the process. A synthetic crystalline zeolite material in which from about 0.5 to 4 weight % of a polar compound, such as water, glycols, amines, ammonia, and alcohols, is incorporated, is employed as the sorbent.

A somewhat similar cyclic and continuous process which may also be operated substantially isothermally in all stages is covered in Ref. 69. A method for regeneration of deactivated molecular sieves is covered in Ref. 70.

One form of the Isosieve process employs two separate vessels containing molecular sieves so that at a given time one vessel is functioning as an adsorber, while the sieves in the other are being desorbed. The process operates in the vapor phase and desorption is accomplished by reducing the pressure in the vessel to be desorbed [70]. References 71-76 give additional information concerning the Isosieve processes.

The Texaco process employs molecular sieves to separate linear paraffins from a variety of petroleum fractions by selective vapor phase adsorption. The first commercial unit started operating at Pointe-à-Pierre, Trinidad, in 1965 [77]. The importance of optimizing the desorption step is discussed in Ref. 77.

References 78-80 to patents issued to Texaco are illustrative of some modifications of adsorption and desorption methods in vapor phase separation processes. Reference 77 relates to an improved method for the regeneration of molecular sieves such as the Linde 5A Sieve.

The Esso Research and Engineering Company has a great many patents covering molecular-sieve-separation processes. A number of selected examples are covered in Refs. 82-87.

Reference 82 describes a process for contacting a vaporized stream of a mixture containing linear paraffins and other hydrocarbons in an adsorbing zone of a crystalline alumina-silicate for less than 4 min. This may be followed quickly by removal to a desorbing zone operated at elevated temperature with a purge gas or at reduced pressure. The desorbed molecular sieve is then recycled to the adsorption zone. Reference 83 covers an improved process for separation of linear paraffins of greater than C_{10} chain length from admixture with nonlinear hydrocarbons by contacting in vapor phase with a molecular-sieve adsorbent of a selective pore size. This is followed by a desorption step in which the displacing medium is a small polar molecule such as ammonia. The sieves are then contacted again with feed and the adsorption and desorption steps are repeated over a number of cycles.

A process for the separation of linear paraffins and linear olefins from admixture with branched-chain and cyclic hydrocarbons with the aid of molecular sieves is described in Ref. 84. The sieve pores are of such a size as to exclude the branched-chain and cyclic components of the feed. The olefins are then desorbed from the sieves with a linear olefin differing by 1 or 2 carbon atoms from the adsorbed olefin. The linear paraffins are desorbed from the sieves with a linear paraffin differing by 1 or 2 carbon atoms from the adsorbed linear paraffin.

Other modifications of the molecular sieve process are covered by Refs. 85 and 86. Conditions of desorption and sequence of processing steps differ somewhat from other processes. The preferred displacing medium is usually ammonia although amines, carbon dioxide, lower alcohols, and glycols are also mentioned.

Reference 87 covers a process for desorbing n-paraffins with steam from a sieve containing a naturally occurring molecular-sieve material called erionite. It consists of zeolitic mixed aluminum silicate containing sodium, potassium, and calcium. According to the invention [87], it was found that erionite, particularly one in which the sodium and potassium ions have been at least partially replaced by hydrogen ions, is capable of being desorbed by steam without loss of adsorbent capacity. Sieves such as Linde 5A are not steam stable and collapse in the presence of steam, thus losing their ability to adsorb. Thus, an erionite material makes possible the use of steam as the displacing agent which is more economical than other displacing agents such as ammonia, amines, etc.

The Exxon Ensorb process for displacing n-paraffins from molecular sieves is described in Ref. 88.

References 89-91 represent modifications of the molecular-sieve separation process of n-paraffins developed by the British Petroleum Company, Ltd.

The process of Ref. 89 consists of three stages, all operated in the vapor phase. The first stage is adsorption on a molecular-sieve bed. The bed is then purged in a second stage to remove both surface and interstitially held material by treatment with a stream of n-pentane which is passed through the bed in the direction opposite to that in which the C_9 and higher feed (diluted with n-pentane) was passed in the first stage.

The n-paraffins are then desorbed in a third stage by continuing the flow of n-pentane in the same direction as in the purge stage. The second and third stages are distinguished by switching the effluent at the beginning of the third stage to a product-recovery system. The effluent from the second stage containing some linear paraffin is recycled to prevent loss of such linear paraffin and to ensure good extraction efficiency in the process.

References 90 and 91 cover somewhat similar processes in which the purging and desorbing stages are effected by successively reducing the pressure from the purging to the desorbing stages.

Examples of yet other modifications of the continuous cyclic separation process are described in two patents issued to Shell Oil Company [92, 93].

All molecular-sieve adsorption processes for separating n-paraffins from kerosene, gas oil, and naphtha-type petroleum feeds have six common operating variables: phase, means of sulfur removal, degree of adsorption-desorption, space velocity, temperature, and purge material. Comparison of the commercial processes is given in Ref. 94.

2. Urea-Adduction Separation Method

The urea-adduction method is an alternative method to that of the molecular-sieve-adsorption process for separation of n-paraffins from various petroleum fractions in which they are contained in admixture with branched-chain and cyclic hydrocarbons. It is based on the discovery made by F. Bengen in Germany in 1940 [95] that urea forms relatively stable crystalline adducts with linear paraffins, but does not with branched-chain and cyclic hydrocarbons. These crystalline complexes are commonly known as clathrates and can be separated from other components of the feed by filtration. There are many literature and patent references relating to the urea-adduction method, e.g., [95-102].

One general procedure for effecting adduction by urea is by addition of the hydrocarbon feed to a solution of urea in an activator with constant stirring. Examples of activators are methyl ethyl ketone, acetone, or preferably methanol.

About 3.5 parts of urea per one part of the hydrocarbon to be adducted is preferable. The activator is required in sufficient quantity to keep any free urea existing in the adduction mixture soluble at a temperature as low as 25°C. After complexing has occurred, usually after one or not more than two hours, the mixture is filtered and washed with a suitable hydrocarbon solvent such as butane, pentane, hexane, etc. The washed crystals can then be decomposed in water at 80-90°C to yield the n-paraffins.

Urea-adduction processes were originally developed to produce low-pour-point oils by dewaxing. An example is the Edeleanu process. The first continuous plant using this process is that of the Deutsche Erdöl A.G., which has been operating in Heide, West Germany, since 1955 with a distillate throughput of 200 tons per day [96]. A considerably larger plant using the same process is operating in the Shell Refinery in Pernis, Holland.

Reference 97 presents an excellent summary of processes for dewaxing with urea.

There are three general processes for dewaxing with urea. One process employs the use of solid crystalline urea, the second uses a urea solution of low concentration and a highly concentrated urea solution is used in the third [98]. Each type of urea-adduction process can be operated in a continuous cycle and involves four steps of varying complexity.

    a. Formation of the crystalline adduct by agitation of the hydrocarbon feed with either dissolved or solid urea.

    b. Separation of the reaction product into solid crystalline adduct and liquid phases.

    c. Decomposition of the crystalline adduct to urea (in the form of a solution or as a solid) and the adducted linear hydrocarbons.

    d. Purification of the product streams and recovery of urea and solvent.

Reference 98 relates to a process for separating linear-chain organic compounds from admixture with other organic compounds by the urea-adduction method. The improvement in this process lies in bringing such mixture into contact with unsaturated urea solution in high-purity methanol contained in a large retention capacity crystallizer, while heat is extracted from the mixture. This causes formation of relatively large rod-shaped or cylindrical crystals of the urea adduct. Such large crystals are said to possess rather high purity and are more easily separated in a slurry from the nonadductable portion of the feedstock than are smaller urea-adduct crystals. One special advantage of this process involves separation of n-paraffins in the preferred range of C_9-C_{17}. This range is of particular importance in the manufacture of biodegradable detergents.

Reference 99 covers an improved process for preparation of a bed of solid urea dispersed throughout an inert carrier in such a manner that high efficiency and ease in loading of the carrier are attained. The process of this invention is said to result in deposition of greater quantities of urea on the inert carrier under less extreme conditions, and allows the use of less expensive and more easily handled solvents than conventional methods.

Reference 100 covers a process designed to improve physical separation of the urea adducts from the nonadducted compounds with which they are intimately mixed since by many other techniques, the adducts have been slushy or semisolids which could not be economically separated. The adducts formed by the technique of this process are hard crystals and can be handled as solids mixed with unreacted urea while the liquid nonadducted materials may be readily drained or washed from the reactor.

An n-paraffin separation process employing the urea adduct has been developed and commercialized in Japan. It is called the Nurex process and was developed by Nippon Mining Co., Ltd., in collaboration with Chiyoda Chemical Engineering and Construction Company, Ltd. The Nurex process started operations in 1967 in a 40,000 ton-per-year plant at the Mizushima refinery of Nippon Mining [101].

In the Nurex process urea is used in the solid state and is continuously recycled between the adduction and eduction (decomposition) steps. The features of the Nurex process are stated to be as follows [101]:

a. Use of solid urea for adduction.
b. Application of a wide range of hydrocarbons and no necessity of process change with variation of feedstocks.
c. No necessity of feed preparation such as desulfurization.

A flow diagram of the four sections of the Nurex process and a description of its operation is given in Refs. 101 and 102.

B. Detergent-Range Monoolefins

Olefins do not occur naturally in petroleum. The broad class includes acyclic monoolefins, diolefins, and cyclic olefins. In this section only those acyclic monoolefins which are widely used as raw materials for the production of anionic surfactants are discussed. These generally encompass a carbon range of C_8 to C_{18}.

1. Linear Internal Olefins

Linear internal olefins are straight-chain hydrocarbons containing one double bond per molecule which is located randomly in the carbon chain. For example, linear dodecene (mixed isomers) contains predominantly the 2-, 3-, 4-, and 5-, isomers and a small amount of the 1- or α-isomers. Such olefins may be produced directly from n-paraffins by three methods, the first two listed being more important:

Chlorination-dehydrochlorination of n-paraffins
Catalytic dehydrogenation of n-paraffins
Dimerization processes

Formation of linear internal olefins by the foregoing methods is discussed below.

a. Chlorination-Dehydrochlorination of n-Paraffins

The chemical reactions involved are illustrated in Eqs. (10) and (11), as follows:

Chlorination

$$CH_3(CH_2)_x CH_3 \xrightarrow{\ Cl_2\ } CH_3(CH_2)_y CHCl(CH_2)_2 CH_3 + HCl \qquad (10)$$

Linear paraffins   Chloro-n-paraffins (C_{10}-C_{18})
(e.g., C_{10}-C_{18})   (mixed primary and secondary)

Dehydrochlorination

$$CH_3(CH_2)_yCHCl(CH_2)_2CH_3 \longrightarrow CH_3(CH_2)_yCH=CHCH_2-CH_3 + HCl$$

Linear internal olefins
$(C_{10}-C_{18})$ (11)

A process for converting C_7-C_{15} n-paraffins to the corresponding molecular-weight-range monoolefins is described in Ref. 103. This process involves photochlorinating the n-paraffin and dehydrochlorinating the resultant chloro-n-paraffin to produce monoolefin, illustrated by Eqs. (10) and (11). The monoolefin so produced is predominantly linear internal olefin containing some linear α-olefin, but may be recycled in an isomerization zone to produce the selected type olefin whether it be the internal or α-type.

The photochlorination step is carried out in the presence of uv radiation of approximately 2500-6000 Å at a temperature of approximately 40-230° F. Several photochlorination zones are used. The temperature is maintained relatively low and the amount of chlorine present in each zone is kept considerably below the stoichiometric ratio to minimize production of di- and polychlorohydrocarbons.

After fractionation to recover unconverted n-paraffin and such higher chlorinated hydrocarbons as were formed, the chloro-n-paraffins are then passed to the dehydrochlorination operation wherein HCl is split off in the presence, of high-surface oxidized carbon or diatomite in conjunction with a ceramic binder at a temperature of approximately 750-850° F [103]. The α-olefin formed may then be separated by fractionation from the internal olefins. The latter are then passed to an isomerization zone wherein they are isomerized to the desired α-olefin. This isomerization may be accomplished by the hydroboration-displacement technique described in Refs. 103 and 104.

When internal olefins are the desired products, the separated α-olefin may be isomerized in a packed column, the packing of which may contain catalysts such as kaolin, diatomaceous earth, bauxite, activated alumina, silica gel, or the like. These materials or inert support such as pumice impregnated with a slight amount of sulfuric or phosphoric acid or with acid phosphate, sulfate, etc., may also be used [103].

Another dehydrochlorination process is described in Ref. 105. In this case $C_{10}-C_{15}$ chloro-n-paraffins are converted principally to the corresponding internal olefins by contacting with a solid acidic dehydrochlorination catalyst composed at least partially of alumina, at a temperature range of 425 to 550° F at super atmospheric pressure below 100 psig. An inert gas such as nitrogen is introduced into the dehydrochlorination reactor in an amount necessary to reduce the partial pressure of HCl to less than 35% of the total pressure.

Thermal halogenation of straight-chain paraffins of at least nine carbon atoms, followed by catalytic dehydrohalogenation in the presence of an aromatic diluent is described in Ref. 106. The linear internal olefin-n-paraffin mixture so obtained is suitable for producing biologically soft detergent alkylate by alkylation of benzene.

b. Catalytic Dehydrogenation of n-Paraffins

The chemical reaction involved in this process is illustrated by Eq. (12).

$$CH_3(CH_2)_x CH_3 \longrightarrow CH_3(CH_2)_y CH{=}CH(CH_2)_z CH_3 + H_2 \qquad (12)$$

Linear paraffins   Linear internal olefins (C_{10}-C_{18})
(e.g., C_{10}-C_{18})

References 107-113 describe a process for producing linear internal olefin by catalytic dehydrogenation of linear paraffinic hydrocarbons with various catalyst compositions and under operating conditions particularly suited to the process claimed.

In Ref. 107 it is pointed out that some catalytic composites are so active that certain undesirable side reactions (e.g., cracking, diene formation, or isomerization) are promoted even at low temperatures. Others are insufficiently active at low temperatures to promote an appreciable degree of dehydrogenation. The life of certain catalysts is too short to be practical for an economic process. A preferred catalyst composition may comprise as a carrier an alkylized alumina which is combined with a Group VIII metallic component, e.g., platinum. Another catalytic component used is an "attenuator," the function of which is to poison the platinum to the extent that its residual cracking activity is almost completely curtailed and its tendency to promote side reactions, particularly cyclization, virtually eliminated. The attenuator is selected from the group consisting of arsenic, antimony, and bismuth.

Dehydrogenation may, for example, be effected with an approximately C_{10}-C_{20} mixture of n-paraffins by reacting it with hydrogen and approximately 400-3000 ppm of water in contact with the catalyst under dehydrogenation conditions, such as a pressure above about 10 psig and a temperature range of 750-1110° F.

Reference 108 describes a process applicable to catalytic dehydrogenation of about C_3-C_{20} linear paraffins. High yields of desired corresponding linear monoolefins are obtained under specified reaction conditions. A catalyst composite of crystalline alumino-silicate chemically combined with a metal subfluoride vapor is employed.

Reference 109 relates to a conversion process for dehydrogenation of about C_3-C_{20} paraffinic hydrocarbons in the presence of hydrogen and 0.5-2.0 moles of benzene per mol of paraffinic hydrocarbon. The catalyst and dehydrogenating conditions are about the same as described in Ref. 107, but no water is present.

Use of a dehydrogenation catalyst composite similar to that described in Ref. 107 is described in Ref. 110. The improvement is in using the catalyst in a particular size with a maximum dimension not exceeding one-thirty-second inch.

The process of Ref. 111 covers catalytic dehydrogenation of C_{10}-C_{18} linear paraffins to the corresponding carbon number linear internal olefins without skeletal rearrangement. The catalyst composite is arsenic-attenuated platinum on lithiated alumina in a fixed bed. Dehydrogenation is accomplished at pressures of 10 psig or higher and an 8:1 hydrogen to hydrocarbon mole ratio in a temperature range of 400 to 600° C.

All catalytic dehydrogenation processes employing for example the type catalyst component described in Ref. 107 result in formation of at least some diolefins of the same carbon numbers as the desired linear internal monoolefins. The process of Ref. 112 includes a method for eliminating the conjugated diolefin from admixture with the monoolefin. This is accomplished by treating the mixture with ethylene under conditions affecting its adduct formation with the conjugated diolefin. The adduct may be separated from the monoolefins by fractional distillation.

Universal Oil Product Company's Olex process is a combination catalytic dehydrogenation and selective extraction described in Ref. 113. Briefly, the process involves less than complete catalytic dehydrogenation of linear paraffins to linear internal olefins and hydrogen, followed by separation of a small amount of light ends by fractionation. The unconverted linear paraffin-monoolefin mixture is separated by selective adsorption in the liquid phase in a fixed bed of selective adsorbent. The desorbent stream is a hydrocarbon boiling lower than the feedstock and is recycled continuously as an internal stream to the extraction section.

For example, when employing the Olex process with a C_{11}-C_{14} linear paraffin feedstock containing 98.8 weight % linear paraffins, an extract was made containing 93.5% linear olefins, 87% of the linear paraffins appearing in the extract as linear olefins. About 97% of the linear olefins in the extract were shown to be monoolefins by mass spectrometry. Thus this process provides a means of producing highly concentrated linear internal monoolefins.

Other dehydrogenation catalysts and operating conditions for converting n-paraffins of various molecular-weight ranges to the corresponding carbon numbered linear internal monoolefins are described in Refs. 114-121.

c. Dimerization Processes

References 122-125 relate to processes for preparing detergent-range linear-type olefins suitable for alkylation of benzene to produce biologically degradable detergent alkylate.

References 122-124 describe two-stage processes for catalytically dimerizing normally gaseous monoolefin hydrocarbons to a C_8-C_{16} olefin product which is relatively high in content of linear isomers and homologues.

Reference 125 relates to a catalytic polymerization process for producing dimers, trimers, etc., of ethylene, propylene, 1-butene, 1-pentene, 1-hexene, or their mixtures. For example, if the liquid dimer product olefins are of high enough molecular weight, they may be used directly for alkylation of aromatics which then may be converted to biodegradable anionic surfactants.

2. Linear α-Olefins

As the name implies, the carbon chain structure of the linear α-olefin molecule is straight and the only double bond is located in the α- or 1-position. A typical example is 1-dodecene.

Linear α-olefins may be produced by the following methods:

Polymerization of ethylene (growth reaction)
Catalytic polymerization of ethylene
Cracking of suitable paraffinic petroleum fractions
Dehydration of linear alkanols

a. Polymerization of Ethylene (Growth Reaction)

This method for synthesizing linear α-olefins was discovered in the early 1950s by Karl Ziegler of the Max Planck Institute in West Germany. He found that straight-chain alkyl groups could be polymerized from ethylene by a method involving alkylaluminum compounds as intermediates.

Later a process was developed by the Continental Oil Company based on Ziegler's discovery. The Company's registered trademark for the olefin products is Alfene; details of the process, then in the semiworks stage, are disclosed in Ref. 126. (See also Sec. VI,A below relating to Continental Oil Company's Alfol alcohol process, the early processing stages of which are the same as those employed in the Alfene olefin process.)

In the late-1960s Gulf Oil Company was the only manufacturer of linear α-olefins employing the ethylene polymerization method. Although the olefin products are of similar high quality as those made for example by the Alfene olefin process, Gulf's process is believed to be one in which the alkylaluminum compound is used catalytically rather than as a reactant.

The Ethyl Corporation has a process for the production of linear α-olefins [127]. The production scheme is believed to be based on polymerization of ethylene in a growth-reaction-type operation of its own development.

The processing scheme described here involves ethylene, aluminum, and hydrogen as the basic raw materials. There are five steps: first, the preparation of triethylaluminum; second, the controlled addition of ethylene to triethylaluminum which has been named the "growth reaction"; third, the "displacement reaction" by which the linear alkyl chains of the growth product attached to the aluminum atoms are displaced thermally with ethylene or by catalytic reaction with ethylene; fourth, the recovery of triethylaluminum by formation and subsequent decomposition of a complex; and fifth, fractionation of the olefins into the desired fractions.

The simplified flow diagram shown in Fig. 3 illustrates a processing scheme for producing linear α-olefins by polymerizing ethylene in a growth reaction. It is similar to the Alfene olefin processing scheme illustrated and described in Ref. 126, except that in Fig. 3 the triethylaluminum is shown as being recovered for recycle by formation of a complex and subsequent decomposition.

In the first step, preparation of triethylaluminum may be accomplished by reaction of diethylaluminum hydride with ethylene or by direct reaction of aluminum metal with hydrogen and ethylene.

The conventional preparation of diethylaluminum hydride involves the reaction of high-purity aluminum with triethylaluminum and hydrogen at elevated temperature. This method is described by Karl Ziegler in Ref. 128 and illustrated by Eq. (13).

$$Al + 3/2\ H_2 + 2\ Al(C_2H_5)_3 \longrightarrow 3\ Al(C_2H_5)_2H \qquad (13)$$

$$\text{Triethylaluminum} \qquad\qquad \text{Diethylaluminum} \\ \text{hydride}$$

The diethylaluminum hydride is then reacted with ethylene to produce triethylaluminum, as shown by Eq. (14).

$$3\ Al(C_2H_5)_2H + C_2H_4 \longrightarrow 3\ Al(C_2H_5)_3 \qquad (14)$$

$$\text{Diethylaluminum} \quad \text{Ethylene} \qquad \text{Triethylaluminum} \\ \text{hydride}$$

The net reaction of the foregoing two reactions is expressed in Eq. (15).

$$Al + 3/2\ H_2 + 3\ C_2H_4 \longrightarrow Al(C_2H_5)_3 \qquad (15)$$

The formation of diethylaluminum hydride, as described in Ref. 128 and illustrated by Eq. (13), is rather slow. Instead of aluminum, the use of aluminum alloys containing catalytic elements such as titanium, zirconium, niobium, vanadium, hafnium, scandium, and ytterbium increases the reaction rate considerably [129]. Such alloys may, for example, contain at least 98.0% by weight of aluminum and at least 5 ppm of a catalytic element such as those named above. Alloys of this type containing about 0.02-2.0% by

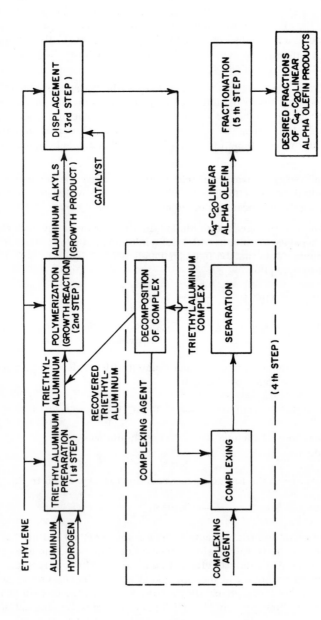

FIG. 3.   Preparation of linear α-olefins by polymerization of ethylene.

weight of the catalytic agent also improve the preparation of triethylaluminum by the direct reaction of aluminum with hydrogen and ethylene.

Reference 130 covers a process similar to that described in Ref. 128 except that aluminum need not be alloyed with catalytic elements to increase its activity. Certain halides of catalytic elements, e.g., titanium tetrachloride, vanadium tetrachloride, etc., by simple admixture with aluminum powder were found to increase the reaction rate for the formation of the dialkylaluminum hydride. Both Refs. 129 and 130 are broader in scope than the preparation of only diethylaluminum hydride.

Manufacture of triethylaluminum and other alkylaluminum compounds by direct reaction of aluminum metal with hydrogen and ethylene is described in Refs. 131 and 132. A specific embodiment of this invention is a process for preparing triethylaluminum by contacting particulated aluminum with sufficient triethylaluminum to wet metal surfaces in a reaction zone. The zone is then heated to a temperature of between 30-130° C under 10-300 atm pressure of a gaseous mixture of hydrogen and ethylene. Equation (16) is illustrative of this direct process.

The second step (see Fig. 3) is the controlled addition of, for example, ethylene to triethylaluminum to form higher-molecular-weight linear trialkylaluminum compounds [133]. This is illustrated by Eq. (16), commonly called the growth reaction.

$$Al(C_2H_5)_3 \; + \; n = C_2H_4 \quad \longrightarrow \quad Al \begin{array}{l} -CH_2CH_2R \\ -CH_2CH_2R_1 \\ -CH_2CH_2R_2 \end{array} \qquad (16)$$

Growth product

This reaction is usually carried out at a temperature range from 100 to 125° C, a pressure varying from about 10 to 300 atm and in the presence of a solvent. It is a highly exothermic reaction. Reference 134 describes a process wherein the heat of reaction can be effectively controlled.

References 135-139 relate to improvements over the prior art concerning the production of higher-molecular-weight alkylaluminums, the alkyl groups of which are of straight-chain configuration.

Reference 140 describes a process for producing alkylaluminums of higher-than-average molecular weight than that which would be produced statistically according to the Poisson distribution equation:

$$P_n = \frac{m^n e - m}{n!}$$

wherein P_n represents the probability that a certain hydrocarbon radical will be formed by n additions of ethylene to the aluminum-ethyl bond originally present and m is the mean number of additions of ethylene per growing chain.

An example of the statistical distribution of alkyl chain lengths provided
by the conventional growth reaction [Eq. (17)] at about m = 5, is illustrated
in Table 7.

References 141-144 also relate to methods by which the chain-length
distribution of the alkyl groups attached to the aluminum can be controlled to
a considerable extent to the desired number of carbons.

In the growth reaction involving triethylaluminum and ethylene to form
higher-molecular-weight alkylaluminums, olefins are a by-product and are
used to produce additional quantities of the desired growth product [145]. The
high-molecular-weight trialkylaluminums can be oxidized to aluminum alkox-
ides and hydrolyzed to linear primary alcohols.

The third step in the production of linear α-olefins by polymerization of
ethylene (see Fig. 3) is commonly referred to as the "displacement reaction."
The alkyl groups of the growth product [see Eq. (16)] may be displaced either
catalytically or thermally, as illustrated by Eq. (17).

TABLE 7

Typical Alkyl-Group Distribution Characterized
by the Poisson Equation

Alkyl group	Weight percent
C_2	0.04
C_4	0.23
C_6	3.37
C_8	11.79
C_{10}	20.42
C_{12}	22.63
C_{14}	18.20
C_{16}	11.61
C_{18}	6.21
C_{20}	3.47
C_{22}	1.25
C_{24}	0.49
C_{26}	0.18
C_{28}	0.05

$$Al \begin{cases} CH_2CH_2R \\ CH_2CH_2R_1 \\ CH_2CH_2R_2 \end{cases} + 3\ C_2H_4 \xrightleftharpoons{Ni} 3\ Al(C_2H_5)_3 + \begin{array}{l} R\ -CH=CH_2 \\ R_1-CH=CH_2 \\ R_2-CH=CH_2 \end{array} \quad (17)$$

| Growth product | Ethylene | Triethyl-aluminum | Linear α-olefins |

This reaction is actually not as simple as shown above because the displacement products, triethylaluminum and α-olefins, tend to undergo a reverse displacement reaction. Also, under the conditions of the reaction there is a tendency for the α-olefins to isomerize at atmospheric pressure. Both the reverse displacement reaction and the tendency of the α-olefins to isomerize have been shown to be accelerated by the reducing-type catalyst employed in the displacement reaction.

Reference 146 describes a means of inhibiting the undesirable reverse displacement reaction and isomerization tendency. In this process, the preferred displacement catalyst is nickel or a nickel compound which reacts with the trialkylaluminum compound. Examples include finely divided metallic nickel, Raney nickel, nickel acetylacetoacetate, and nickel naphthanate. An acetylene alcohol is also added to inhibit reverse displacement and isomerization. Examples of such alcohols are propargyl alcohol, 2-methyl-3-butyne-2-ol, butyne diol, and 1-butyne-1-ol.

When the preferred nickel catalyst is used, its amount may vary from about 0.001 to 0.1% based upon the weight of the growth product present. Generally, not more than 100 parts of the acetylene alcohol is added per part of nickel.

The displacement reaction, as described in Ref. 146 is carried out at a temperature preferably of 90–125° C.

Reference 147 is an example of a process in which the displacement reaction is carried out in the absence of a catalyst. Ordinarily, noncatalytic displacement requires such high temperatures that substantial quantities of internal and branched olefins are produced, which are usually undesirable.

According to the process of Ref. 147, the alkylaluminums (growth product) in finely divided state (atomized) are reacted preferably with a large excess of ethylene in the presence of the diluent ordinarily used in the growth reaction. Other low-molecular-weight α-olefins, such as propylene or 1-butene, may be substituted for ethylene if desired. The reaction is carried out in a preferred temperature range of approximately 400 to 550° F and at pressures varying from 20 to 100 psia, for about 3 to 7 min. Under these conditions, the lower-molecular-weight olefins are preferentially displaced over higher-molecular-weight olefins. This makes it possible to recycle the undisplaced growth product to the reaction system, thereby increasing the quantity of higher-molecular-weight olefins, when desired. A very important feature of this process is atomization of the growth product.

When the growth product is displaced with ethylene, the resulting mixture of linear α-olefins covering a molecular weight range of, for example, from C_4 to C_{20} also contains triethylaluminum. It is desirable from an economic standpoint to recover the triethylaluminum for use in the growth reaction. Since its boiling point is so close to 1-dodecene, it is impractical to separate the two by ordinary methods of fractional distillation. A number or methods have been developed to solve this problem by formation of complexes of the trialkylaluminum compound, which can then be separated from the olefins and decomposed to recover the complexing agent and triethylaluminum [148-152]. This is the fourth step in Figure 3. The fifth step is fractionation of the olefin product.

In Refs. 148 and 149 preparation of alkali metal cyanide complexes of the general formula $Na_xK_y[n Al(C_2H_5)_3CN]$ are described and also claimed as new composition of matter. In this formula $x + y = 1$ and $n = 1.5$ to 2. The thermal (120–180° C) decomposition reactions can be illustrated by Eqs. (18) and (19):

$$Na_{0.5}K_{0.5}[2 Al(C_2H_5)_3CN] \longrightarrow$$

$$Na_{0.5}K_{0.5}[1.5 Al(C_2H_5)_3CN] + 0.5 Al(C_2H_5)_3 \qquad (18)$$

$$Na_{0.5}K_{0.5}[1.5 Al(C_2H_5)_3CN] \longrightarrow$$

$$0.5 NaCN + 0.5 KCN + 1.5 Al(C_2H_5)_3 \qquad (19)$$

In addition to complexing with alkali metal cyanides, both Refs. 148 and 149 also describe other complexing agents for achieving separation and recovery of the trialkylaluminum.

References 150-153 relate respectively to separation techniques employing other trialkylaluminum complexes with, for example, a tetraalkylammonium halide, phosphonium compounds, alkali metal azides, and polymers containing a Lewis base group with elements such as oxygen, sulfur, selenium, nitrogen, phosphorous, or arsenic.

A process for producing α-olefins by catalytically displacing a trialkylaluminum growth product with a monoolefin containing about 2 to 4 carbon atoms is described in Ref. 154. This produces a mixture of linear α-olefins containing, for example, 2 to 40 carbon atoms and a trialkylaluminum containing 2 to 4 carbon atoms. This mixture is then oxidized with an oxygen-containing gas to convert the trialkylaluminums to the corresponding aluminum trialkoxides without affecting the C_2-C_{40} linear α-olefins. The trialkoxides and olefins may then be separated by fractionation. Alternate procedures are described.

A process for the preparation of C_4-C_{20} linear α-olefins by a displacement technique in which higher olefins are displaced from lower-molecular-weight trialkylaluminums with butylene and recycling the latter to the process, is described in Ref. 155.

Reference 156 describes a process for controlling the molecular size of alkyl groups on trialkylaluminums by subjecting triethylaluminum to growth followed by displacement with ethylene. The olefins are then separated from the triethylaluminum formed in the displacement zone by complexing, for example, with tetramethyl ammonium chloride. The complex is broken and the triethylaluminum is passed to a reverse displacement zone. The olefins from the first displacement zone are separated into a light and heavy olefin fraction, the light fraction is used in the reverse displacement zone. The resultant trialkylaluminums are then passed back to the growth-reaction zone. By a particular sequence of steps the alkyl carbon number can be controlled and in recovering olefins no net consumption of triethylaluminum is involved.

A process has been described [157] for recovering olefins from trialkylaluminums by displacement with ethylene, propylene, or 1-butene, the improvement consisting of presaturating the trialkylaluminums with the low-molecular-weight olefin before subjecting the trialkylaluminums to displacement conditions. Presaturation before displacement is carried out under a pressure range of 150 to 1000 psig.

A procedure for displacing alkyl residues of selected molecular weight by a process permitting high displacement percentages and small recirculating streams and also avoiding high-temperature conditions employed in certain prior-art displacement techniques is covered in Ref. 158.

In processes for preparing liquid α-olefins by the growth reaction, a small but troublesome amount of solid polyethylene may be formed which deposits on reactor surfaces. This interferes with heat transfer and may cause plugging, thus necessitating frequent shutdowns for removal of such deposit. References 159-161, respectively, cover the use of phenothiazone, an organic cyanide such dodecyl cyanide, and benzothiazole.

A process in which linear α-olefins, linear primary alcohols, and alumina or alum are produced is described in Ref. 162. The processing scheme includes a first-growth reactor, followed by a catalytic displacement step. Following catalytic displacement, the next step is complexing with sodium cyanide as described above [149]. The olefins from that step are separated and collected as part of the olefin product and the $NaCN-2$ $Al(C_2H_5)_3$ complex is grown in a second growth reactor with ethylene and additional triethylaluminum. From there the "grown" complex may be divided. Part may be passed to a thermal displacement step such as the one described above [147]. The other part of the grown complex may be passed to an oxidation step and converted to liquid trialkoxyaluminums, followed by filtration to separate sodium cyanide which may then be reused in complexing. Olefins separated from the thermal displacement step may be combined as product with those obtained from the complexing and separation step immediately following the catalytic displacement step. The trialkoxyaluminums may then be hydrolyzed with either water or sulfuric acid solution to yield linear primary alcohols and alumina or alum solution.

Reference 165 covers three alternate processing schemes all of which involve the growth reaction, but only one of which involves complexing.

Linear primary alcohols are produced and, additionally, linear α-olefins may be produced, if desired. Features of this invention include a method for utilizing low-molecular-weight olefins to produce linear primary alcohols; a method to alter the distribution curve of a conventional growth process; and a method to increase the proportion of higher-molecular-weight alkyl groups in a growth process at the expense of lower-molecular-weight groups. Operation of the processing schemes is explained and illustrated by three separate diagrams in Ref. 163.

In Ref. 164 S. A. Miller reviews the chemistry involved in the production of Alfol alcohols (linear primary alcohols) and Alfene olefins (linear α-olefins). In addition to flow sheets showing the processing steps in the Alfol alcohol and Alfene olefin processes, one flow diagram of a Russian pilot plant is included. About 60 patent and literature references are cited in that publication.

b. Catalytic Polymerization of Ethylene

Ethylene can be polymerized at elevated temperatures and pressures in the presence of a catalyst to yield a product containing a mixture of α-olefins which contain 4 to 20 carbon atoms or even higher. Examples of such processes are covered in Refs. 165-170.

Reference 165 covers a continuous process for polymerizing ethylene in the presence of hydrocarbon compounds of aluminum type metals selected from Group III-A, such as aluminum, gallium, and indium. An example is triethylaluminum. The polymerization is carried out at an elevated temperature and pressures ranging above 700 psi. Normally the olefin product amounts to more than 50 parts by weight per 1 part by weight of the metal content of the catalyst expressed as triethylaluminum.

Reference 166 also relates to a continuous catalytic polymerization process. A few examples of catalysts suitable for this process are: $Be(C_2H_5)_2$, $Li(C_2H_5)$, $Hal(CH_3)_2$, $Al(C_2H_5)_3$, $Al(C_3H_7)_3$, $NaHAlH_4$, and $Al(C_2H_5)Cl$. The catalyst can be used as such but is usually employed with about 70-99% of its weight of an inert hydrocarbon diluent. The amount of the catalyst employed is not critical but can vary from 0.01 to 0.001 mole per mole of ethylene. The reaction temperature is preferably from about 180 to 220° C. The pressure must be kept high enough at all times (about 140 kg per sig) to maintain the ethylene and the contents of the reactor in a single phase. The product is principally a mixture of α-olefins of carbon chain lengths from C_4 upward.

The continuous process described in Ref. 167 specifies that the catalyst is an ether-free organo-aluminum compound. The reaction temperature and pressure ranges must be high enough so that very high reaction rates are obtained. Curves are given in Ref. 167 which represent the ranges of optimum residence times as a function of pressure. The curves cover operating temperatures at 25° intervals of from 225 to 350° C. The products are predominantly even-carbon-numbered linear primary C_4-C_{20} α-olefins.

In the process described in Ref. 168, ethylene is converted to higher (about C_{10} to C_{20}) α-olefins which are principally linear, by treatment with a catalyst in a hydrocarbon diluent at a pressure of from at least 50 psi, but preferably not more than about 300 psi above the vapor pressure of the diluent at reaction temperature. The catalyst may be prepared, for example, by mixing cerous chloride and butyllithium at the reaction temperature and pressure specified.

Reference 169 relates to polymerization processes in which catalysts are formed by mixing a chromium halide and an organolithium compound. Reference 170 describes a product formed by mixing at least one rare earth metal halide and at least one alkyl magnesium halide.

In the process discussed in Ref. 169 the contact temperature of the catalyst with ethylene is at least 175° C and the product olefins contain about 4 to 10 carbons.

According to Ref. 170 the contacting temperature range is between 150 and 250° C; the olefin product range is C_4 to C_{20}.

An unexpectedly high selectivity towards linear α-olefins is achieved by polymerizing ethylene in the presence of about 0.01 to 0.0001 mole of, for instance, triethylaluminum catalyst per mole of ethylene in a tubular reaction zone in which the amount of polymer increases throughout its length [171]. The reaction temperature is about 180 to 240° C and the pressure is 1500 to 2000 psig. The catalyst is preferably employed with 70-99% of its weight of an inert hydrocarbon solvent.

References 172 and 173 relate to continuous processes for producing linear α-olefins by a combination of catalytic polymerization of ethylene and the growth and displacement procedures previously discussed [see Eqs. (16) and (17)]. By the techniques of these processes it is possible to control to a considerable extent the spread of molecular weights of the α-olefin product and to narrow it down so as to produce a larger amount of the carbon chain lengths desired. This cannot be done by catalytic polymerization alone.

A major problem in all of the α-olefin manufacturing processes described above has been the wide molecular-weight-range of olefin products obtained. Olefins varying from C_4 to C_{30} are produced in all cases.

According to recent patents, Shell Oil has developed a new process for the selective production of detergent-range linear olefins via ethylene oligomerization. In this process, both the lower- and higher-molecular-weight olefin coproduct fractions are converted to the desired detergent range by a combined isomerization and disporportionation scheme.

Two versions of this process are disclosed. In one case [174], the process consists of five steps.

a. Ethylene oligomerization
b. α-Olefin separation

    c. Isomerization and disproportionation of lower molecular weight
       α-olefins

    d. Isomerization of higher molecular weight α-olefins

    e. Ethenolysis

In this scheme ethylene is first oligomerized by a catalytic polymerization to a broad range of α-olefins (C_4-C_{40}). The α-olefins are then separated into light, intermediate, and heavy fractions. The higher-molecular-weight fraction (ex—C_{20+}) is isomerized to internal olefins, then subjected to an ethenolysis step, and disproportionated with ethylene to give additional α-olefins. The lower-molecular-weight fraction is fed to the isomerization-disproportionation step and converted to ethylene and intermediate and high-molecular-weight internal olefins. This mixture is also subjected to ethenolysis for conversion to α-olefins. Thus detergent range α-olefins are the sole product.

In subsequent patents [175, 176] a modification of the above process is disclosed. In this scheme, the desired detergent-range α-olefins are first separated from the ethylene oligomerization product mixture. The remaining lower- and higher-molecular-weight α-olefins are isomerized and mutually disproportionated to internal olefins to give a mixture of internal olefins of both odd and even carbon number. The detergent range is separated and the lower and higher internal olefin fraction recycled back to isomerization. Thus the net product of this process consists of an even carbon number α-olefin stream and a mixture of even and odd carbon number internal olefins, both streams being in the desired detergent range, for example, C_{10}-C_{20}.

The Shell Oil Co. is reportedly constructing a commercial plant utilizing the above technology for the manufacture of linear olefins. A number of Shell patents have been published covering the various steps of the olefin process.

The ethylene oligomerization step is described in several patents [177-183]. The catalyst claimed in the earlier references is a nickel chelate compound obtained by reaction of a zero valent nickel compound, such as bis-1,5-cyclooctadiene nickel, with phosphine compounds [177-179], or with fluorine compounds [180]. Later patents [181-183] disclose preparation of the catalyst in a polar solvent in the presence of ethylene contacted with a divalent nickel salt, a boron hydride reducing agent, and a phosphorous containing ligand.

Ethylene oligomerization to α-olefins with the above type catalyst system is conducted at temperatures in the 50-90°C range, and typically at pressures of 400-1500 psig.

α-Olefins of 93% and higher purity are reported by these processes.

Isomerization. Suitable isomerization catalysts cited by Shell for converting α to internal olefins [174, 176] include supported phosphoric acid, bauxite, iron oxide, and magnesium oxide. More recent patents disclose the use of potassium carbonate [184] and sodium potassium carbonate [185] on alumina. Isomerization is carried out at 20-100°C in the liquid phase.

Isomerization-disporportionation. Some catalysts suitable for simultaneous isomerization and disproportionation include rhenium oxide and potassium oxide on alumina [174, 186], molybdenum trioxide, cobalt oxide, and magnesium oxide on alumina [174, 187], and molybdenum trioxide, cobalt oxide, palladium oxide, and potassium oxide on alumina [188].

Suitable reaction temperatures vary from 25 to 300° C, and reaction pressures from 1 to 80 atm, depending on the catalyst used and the composition of the feed.

Ethenolysis of internal olefins to α-olefins. Ethenolysis is carried out in the liquid phase in the presence of a conventional disproportionation catalyst. According to the basic Shell patent [174], the preferred ethenolysis catalysts are rhenium oxides on alumina support.

c. Cracking of Paraffin Wax

It has been known for at least a century that unsaturated hydrocarbons can be produced by thermally cracking petroleum fractions containing paraffin hydrocarbons. However, thermal cracking was not commercialized until fairly early in the twentieth century.

About one-half of the total crude capacity of U.S. oil refineries was devoted to thermal cracking processes just before World War II, when catalytic cracking processes began to replace many of the older thermal processes. The catalytic process was developed to meet the heavy and increasing demand for high-octane gasoline.

Low-boiling olefins such as ethylene to C_4-olefins are produced in refineries today in large volume by thermally cracking naphtha and low-boiling paraffins. Also, suitable heavier paraffinic fractions of petroleum, in which the normal paraffin content usually has been maximized by, for example, a molecular-sieve adsorption or urea-complexing process, are thermally cracked to produce an olefin product containing both odd and even numbers of carbons from about C_5-C_{20}. These olefins predominate in the linear α-homologues. Those falling in the range of about C_{11}-C_{18}, and more particularly from about C_{14}-C_{18}, are of most interest as surfactant intermediates.

Although higher linear α-olefins may be produced by catalytic cracking of suitable paraffinic hydrocarbon feedstocks, the catalytic process is practiced to a minor extent. The yield of olefins above about C_{10} is limited and the purity of the straight-chain α-olefins is lower than in thermal processes. Therefore, the catalytic process will not be covered here.

The following paragraphs of this subsection are devoted principally to descriptions of thermal processes for cracking suitable high-boiling paraffinic fractions of petroleum and processes for purification of olefins so produced to maximize the content of straight-chain α-monoolefins. A number of patent references illustrate methods for producing predominantly linear α-olefins having alkyl chains of from about C_4-C_{20} [189-192].

The vapor-phase thermal cracking of selected petroleum-derived waxes at about 1000-1050° F at atmospheric pressure for at least 7 sec and recycling of higher-molecular-weight olefins yields predominantly C_5-C_{20} n-α-olefins [189]. The yield of that range olefin product is claimed to be in excess of 60%. Another feature of the process is that of selectively producing a C_{10}-C_{20} olefin product containing in excess of 95% of the α-type with less than 2% internal olefins. This is accomplished by selecting as feedstock only those hydrocarbon waxes having a rather narrow melting point-refractive index relationship.

A process to produce predominantly linear α-olefins by vapor-phase steam cracking petroleum wax is described in Ref. 190. Best results are obtained with a paraffinic wax having a melting-point range of about 130-140° F and an oil content preferably of less than about 1% by weight. The molten feedstock and at least 2% (preferably about 3-40%) by weight of steam are mixed and fed to a cracking zone at about 900-1200° F. Careful control of conditions in the cracking reactor must be maintained in order to achieve a satisfactory cracking rate and to minimize side reactions. The temperature must be kept at least at 1000° F but should not exceed 1250° F with about 1000-1175° F being preferred. A pressure range of 1 to 3 atm is satisfactory. In general, a residence of 0.2-6.0 sec is satisfactory with 0.3-3.0 being pre-ferred. The resulting product is cooled in less than 2 sec to about 600-800° F and passed to a flash tower to remove C_{12}-C_{20} and lighter hydrocarbons. The remainder of the cooled product is then removed and a portion is further cooled to 300-500° F and recycled to the flash tower. This constitutes the sole means by which the cracked products from the cracking reactor are cooled before their entry into the flash tower.

Reference 191 relates to another process of steam cracking paraffinic feedstocks to produce predominantly linear α-olefins suitable for alkylation of aromatics to produce detergent alkylate. The salient feature of this proc-ess lies in recycling the liquid olefinic product of below C_{11} to the cracking zone to increase the yield of C_{11} and higher olefin products. These may be separated into fractions as desired.

Linear α-olefins produced by thermally cracking such feedstocks as petroleum wax, petrolatum, or gas oil will be contaminated to a greater or lesser extent, depending on the specific process used, with certain undesir-able impurities. Such impurities are nonlinear and cyclic monoolefins, aromatics, and linear dienes. Urea extraction separates most of these but is not particularly effective in separating dienes.

References 192-194 describe methods of purifying olefins obtained by thermally cracking paraffinic feedstocks. The process of Ref. 192 relates to a two-stage treatment for refining olefins obtained by thermal cracking of paraffinic feedstocks so as to markedly reduce the percentage of all impuri-ties listed in the preceding paragraphs, including linear dienes. The first-stage treatment generally consists of a continuous operation in which the olefins from the cracking zone are being percolated through a fixed bed of a nonacidic catalyst such as silica gel. The preferable temperature range is 50-150° F at pressures from atmospheric to 100 psig. At least 70% or more of the nonconjugated dienes are thus converted to monoolefinic impurities which may then be substantially removed by urea adduction by processes such as those disclosed in Ref. 192.

A process for separating α-olefins from mixtures with nonolefinic hydrocarbons and certain branched-chain olefins is described in Ref. 193. The basis of the process lies in the preferential reaction of linear α-olefins with dialkyl-aluminum hydride. The latter is converted to trialkylaluminum which is separated from non-α-olefin materials, recovered, and heated to a temperature at which it decomposes into dialkylaluminum hydride and linear α-olefins. The olefins are then separated from the hydride and the latter recycled to react again with **α-olefins** admixed with other hydrocarbons and the process is repeated again and again.

Reference 194 relates to a process for selectively purifying and separating linear α-olefins from mixtures with other hydrocarbons from which they cannot be effectively separated by ordinary means. Another feature of this process is that the separation is made under conditions which inhibits isomerization of the α-position of the double bond to internal position. The separation is accomplished by employing a sorbent material, such as natural or synthetic faujasites, which are pretreated with steam and dried. Mineral faujasites are large-pore crystalline alumino-silicate zedites. Synthetically produced faujasites are available and are preferred. The pore size is critical and pore diameters of preferably 8-15 Å units are particularly effective for cracked-wax olefin feeds. The branched and cyclic monoolefins, diolefins, and aromatics are adsorbed on the sorbent and a stream of purified C_6-C_{30} linear α-olefins pass through and can be separated by fractionation into the desired fractions. The other hydrocarbons may be desorbed from the faujasite with steam or ammonia.

Reference 195 discloses a cracking process for producing linear-type olefins suitable for alkylation of benzene to produce soft-detergent alkylates. The feedstock employed for cracking may be of any origin. For example, such feedstocks may contain 30-35% aromatics and the total content of nonlinear hydrocarbons including aromatics, naphthenes, and isoparaffins may be higher than 85% by weight. The cracking conditions are chosen in such a manner as to maximize the content of linear olefins in the product. Purification is effected in two steps: First, a stream rich in C_8-C_{15} olefins is separated from the cracked product by fractionation. The C_8-C_{15} fraction is then treated with urea to selectively form adducts of urea with the olefins. The adducts are then separated and decomposed to yield the linear olefin product of ultimate purity accorded by the whole process. The urea may be recovered.

d. Catalytic Dehydration of Linear Primary or Secondary Alcohols

Linear primary or secondary alcohols may be converted to linear α-olefins by catalytic dehydration as illustrated by Eqs. (20) and (21).

$$RCH_2CH_2CH_2OH \xrightarrow[\text{Heat}]{\substack{\text{Catalyst} \\ \text{(e.g., alumina)}}} RCH_2CH = CH_2 + H_2O \qquad (20)$$

Linear primary   Linear α-olefin
   alcohol

$$RCH_2CHOHCH_3 \xrightarrow[\text{Heat}]{\substack{\text{Catalyst} \\ \text{(e.g., thoria)}}} RCH_2CH = CH_2 + H_2O \qquad (21)$$

Linear secondary   Linear α-olefin
   alcohol

Linear α-olefins of about C_8-C_{18} molecular-weight range can be made from linear primary alcohols of the corresponding chain lengths by catalytic vapor or liquid-phase dehydration illustrated by Eq. (20). Catalysts such as γ-activated alumina [196, 197] and phosphoric acid [198] are reported as being very efficient dehydration catalysts when used under specified conditions. Olefins of more than 95% purity as the linear α-type have been obtained.

About 3-million pounds per year of linear α-olefins were formerly manufactured by the Archer Daniels Midland Company by catalytic dehydration of linear primary alcohols (mainly C_8-C_{18}) produced from natural fats and oils. The relatively high cost of such alcohols and the small plant capacity resulted in unfavorable economics as compared with newer and much larger plants employing processes based on cracking of paraffin wax and the polymerization of ethylene. Operation was therefore discontinued.

Secondary alcohols such as 2-butanol, 2-hexanol, and up to 2-octadecanol may be converted to linear α-olefins by selective catalytic dehydration. A specially prepared thoria catalyst is generally preferred, although other oxides of the lanthanide series of rare-earth elements may be employed. The alcohol must be one that is not too difficult to vaporize at or below the preferred temperature range of the reaction which is 350-450°C.

3. Branched-Chain Olefins

   a. Polymerization of Propylene and Higher Olefins

From a historical standpoint the trimer, tetramer, and pentamer of propylene are of primary importance for the manufacture of intermediates for synthetic surfactants. But since surfactants based on such branched-chain monoolefins are not biodegradable, their consumption has greatly decreased since 1965.

This subsection will include a brief review of selected processes by which the aforementioned highly branched monoolefins may be produced.

As previously mentioned (Sec. II,A), a propylene-propane stream resulting from the catalytic cracking process was catalytically polymerized to produce polymer gasoline. In some instances the propylene-propane stream may have been mixed with a butylene-butane stream as feedstock to produce polymer gasoline, but only the C_3 stream is considered here.

A widely used process for the catalytic polymerization of propylene employs a phosphoric acid catalyst deposited on kieselguhr. A process employing such a catalyst (manufactured by Universal Oil Products Co.) is described in Ref. 199. A process flow diagram shows the arrangement for the polymerization of propylene-butylene mixtures to polymer gasoline [199]. When producing propylene dimer (C_6 olefins), trimer (C_9 olefins), and tetramer (C_{12} olefins), a slight operation modification and a rerun column are required. The debutanizer shown in the diagram [199] is not required. The feed stream is contacted with the catalyst at 350-435° F at 400-1200 psig.

In this type of process the selectivity to form higher-molecular-weight tetramer and pentamer leaves something to be desired. Dodecene and tridecene were the principal propylene polymers used to produce detergent alkylates but were not actually pure C_{12} and C_{13} olefins. Rather they were fractions predominating in dodecene mixed with lower (e.g., C_9) and higher (e.g., C_{15}) propylene polymers in such amounts as to average the molecular weight of dodecene or tridecene, respectively.

A process relating to the catalytic polymerization of propylene under conditions which yield higher than previously attainable selectivity of polymers not exceeding C_{15}, especially dodecene and pentadecene, is described in Ref. 200. The catalyst is a tungsten containing heteropoly acid deposited on a preferred support of silica gel or silica alumina. The temperature of the polymerization reaction is below 300° F. A refinery propane stream is suitable as feedstock for the process. The maximum yield of dodecene is obtained between 250 and 285° F, while the maximum yield of pentadecene is obtained at slightly higher temperatures but still less than 300° F. In one preferred embodiment of the invention, the dodecene and pentadecene fractions are said to be substantially free of diolefins. The polymerization temperature in this case is reported to be between 250 and 275° F.

References 201-205 relate to liquid-phase catalytic processes for polymerization of C_3- to C_{12}-monoolefins, and especially C_3- to C_4-olefins with high selective yields to the dimer, trimer, tetramer, pentamer, hexamer, and higher-molecular-weight homologues with high conversions of olefin. A different catalyst and somewhat different operation conditions are employed in each case.

According to the process of Ref. 201 for polymerizing propylene, the catalyst consists essentially of 3-15% by weight of zinc fluoride supported on activated alumina. The polymerization is carried out between 70 and 180° F at a pressure of 200-800 psig.

For polymerizing propylene the catalyst of Ref. 202 is 3-15% by weight of boric oxide supported on activated alumina. The reaction conditions are the same as those stated above for Ref. 201.

Again, the process of Ref. 203 is similar to those of Refs. 201 and 202. The reaction temperature range is the same for propylene, but the pressure is between 200 and 1000 psig. The catalyst in this case may, for example, consist of essentially an activated alumina support, 2-20 weight % of zinc fluoride or boric oxide and between 0.1 and 20 weight % of a metal or oxide of platinum or nickel.

Reference 204 covers a process for polymerizing (dimerizing) a wide range of either linear or branched-chain monoolefins. For example, a feed-stock of catalytic polymer gasoline consisting generally of branched-chain C_5-C_{12} olefins or a dodecene fraction thereof may be used. A preferred catalyst may be from 0.15 to 0.3 parts per part of the olefin of 95-100% sulfuric acid. The olefin dimers so produced are used to alkylate benzene. The higher-molecular-weight alkylate is then sulfonated to produce oil-soluble sulfonates.

V. PROPERTIES OF HIGHER ACYCLIC HYDROCARBONS

A. n-Paraffins

The series of n-paraffins occurring in crude petroleum starts with ethane (C_2) and includes homologues containing both odd and even numbers of carbon atoms up to at least C_{40}. The range covered here is narrowed to C_8-C_{20} which is still somewhat broader than necessary to include the spread most important to the manufacture of anionic surfactants.

Each individual n-paraffin in the C_8-C_{16} range is a water-white liquid at room temperature. Those of higher-molecular-weights are waxy white solids.

From the standpoint of fire hazard, the lower members have relatively low flash points as noted in Table 8. Although the fire hazard decreases with increasing molecular weight, all liquid n-paraffins should, as a general precaution, be kept away from heat and open flame. Also care should be taken to avoid breathing the vapors for prolonged periods.

High-purity individual n-paraffins are not generally used commercially in large volumes. Tables 9A and 9B show some physical properties of a C_{10}-C_{14} n-paraffin mixture representative of a composition and quality produced in a UOP Molex unit as the first step in the manufacture of LAB (linear alkylbenzene).

The most important application for high-purity n-paraffins is as starting materials for the manufacture of a number of different surfactant materials. These are listed in Fig. 4 which also indicates the routes by which the surfactant materials are derived from n-paraffins.

The sodium alkanesulfonates shown in Fig. 4 are produced by Hoechst in West Germany [207]. This is an anionic surfactant having detergent properties comparable to linear alkylbenzenesulfonates.

TABLE 8

Physical Properties of C_8-C_{20} n-Paraffins

n-Paraffin	Formula	Bp at 1 atm, °C[a]	Flash pt, °F[a]	Ref. index n_D^{25}	Density at 25°C, g/ml	Freez. pt, °C[b]
Octane	C_8H_{18}	125.7	60	1.39505	0.6985[c]	-56.8
Nonane	C_9H_{20}	150.8	88	1.40311	0.7138[c]	-53.5
Decane	$C_{10}H_{22}$	174.1	115	1.40967	0.7263[c]	-29.7
Undecane	$C_{11}H_{24}$	185.9		1.41500	0.7366[c]	-25.6
Dodecane	$C_{12}H_{26}$	216.3	165	1.41949	0.7452[c]	- 9.6
Tridecane	$C_{13}H_{28}$	235.4		1.4234	0.7528	- 5.4
Tetradecane	$C_{14}H_{30}$	253.6	212	1.4268	0.7593	+ 5.9
Pentadecane	$C_{15}H_{32}$	270.6		1.4298	0.7650	9.9
Hexadecane	$C_{16}H_{34}$	286.8		1.43250	0.770[c]	18.2
Heptadecane	$C_{17}H_{36}$	301.8		1.4348	0.7745	22
Octadecane	$C_{18}H_{38}$	316.1		1.4369[d]	0.7785[d]	28.2
Nonadecane	$C_{19}H_{40}$	329.7		1.4388[d]	0.7821[d]	32.1
Eicosane	$C_{20}H_{42}$	342.7		1.4405[d]	0.7853[d]	36.8
Reference		205	206	205	205	205

[a]Closed cup.
[b]In air at 1 atm.
[c]Rounded from figures in Ref. 205.
[d]For the undercooled liquid below the normal freezing point.

TABLE 9A

Typical Physical Properties of a C_{10}-C_{14} n-Paraffin Mixture
from a Commercial-Scale U. O. P. Molex Unit

Property	Value
Purity as n-paraffins, %	99.0
Aromatics content, %	0.1
Bromine number	0
Specific gravity at 15° C	0.7499
Color, Saybolt	+30
Freezing point, ° C	−20

TABLE 9B

Molecular-Weight Distribution[a] of the n-Paraffin
Mixture in Table 9A

n-Paraffin	Percent by weight
$C_{10}H_{22}$	10.6
$C_{11}H_{24}$	37.8
$C_{12}H_{26}$	31.4
$C_{13}H_{28}$	19.8
$C_{14}H_{30}$	0.4

[a]By gas chromatography.

Since higher acyclic monocarboxylic acids (synthetic fatty acids) are not produced commercially from n-paraffins in the United States, they are not included in Fig. 4. The technology for producing such acids by oxidation of n-paraffins has been developed and is practiced in 20 or more commercial-scale plants in the Soviet Union and several in other East European countries.

n-Paraffins also serve as chemical intermediates for commercial-scale production of other products used outside the surfactant field.

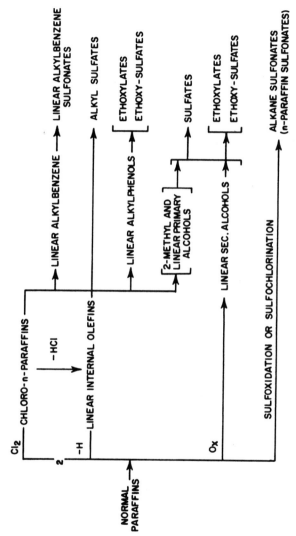

FIG. 4. Principal surfactants derived from n-paraffins in the United States.

B. Detergent-Range Monoolefins

1. Linear Olefins

Typical physical properties of linear α-olefins produced by the Chevron Chemical Company by its wax cracking process are presented in Table 10. Some of the linear α-olefins produced by the Gulf Oil Company via its ethylene growth process are described in Table 11.

Linear α-olefins in the C_8-C_{18} range are liquids at ordinary room temperature. Above C_{18} they are solids. The C_8 linear olefin (e.g., 1-octene) has a flash point of below 80° F and, therefore, requires a red I.C.C. warning label for flammable materials. The hazard from fire decreases for higher homologues, but all linear olefins in the C_8-C_{20} range should, as a general precaution, be kept away from heat and open flame. Care should also be taken to avoid breathing the vapors for prolonged periods.

Linear olefins of both the α- and internal types are versatile chemicals. An excellent survey of more important reactions of olefins with emphasis on more recent developments of commercial or potential significance (excepting the oxo reaction and polymerization) is given in Ref. 210.

The largest volume consumption of linear olefins is in the production of intermediates for synthetic detergents and plasticizers, particularly mixed 2-methyl and linear primary alcohols via the oxo route and LAB (linear alkylbenzene). This type oxo alcohol may be 70-80% linear and is usually produced from either linear α-olefins or internal olefins. Between 20 and 30% may be the 2-methyl branched isomers. It is the C_7-C_{10}-range alcohols which serve as intermediates for plasticizers. The C_{11}-C_{18}-range alcohols are used as intermediates for the manufacture of synthetic surfactants such as oxyethylated alcohol nonionics, ether-alcohol sulfates and alcohol sulfates.

Besides the well-established, large-volume applications of linear olefins mentioned above, there are many of lesser current importance under development. Only a few will be discussed here, but mention is again made of Ref. 210, as a good source of more information about reactions of olefins. The technical literature of linear-α-olefin manufacturers [208, 211] also presents valuable information and references pertinent to many applications in various stages of development.

In the surfactant field, one potentially important use of linear α-olefin is its sulfonation with sulfur trioxide to produce an olefin sulfonate. Because of objectionable impurities inherent in linear α-olefins made by wax cracking, only the high-purity linear α-olefins produced by controlled ethylene polymerization yield a sulfonate of acceptable quality without bleaching with an agent such as sodium hypochlorite.

α-Olefin sulfonates are rapidly and practically completely degraded as demonstrated by the river die-away and shake-culture tests.

TABLE 10

Typical Physical Properties of Chevron Linear α-Olefins [208]

Property	C_6-C_9	C_{10}	$C_{11}-C_{14}$	$C_{15}-C_{18}$	$C_{18}-C_{20}$	$C_{15}-C_{20}$
Straight-chain mono-α-olefins, wt. %	89	90	89	91	86	88
Diolefins, wt. %	4	5	6	8	4	5
Paraffins, wt. %	3	2	1	2	9	5
Appearance	Clear and bright and free of sediment					
Color, Saybolt	+18	+17	+14	+ 7	-16	-12
Density (20°/4°C), g/ml	0.713	0.751	0.770	0.783	0.797	0.787
(60°/60°F), lb/gal	5.95	6.27	6.42	6.57	6.68	6.60
Flash point, TOC, °F	30	103	162	260	330	280
Pour point, °F	--	--	-20	+40	+70	+55
Bromine number, g/100 g	165	118	98	73	57	67
Water content, ppm	130	130	130	80	40	50
Sulfur content, ppm	5	8	10	15	15	15
Carbon number distribution, wt. %						
C_5	2					
C_6	39					
C_7	24					
C_8	17					

TABLE 10 (Continued)

Property	C$_6$-C$_9$	C$_{10}$	C$_{11}$-C$_{14}$	C$_{15}$-C$_{18}$	C$_{18}$-C$_{20}$	C$_{15}$-C$_{20}$
Carbon number distribution, wt. % (Continued)						
C$_9$	16	4				
C$_{10}$	2	95				
C$_{11}$		1	27			
C$_{12}$			24			
C$_{13}$			24			
C$_{14}$			23	1		1
C$_{15}$			1	29		17
C$_{16}$				28		18
C$_{17}$				27	1	17

C_{18}			14	23	17
C_{19}			1	37	15
C_{20}				30	12
C_{21}				9	3
Average molecular weight	100	140	174	228	269	244
Distillation[a], °F	D-86[a] @ 760 mm Hg			D-1160[a] @ 10 mm Hg		
Start	148	314	169	286	352	288
5%	161	325	178	293	360	295
95%	284	328	246	343	395	367
End point	315	343	356	350	406	406

[a]ASTM method.

63

TABLE 11

Typical Physical Properties of Gulf Linear α-Olefins [209]

Product characteristics	Method	C_8	C_{10}	C_{12}	C_{14}	C_{16}	C_{18}	C_{20}
Specific gravity, 60°/60° F	D 1298[a]	0.719	0.745	0.763	0.776	0.785	--	--
Flash point, °F		<60	120	180	225	270	310	320
Water content, wt. %	Gulf	0.01	0.01	--	--	--	--	--
Linear α-olefins, wt. %	Gulf	96.1	95.2	94.0	93.0	92.0	90.8	88.8
Monoolefins, wt. %	Gulf	98.6	98.6	98.6	98.6	98.6	98.6	98.6
Saturates	Gulf	1.4	1.4	1.4	1.4	1.4	1.4	1.4
Carbon number, wt. %	Gulf	99.7	99.4	99.3	98.9	98.5	98	96.0
Distillation	D 1078[a]							
5%, °C		118	164	205	240	270	--	--
95%, °C		128	175	220	255	300	--	--
Freezing point, °C	D 1015[a]	--	--	--	--	4	18	29

[a]ASTM method.

α-Olefins of most interest for sulfonation are usually various blends of two or three homologues in the C_{14}-C_{20} range.

Another readily biodegradable anionic surfactant may be produced from linear α-olefins by addition of an alkaline bisulfite under free-radical conditions. These are primary alkanesulfonates produced from a suitable blend of linear α-olefins, e.g., the C_{14}-C_{16} olefins. They possess rather low solubility characteristics as compared with α-olefin sulfonates. The primary alkanesulfonates appear more attractive in syndet toilet bars rather than in laundering and dishwashing products [211].

2. Branched-Chain Olefins

Tripropylene and tetrapropylene are prime examples of highly branched-chain monoolefins which, until the advent of biologically soft synthetic detergents in 1965, were widely used as chemical intermediates for the manufacture of synthetic detergents in the United States. Typical physical characteristics of a commercial grade of both the trimer and tetramer are given in Table 12. Listed suppliers include [213]:

Propylene tetramer	Propylene trimer
Gulf Oil Company[a]	Gulf Oil Company[a]
Arco Chemical Company	Arco Chemical Company
Continental Oil Company	Sun Oil Company
Enjay Chemical Company	Sun Oil Company, DX Div.
Sun Oil Company	
Sun Oil Company, DX Div.	
Texaco, Inc.	

[a]Discontinued in 1969.

TABLE 12

Typical Physical Properties of Tripropylene
and Tetrapropylene [212]

Property	Tripropylene[a]	Tetrapropylene[a]
Gravity, °API	59.4	51.7
Specific gravity, 68/60° F	0.739	0.770
Refractive index, n_D^{20}	1.4226	1.4366
Bromine number	125	100

TABLE 12 (Continued)

Property	Tripropylene[a]	Tetrapropylene[a]
Distillation		
I.B.P., °F	276	369
Dry point	282	387
10%	277	372
50%	278	376
90%	280	383
95%	281	385

[a]Inhibitor added at time of manufacture.

The largest volume of these branched-chain olefins in the surfactant field was used as alkylating agents for phenol and benzene to produce alkylphenols and dodecylbenzene (detergent alkylate), see Secs. II and III.

VI. HIGHER LINEAR ALCOHOLS

In this section several production processes for the higher alcohols most valuable in the manufacture of synthetic anionic surfactants are described. Such alcohols are saturated, linear or predominantly linear, primary or secondary alcohols within the C_{10}-C_{18} range.

Several processes by which linear primary alcohols, called fatty alcohols, are derived by conversion of natural fats and oils are well established and are described in Chapter 5 of this volume. The processes described in this section are those which are based only on petroleum-derived raw materials. The most important ones are:

Linear primary alcohols by hydrolysis of alkoxyaluminum compounds
Linear and 2-methyl branched alcohols by hydroformylation of linear olefins
Linear secondary alcohols by oxidation of n-paraffins

These processes are covered in Secs. VI,A, B, and C below. Two other processes of industrial importance are the Aldol condensation and Aldex processes. Since the alcohols produced by those processes are of branched-chain configuration, they are not treated in this chapter.

A. Linear Primary Alcohols by Hydrolysis of Alkoxyaluminum Compounds

Ethylene is the basic petroleum-derived raw material most often used in this process. The first and second steps are the conversion of ethylene to triethylaluminum, followed by reaction with more ethylene to form a mixture of trialkylaluminum compounds which are called "growth products." Preparation of triethylaluminum and trialkylaluminum compounds are illustrated by Eqs. (18)-(21). The preparation of trialkylaluminum compounds in which straight alkyl groups of C_2-C_{28} attached to the aluminum follows a Poisson distribution as illustrated in Table 7. Various methods of preparation of trialkylaluminums and ways to improve control of ordinary distribution of molecular weight are described in Refs. 128-164 and the test relating thereto.

If, instead of subjecting the alkylaluminum compounds to the displacement reaction, they are oxidized, a mixture of alkoxyaluminum compounds is formed, corresponding to the alkyl chain length distribution of the alkyl aluminum compounds oxidized. This oxidation reaction may be illustrated by Eq. (22):

$$Al \begin{cases} CH_2CH_2R \\ CH_2CH_2R_1 \\ CH_2CH_2R_2^1 \end{cases} + 1\,1/2\,O_2 \longrightarrow Al \begin{cases} OCH_2CH_2R \\ OCH_2CH_2R_1 \\ OCH_2CH_2R_2^1 \end{cases} \quad (22)$$

Alkylaluminum compounds   Alkoxyaluminum
(Growth product)   compounds

The linear primary alcohol homologues are produced by hydrolysis as illustrated by Eq. (23):

$$Al \begin{cases} OCH_2CH_2R \\ OCH_2CH_2R_1 \\ OCH_2CH_2R_2^1 \end{cases} + 3\,H_2O \text{ (or 1.5 } H_2SO_4) \longrightarrow$$

Alkoxyaluminum
compounds

$$\begin{array}{l} RCH_2CH_2OH \\ R_1CH_2CH_2OH \\ R_2^1CH_2CH_2OH \end{array} + Al(OH)_3 \text{ [or } 1/2\,Al_2(SO_4)_3] \quad (23)$$

Linear primary
alcohols

By way of historical note, this route to linear primary alcohols was also discovered by Karl Ziegler; the Continental Oil Company obtained a license under certain Ziegler patents shortly thereafter. By late 1961, after a six-year year research and development effort, the first commercial-scale plant operation of this type in the world was initiated in Westlake, Louisiana [214].

An Alfol alcohol process is also located in Brünsbutel, West Germany. A flow diagram for the Alfol process is given in Fig. 5.

An alcohol process also based on the aluminum chemistry route was developed by Ethyl Corporation [215].

Trialkylaluminum compounds (e.g., growth product) are probably oxidized to trialkoxyaluminum compounds in three stages, as described in Refs. 216 and 217, instead of in one stage as illustrated by Eq. 22. These three stages may be illustrated by Eqs. (24)-(26):

$$Al(CH_2CH_2R)_3 + 1/2\ O_2 \longrightarrow$$

$$(RCH_2CH_2)_2AlOCH_2CH_2R \qquad (24)$$

$$(RCH_2CH_2)_2AlOCH_2CH_2R + 1/2\ O_2 \longrightarrow$$

$$RCH_2CH_2Al(OCH_2CH_2R)_2 \qquad (25)$$

$$RCH_2CH_2Al(OCH_2CH_2R)_2 + 1/2\ O_2 \longrightarrow$$

$$Al(OCH_2CH_2R)_3 \qquad (26)$$

The oxidation in Eqs. (24) and (25) take place very rapidly, while the reaction in Eq. (26) proceeds at a much slower rate. According to conditions described in Ref. 216, oxidation is effected initially with a gas such as nitrogen mixed with a small amount of air. As the oxidation proceeds, more oxygen is gradually added until pure oxygen is present in the final stage. The temperature of this noncatalytic oxidation may vary widely, but the preferred range is stated to be between 30 and 60° C.

The rate at which the reaction in Eq. (26) takes place may be increased by adding to the reaction mixture a quantity of trialkoxyaluminum compound [217].

In a process such as this for making linear primary alcohols, the oxidation of trialkylaluminum compounds, i.e., growth product, may be accompanied by the formation of undesirable by-products which in the hydrolysis step are not convertible to the desired alcohols or which are difficultly removed therefrom. This results both in serious contamination problems and in some cases unacceptable losses. Examples of such undesirable by-products are paraffin hydrocarbons, olefins, carbonyl compounds, and small quantities of oxidation products which are converted to polyols in the hydrolysis step.

A great deal of work has been done to develop methods for improving the control of the oxidation reaction so as to maximize formation of the trialkoxyaluminum compounds and, to the largest extent possible, eliminate the undesirable by-products. Such by-products may be the result of either or both underoxidation or overoxidation.

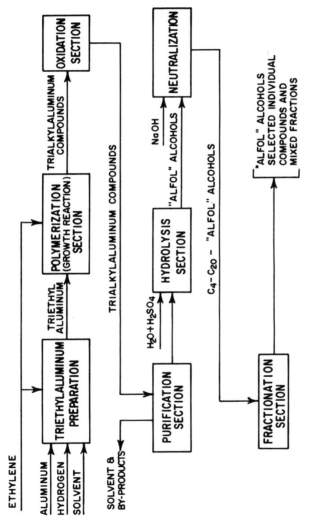

FIG. 5. Alfol alcohol process, simplified flow diagram

Examples of methods designed to improve the oxidation step are covered in Refs. 207-209.

A method of completing the oxidation reaction in the presence of a small percentage of organometal catalyst, not prior to but when the composition of alkylaluminum alkoxide mixture reaches 50-70% alkoxyaluminum compounds, is described in reference 218. Such catalysts are organo compounds of metals of Groups III-B, IV-V, and V-B of the Periodic Table. Titanium with an organic constituency of alkoxy radicals is an example. When the reaction temperature is maintained at about 70° C and other conditions are properly controlled, side reactions are greatly reduced.

The next step in the production of linear primary alcohols is the conversion by hydrolysis of the trialkoxyaluminum compounds to alcohols corresponding in alkyl chain lengths to the alkyl chains of the trialkoxyaluminum compounds as illustrated in Eq. (23). A process for accomplishing this, while at the same time avoiding inherent gelling problems, is described in Ref. 219. Hydrolysis is preferably conducted with dilute aqueous sulfuric acid. The formation of gels is avoided by adding the trialkoxides to a recirculating stream containing the dilute sulfuric acid immediately before a vigorous mixing stage. In this process aluminum sulfate (alum) solution is obtained as a by-product. Because of freezing considerations during storage and shipping the concentrated alum solution is usually adjusted to 27.8 weight % as $Al_2(SO_4)_3$ [214]. If alumina instead of alum is the desired by-product, water instead of sulfuric acid may be used as the hydrolyzing agent. When sulfuric acid is used in the hydrolysis step, the alcohols are freed of any residual acid by neutralization with a base such as ammonium or sodium hydroxide. This is followed by a hot-water wash to remove inorganic sulfate from the alcohols before fractionation [214].

Most schemes designed to reduce the formation of by-product during the oxidation of trialkylaluminums growth product to trialkoxyaluminum compounds, as previously discussed, do not entirely accomplish the desired results. Certain contaminants, which are not removed completely, show up in the linear primary alcohol product mixture in objectionable amounts. Some of these, as previously mentioned (especially polyols), are the impurities which form highly objectionable odors when a particular alcohol or blend of alcohols is further processed to derivatives which serve as plasticizers or synthetic detergents. These polyols and certain carbonyl and hydrocarbon impurities cannot be separated completely from the alcohol products by ordinary means such as fractionations. A number of methods have been developed for purifying alcohols synthesized from petroleum-derived raw materials, particularly those produced by the oxidation of alkoxyaluminum compounds followed by hydrolysis. Several examples are cited in Refs. 220-222.

The C_6-C_{24} alcohols and particularly the detergent-range alcohols (those having alkyl chains of C_{12}-C_{16}), are purified by the two-step hydrogenation process described in Ref. 220. The two distinct steps performed

in either sequence employ in one case a copper chromite hydrogenation catalyst, while in the other step a nickel catalyst is used. In one modification,
fractionation of the C_6-C_{24} alcohols to obtain the desired cuts follows the
two-stage hydrogenation step. The C_{12}-C_{16} middle cut yields low-odor
sweet-smelling surfactant products when sulfated or oxyethylated.

Odorous polyols such as 1,2- and 1,3-glycols in admixture as impurities with monoalcohols are removed by preferential catalytic dehydration to
monoalcohols [221]. This is followed by a conventional hydrogenation step in
the presence of any nickel catalyst. Distillation to remove light ends follows
the catalytic hydrogenation step. Certain neutral oxides of Group III-B elements, especially thorium, scandium, and yttrium may be used as selective
dehydration catalyst.

Another catalytic dehydration method for the removal of polyols, followed by catalytic dehydrogenation of unsaturated compounds obtained by dehydration is described in Ref. 222. In this case, the dehydration catalyst is
an oxide of aluminum, zirconium, or titanium. A nickel hydrogenation catalyst may be used in the hydrogenation step.

B. Primary Alcohols by Hydroformylation of Olefins (Oxo Process)

The hydroformylation reaction may be defined as the addition of an aldehyde
(—CHO) or carbinol group (—CH_2OH) to a primary or secondary carbon of an
olefin by reaction between the olefin, carbon monoxide, and hydrogen. Conventionally a cobalt catalyst is employed and relatively high temperatures
and pressures are required.

Since much information concerning the oxo process has been published
in the technical and patent literature, it will be treated only briefly here. An
excellent comprehensive general technical treatment and review of the patent
literature relating to the process up to 1957 is given in a monograph by L. F.
Hatch [225]. Included is a bibliography of over 400 references relating to
various aspects of the subject.

Various modifications of the oxo process are widely practiced in commercial-scale plants throughout the world. Table 13 lists the producers in
the United States and Puerto Rico; their total annual plant capacity was over
2.0 billion lb in 1971 [226].

Oxo plants are also located in Europe, Australia, and Japan.

Discovery of the process for producing aldehydes and/or alcohols by the
hydroformylation reaction was the result of work by Otto Roelen of Ruhr-
chemie in Germany before World War II. He originated the term "oxo" to
cover the process and applied for a patent in Germany in 1938. Its issuance
was delayed until 1952 [223]. The first U.S. Patent in his name was issued
in 1943 [224].

TABLE 13

Producers Using the Oxo Process in the United States and Puerto Rico

Producer	Plant location
Dow Badische	Freeport, Texas
Eastman Kodak Co., Texas Eastman Co. Div.	Longview, Texas
Exxon Chemical Co., Inc., Enjay Chemical Co. Div.	Baton Rouge, La.
Tidewater Oil Co., Air Products and Chemicals	Delaware City, Del.
Gulf Oil Corp.	Philadelphia, Pa.
Monsanto Co.	Texas City, Texas
Oxochem	Penuelas, P. R.
Shell Oil Co., Shell Chemical Co. Div.	Houston, Texas; Geismar, La.
Union Carbide Corp., Chemical Div.	Ponce, Puerto Rico; Seadrift, and Texas City, Texas
U.S. Steel Corp., U.S.S. Chemicals Div.	Haverhill, Ohio

Primary alcohols of both linear and branched-chain configurations are produced in molecular-weight ranges from C_3 to C_{20}. Some of the principal ones include the following [226]:

n-Propanol	Isohexanol
n-Butanol	Isooctanol
Isobutanol	Isodecanol
2-Ethylhexanol	Tridecanol
Pentanols	C_{12}-C_{15} linear primary alcohols

By well-known, conventional oxo technology alcohols are produced in a three-stage process. Reactions of the first and third stages are represented by Eq. (27) and (28), respectively.

$$RCH = CH_2 + CH + H_2 \longrightarrow RCHCHO + RCH_2CH_2CHO \qquad (27)$$

$$\overset{CH_3}{\underset{|}{RCHCHO}}$$

Linear olefin   Branched and linear
                                      aldehydes

$$\begin{array}{c} CH_3 \\ | \\ RCHCHO + RCH_2CH_2CHO + H_2 \end{array} \longrightarrow \begin{array}{c} CH_3 \\ | \\ RCHCH_2OH + RCH_2CH_2CH_2OH \end{array} \quad (28)$$

Branched and linear alcohols

The standard catalyst employed industrially in the oxo process is a soluble cobalt carbonyl, e.g., dicobalt octacarbonyl. The actual catalyst may be cobalt hydrocarbonyl which is thought to be formed from the octacarbonyl.

Commonly employed conditions of temperature and pressure to effect the hydroformylation of olefins in the presence of a cobalt carbonyl catalyst are 100-200° C and 150-200 atm. The specific reaction conditions of a given process depend not only on the catalyst and its concentration, but also on the nature of the olefin being used as feed and the proportions of n-iso aldehydes and/or alcohols desired as products.

The olefins employed as feeds vary from ethylene to C_{18} or higher, containing odd or even numbers of carbons in the chain. They may be of highly or moderately branched-chain or straight-chain configuration with the double bond located internally or in the α position. Olefins from processes such as wax cracking, dimerization, or trimerization of butylenes and/or propylene, polymerization of ethylene, catalytic dehydrogenation of n-paraffins and chlorination-dehydrochlorination of n-paraffins can be used as feed stock for the oxo process. Composition of the synthesis gas may vary considerably depending on the ratio of CO to H_2 desired in a specific process.

The products from the first-stage reaction may predominate in aldehyde materials and also contain carbinols and other oxygenated products such as ketones and esters. In order for the carbonyl- and ester-type products contained in the reaction product to be converted to carbinols by hydrogenation in the last stage, it is necessary to separate from the reaction products the hydroformylation catalyst employed in the first stage since it is a complex containing carbon monoxide. The latter will poison the cobalt/nickel hydrogenation catalyst if not removed before stage three. This is done in a separate second stage. The hydrogen stream used in that stage must also be substantially free of carbon monoxide [227-230].

An improved operating scheme by which alcohols substantially free of aldehydes are produced directly in a continuous-unit oxo process is described in Ref. 231. The key to the process is control of the exothermic heat of reaction in, for example, a tubular reactor in which the reactants move progressively without longitudinal mixing. A series of autoclaves may be used in place of the tubular reactor, if desired. The temperatures employed depend on the composition of the olefin feedstock. The overall reaction for the direct process is represented by Eq. (29), which is Eqs. (27) and (28) combined:

$$\text{RCH} = \text{CH}_2 + \text{CO} + 2\ \text{H}_2 \longrightarrow \overset{\overset{\displaystyle \text{CH}_2}{\displaystyle |}}{\text{CH}} \text{CH}_2\text{OH} + \text{RCH}_2\text{CH}_2\text{CH}_2\text{OH} \qquad (29)$$

Linear olefin   Branched and linear alcohols

Reference 232 relates to an improved single-stage oxo process for direct production of alcohols. The reaction is promoted by adding a sulfur or a sulfur compound that under the proper reaction conditions converts cobalt of the catalyst to cobalt sulfide.

Improvement in the preparation of alcohols by the oxo process [233] is accomplished by shortening the induction period encountered in many oxo processes. By this process the reaction is initiated by introducing hydrogen gas into the reaction mixture in the presence of a catalyst which comprises iron pentacarbonyl and tertiary amine.

Reference 234 covers a process to produce linear internal olefins by catalytic dehydrochlorination of monochloro-n-paraffins. The feature is to produce olefins with sufficiently low chloride content to prevent chloride poisoning of the hydrogenation catalyst used in conventional oxo processes for producing alcohols.

Shell developed a low-pressure hydroformylation to produce primary alcohols directly from linear internal or α-olefins. The resulting mixture of alcohols predominates in linear isomers, e.g., 80:20 linear:2-methyl branched-chain structures. The sulfates and ether-alcohol sulfates derived therefrom are biologically soft. The improved catalyst system consists of a tertiary organophosphine-cobalt hydrocarbonyl complex. The reaction conditions of temperature and pressure typically preferred are 150–210° C and generally 400–800 psig [235–237].

Other modifications of Shell type catalyst systems for low-pressure oxo processes are described in Refs. 235–245.

C. Linear Secondary Alcohols by Oxidation of n-Paraffins

Oxidation of paraffins has been the subject of a great deal of research and development work in Germany, the Soviet Union, and later in the United States. It has been known for many years that oxidation of paraffins with oxygen (air) is nonselective. Oxygenates such as alcohols, ketones, aldehydes, peroxides, and carboxylic acids are all formed. Furthermore, scission of carbon-to-carbon bonds occurs which adds to the complexity of the mixture of products obtained.

In the past, secondary alcohols have been produced by catalytic oxidation of n-paraffins utilizing known oxidation catalysts such as potassium permanganate which promotes conversion of the n-paraffins to secondary alcohols. Such a process is not particularly satisfactory since it also promotes formation of undesirable oxygenates which tend to increase substantially as conversion of the n-paraffins increases.

Liquid-phase n-paraffin oxidation processes with improved selectivity to linear secondary alcohols are described in Refs. 246 and 247.

By the procedure of Ref. 246, the oxidation, for example, of a C_6-C_{20} n-paraffin mixture is carried out in the presence of boric acid, boric anhydride, or boric oxide at 160-200° C. Sufficient pressure is maintained in the oxidation zone to keep the alkane essentially in the liquid phase and at the same time to permit water of reaction to be removed continuously therefrom. In order for the product recoverable from the alkyl metaborates formed in the oxidation reaction to consist almost entirely of linear secondary alcohols, the conversion rate of n-paraffins is held to approximately 10-15%. The alkyl metaborates are separated from solution in unoxidized n-paraffin by a stripping operation at about 200° C. Preferred stripping agents are propane or isobutane. The metaborates are then hydrolyzed to yield linear secondary alcohols which may be separated by fractionation into various cuts. These are usually mixtures within the C_{11}-C_{15} range.

The reactions involved in the foregoing process are illustrated by Eqs. (30) and (31):

$$CH_3(CH_2)_xCH_3 + H_3BO_3(O) \longrightarrow \left[\begin{array}{c} CH_3(CH_2)_y \\ CH_3(CH_2)_z \end{array} \underset{H}{\overset{H}{C}}-O- \right]_3 B \qquad (30)$$

$$\underset{\text{n-Paraffin(s)}}{} \qquad\qquad \underset{\text{Alkyl metaborate(s)}}{}$$

$$\left[\begin{array}{c} CH_3(CH_2)_y \\ CH_3(CH_2)_z \end{array} \underset{H}{\overset{H}{C}}-O- \right]_3 B + 3/2\,H_2O \longrightarrow$$

$$3 \begin{array}{c} CH_3(CH_2)_y \\ CH_3(CH_2)_z \end{array} \underset{H}{\overset{H}{C}}-OH + H_3BO_3 \qquad (31)$$

In the method described in Ref. 247 the oxidation reaction is carried out in the presence of a borate ester anhydride such as trialkoxyboroxine, oxybis-(dialkoxyborane) or mixtures of the two.

Several U.S. companies have patents on the boric acid oxidation process of n-paraffins to make linear secondary alcohols [247-248]. Union Carbide has installed a commercial-scale plant. The feedstock consists of n-paraffins produced by its own Iso-sieve separation. Apparently no patents covering the process have been issued. This plant was started up in the spring of 1965 [249]. It is believed that the annual production capacity of the secondary alcohols in about the C_{11}-C_{15} range is about 100 million pounds.

VII. HIGHER ACYCLIC MONOCARBOXYLIC ACIDS

The major source of fatty acids in the United States is tallow. The abundance of tallow and its low cost has discouraged development of petroleum-based routes to synthetic fatty acids. The price of tallow has moved erratically, however, and there is now renewed interest in synthetic fatty acids in the United States. In contrast to the situation in this country, the Soviet Union and East European countries have an abundance of paraffins and a relative shortage of tallow. Oxidation of paraffins to synthetic fatty acids is in commercial practice in the Soviet Union, Rumania, Czechoslovakia, and East Germany [250].

A. Higher Fatty Acids from Paraffin Oxidation

Synthetic fatty acids are manufactured in Eastern Europe by continuous oxidation of paraffins with air. The reaction temperature is about 100° C and a permanganate catalyst is used [251]. The crude product is purified by washing with water, heating with aqueous base, acidifying, and final purification by distillation. Feedstocks are chosen to maximize the yield of $C_{10}-C_{20}$ chain-length fatty acids which are in short supply for soap manufacture. A paraffin hydrocarbon which contains not less than 93% boiling in the range from 320 to 450° C is chosen. A typical product distribution is shown in Table 14 [251].

The crude C_5-C_9 acid fraction is a by-product. Small amounts of this fraction have been imported into the United States for use in synthetic lubricants [252].

TABLE 14

Distribution of Acids from Paraffin Oxidation

Acid	Percent by weight
Formic	4.0
Acetic	2.7
Propionic	1.2
Butyric	1.5
C_5-C_6	2.7
C_7-C_9	9.5
$C_{10}-C_{20}$	61.0
C_{20+}	17.4

B. Higher Fatty Acids by Oxidation of Straight-Chain Alcohols or Olefins

Alcohols can be oxidized to acids by several well-known routes. However, the economics of synthesis of tallow-range acids from oxidation of corresponding alcohols make commercialization unlikely.

Similarly, olefins may be converted to carboxylic acids by several routes. Commercial availability of higher α-olefins from ethylene oligomerization has stimulated research to develop a suitable process for the oxidation of olefins, shown in Eq. (32).

$$RCH_2CH{=}CH_2 \longrightarrow \left[\begin{array}{l} \xrightarrow{\ O_2\ } RCH_2COOH \\ \xrightarrow{\ O_2\ } RCH_2COOH \\ \xrightarrow{\ HNO_3\ } \left[\begin{array}{l} RCH_2COOH \\ RCOOH \end{array} \right. \end{array} \right. \qquad (32)$$

Ozonolysis of an α-olefin to a fatty acid of one less carbon is a high-yield reaction but may be too expensive to compete with fatty acids from tallow. An additional disadvantage is the formation of odd-carbon acids from the even-carbon olefins. Ozonolysis of olefins as a route to acids has been patented [253].

Air oxidation of olefins to carboxylic acids would be the method of choice if the many problems of yield, quality, and process could be solved. A process using bromine catalysis appears to present severe corrosion problems [254]. If these problems can be solved, air oxidation of olefins should provide the best chance of a synthetic fatty acid process that is competitive with tallow-derived acids. For a review of properties and industrial applications of fatty acids, see Ref. 255.

REFERENCES

1. United States International Trade Commission, Synthetic Organic Chemicals, United States Production and Sales of Surface-Active Agents, 1973. U.S. Government Printing Office, Washington, D.C., January 1975, p. 1.
2. United States Tariff Commission, Synthetic Organic Chemicals, United States Production and Sales, 1967. T. C. Publication 295, U.S. Government Printing Office, Washington, D.C., 1969, pp. 3, 13.
3. United States Tariff Commission, Synthetic Organic Chemicals, United States Production and Sales, 1970. T. C. Publication 479, U.S. Government Printing Office, Washington, D.C., 1972, pp. 11, 24.
4. United States International Trade Commission, Preliminary Report on U.S. Production of Selected Synthetic Organic Chemicals, Cumulative 1974. S.O.C. Series c/p-74-12, U.S. Government Printing Office, Washington, D.C., February 5, 1975.

5. C. A. Thomas, U.S. Pat. 2,072,061 (1933), to The Sharples Solvents Corp.
6. L. P. Kyrides, U.S. Pat. 2,161,173 (1939), to Monsanto Chemical Co.
7. L. H. Flett, U.S. Pat. 2,499,578 (1950), to Allied Chemical & Dye Corp.
8. F. Guenther et al., U.S. Pat. 2,220,099 (1940), to General Aniline and Film Corp.
9. G. C. Feighner, J. Am. Oil Chemists' Soc., 35, 520 (1958).
10. K. M. Thompson, U.S. Pat. 2,875,257 (1959), to Atlantic Refining Co.
11. G. C. Feighner and L. D. Lindemuth, U.S. Pat. 3,209,045 (1965), to Continental Oil Co.
12. R. L. Kylander, U.S. Pat. 2,941,015 (1960), to Continental Oil Co.
13. F. P. Williamson and R. C. Bieneman, U.S. Pat. 3,207,800 (1965), to Continental Oil Co.
14. G. C. Feighner and B. L. Kapur, U.S. Pat. 3,316,294 (1967), to Continental Oil Co.
15. G. L. Kapur and D. B. Ratliff, U.S. Pat. 3,365,508 (1968), to Continental Oil Co.
16. G. C. Feighner et al., U.S. Pat. 3,403,194 (1968), to Continental Oil Co.
17. H. N. Moulden, U.S. Pat. 3,355,508 (1967), to Chevron Research Co.
18. R. T. Adams and H. J. Aroyan, U.S. Pat. 3,333,014 (1967), to Chevron Research Co.
19. B. L. Kapur and R. D. Eccles, U.S. Pat. 3,274,278 (1966), to Continental Oil Co.
20. A. McLean, Br. Pat. 968,338 (1964), to British Hydrocarbon Chemicals, Ltd.
21. G. C. Feighner et al., U.S. Pat. 3,391,210 (1968), to Continental Oil Co.
22. W. J. DeWitt et al., U.S. Pat. 3,342,888 (1967), to The Atlantic Refining Co.
23. Br. Pat. 1,066,668 (1967), to The Lummus Co., a Corporation of the U.S.
24. T. R. Liston, U.S. Pat. 3,358,047 (1967), to Allied Chemical Corp.
25. Br. Pat. 1,037,868 (1966), to Chemische Werke Hüls, A.G.
26. E. K. Jones, U.S. Pat. 3,303,233 (1967), to Universal Oil Products Co.
27. R. T. Adams, U.S. Pat. 3,351,672 (1967), to Chevron Research Co.
28. H. S. Bloch, U.S. Pat. 3,169,987 (1965), to Universal Oil Products Co.
29. G. J. McEwan and S. G. Clark, U.S. Pat. 3,248,443 (1966), to Monsanto Co.
30. W. A. Sweeney, U.S. Pat. 3,349,141 (1967), to Chevron Research Co.
31. H. A. Sorgenti, U.S. Pat. 3,352,933 (1967), to The Atlantic Refining Co.
32. J. J. Shook, U.S. Pat. 3,365,509 (1968), to Chevron Research Co.
33. G. J. McEwan and S. G. C. Clark, U.S. Pat. 3,387,056 (1968), to Monsanto Co.
34. R. T. Adams et al., U.S. Pat. 3,433,846 (1969), to Chevron Research Co.
35. H. T. Hutson, Jr., U.S. Pat. 3,372,207 (1968), to Phillips Petroleum Co.

36. O. C. Kerfoot and D. D. Krehbiel, U.S. Pat. 3,401,208 (1968), to Continental Oil Co.
37. Br. Pat. 990,744 (1965), to Universal Oil Products Co. of the U.S.
38. H. S. Bloch, Detergent Age, 3 (8), 33-35, 105 (1967).
39. H. S. Bloch, U.S. Pat. 3,413,373 (1968), to Universal Oil Products Co.
40. D. B. Carson and R. A. Lengemann, U.S. Pat. 3,426,092 (1969), to Universal Oil Products Co.
41. H. R. Alul et al., U.S. Pat. 3,349,144 (1967), to Monsanto Co.
42. C. R. Enyeart, in Nonionic Surfactants (M. J. Schick, ed.), Vol. 1, Marcel Dekker, Inc., New York, 1967, Chap. 3.
43. R. J. O'Neill and T. C. Tesdahl, U.S. Pat. 3,257,467 (1966), to Monsanto Co.
44. K. L. Mai, U.S. Pat. 3,268,595 (1966), to Shell Oil Co.
45. G. B. Arnold and H. D. Kluge, U.S. Pat. 2,789,143 (1957), to The Texas Co.
46. L. Schmerling and J. P. Luvisi, U.S. Pat. 2,916,532 (1959), to Univ. Oil Products Co.
47. C. T. Hathaway, U.S. Pat. 2,800,514 (1957), to General Electric Co.
48. Physical Data Nalkylene Soft-Detergent Alkylates and Derivatives, Tech. Data Bull. 51-2500-865, Continental Oil Co., Petrochemical Dept., Saddle Brook, N.J.
49. Ucane Alkylate, Tech. Broch. F-41143A, Union Carbide Corp., New York, N.Y.
50. Linear Alkylbenzene .... 98% Minimum Biodegradable, Monsanto Co., Inorganic Chemicals Div., St. Louis, Mo., 1966.
51. Linear Alkylbenzenes, Tech. Data Bull., Continental Oil Co., Petrochemical Department Data, Saddle Brook, N.J., Nov. 11, 1969.
52. Detergents and Emulsifiers, 1974 Annual, John W. McCutcheon, Inc., Morristown, N.J.
53. United States Tariff Commission, Synthetic Organic Chemicals, United States Production and Sales of Cyclic Intermediates, 1971 Preliminary. U.S. Government Printing Office, Washington, D.C., March 1973, p. 3.
54. United States Tariff Commission, Synthetic Organic Chemicals, United States Production and Sales of Cyclic Intermediates, 1972 Preliminary. U.S. Government Printing Office, Washington, D.C., April 1974, p. 3.
55. United States Tariff Commission, Washington, D.C., Preliminary Report on U.S. Production of Selected Synthetic Organic Chemicals, Cumulative Totals 1973. S.O.C. Series c/o-73-12, U.S. Government Printing Office, Washington, D.C., February 6, 1974.
56. F. D. Rossini et al., Selected Values of Physical and Thermodynamic Properties of Hydrocarbons and Related Compounds, Carnegie Press, Pittsburgh, Pa., 1953.
57. Alkylphenols, Tech. Bull. CO-50, Rohm and Haas Co., Special Products Department, Philadelphia, Pa.
58. Monsanto's Versatile and Reactive Alkylphenols, Tech. Bull. I-218, Monsanto Co., Inorganic Chemicals Div., St. Louis, Mo.
59. Nonylphenol, Tech. Broch. JO-102, Jefferson Chemical Co., Inc., Houston, Texas, Feb. 11, 1960.

60. Tech. Reps. TMO-8 and TMO-9, Rohm and Haas Co., Special Products Department, Philadelphia, Pa.

61. L. W. Burnette, in Nonionic Surfactants (M. J. Schick, ed.), Vol. 1, Marcel Dekker, Inc., New York, 1967, Chap. 11.

62. E. C. Steinle et al., J. Am. Oil Chemists' Soc., 41, 804-807 (1964).

63. W. J. Haensel, U.S. Pat. 3,422,004 (1960), to Universal Oil Products Co.

64. C. K. Hersh, Molecular Sieves, Reinhold Publishing Corp., New York, 1961.

65. M. J. Sterba, Hydrocarbon Process. Petrol. Refiner, 44 (6), 151-153 (1965).

66. D. B. Broughton and D. B. Carson, Hydrocarbon Process. Petrol. Refiner, 38 (4), 130-134 (1959).

67. Hydrocarbon Process. Petrol. Refiner, 47 (9), 238 (1968).

68. R. C. Wackker and D. B. Broughton, U.S. Pat. 3,306,848 (1967), to Universal Oil Products Co.

69. D. B. Broughton and C. S. Brearley, U.S. Pat. 3,310,486 (1967), to Universal Oil Products Co.

70. F. G. Padrta, U.S. Pat. 3,422,004 (1969), to Universal Oil Products Co.

71. K. A. Scott, Hydrocarbon Process. Petrol. Refiner, 44 (3), 97-100 (1964).

72. G. J. Greismer et al., Hydrocarbon Process. Petrol. Refiner, 44 (6), 147-150 (1965).

73. W. F. Avery and M. N. Y. Lee, Oil Gas J., 60 (24), 121-123 (1962).

74. Ind. Eng. Chem., 54 (5), 13-14 (1962).

75. Chem. Eng., 69 (5), 58-59 (1962).

76. Hydrocarb. Process. Petrol. Refiner, 47 (9), 236 (1968).

77. D. E. Cooper et al., Chem. Eng. Progr., 62 (4), 69-73 (1966).

78. T. A. Cooper and R. A. Woodle, U.S. Pat. 2,886,522 (1959), to The Texas Co.

79. F. A. Clauson, U.S. Pat. 3,010,894 (1961), to Texaco, Inc.

80. D. E. Cooper et al., U.S. Pat. 3,373,103 (1968), to Texaco, Inc.

81. N. D. Carter et al., U.S. Pat. 2,908,639 (1959), to Texaco, Inc.

82. H. A. Ricards et al., U.S. Pat. 2,988,502 (1961), to Esso Research and Engineering Co.

83. W. J. Asher and W. R. Epperly, U.S. Pat. 3,070,542 (1962), to Esso Research and Engineering Co.

84. W. J. Mattox et al., U.S. Pat. 3,160,581 (1964), to Esso Research and Engineering Co.

85. W. H. Mueller, U.S. Pat. 3,306,847 (1967), to Esso Research and Engineering Co.

86. W. R. Epperly and W. J. Asher, U.S. Pat. 3,309,311 (1967).

87. M. G. Lorenz, U.S. Pat. 3,362,904 (1968), to Esso Research and Engineering Co.

88. W. J. Asher et al., Hydrocarb. Process. Petrol. Refiner, 48 (1), 134-138 (1969).

89. R. T. M. Mowell and H. M. C. Smith, U.S. Pat. 3,342,726 (1967), to
     The British Petroleum Co., Ltd.
90. J. N. Turnbull and R. J. H. Gilbert, U.S. Pat. 3,428,552 (1969), to
     The British Petroleum Co., Ltd.
91. R. H. Anstey and R. M. Macnab, U.S. Pat. 3,422,003 (1969), to The
     British Petroleum Co., Ltd.
92. G. F. Eggesten, U.S. Pat. 2,987,471 (1961), to Shell Oil Company.
93. F. G. Helfferich and R. A. Loth, U.S. Pat. 3,451,924 (1969), to Shell
     Oil Co.
94. T. C. Ponder, Hydrocarbon Process. Petrol. Refiner, $\underline{44}$ (1), 141-
     142 (1965).
95. F. Bengen, Ger. Pat. Appl. O.Z. 12438 (Mar. 18, 1940); Ger. Pat.
     869,070 (1953).
96. H. Franz, Hydrocarbon Process. Petrol. Refiner, $\underline{44}$ (9), 183-185
     (1965).
97. A. Hoppe, in Advances in Petroleum Chemistry and Refining (K. A.
     Kobe and J. J. McKetta, eds.), Vol. VIII, Interscience Publishers,
     New York (1964), pp. 193-234.
98. G. B. Dellow, U.S. Pat. 3,328,313 (1967), to Shell Oil Co.
99. C. H. Middlebrooks, U.S. Pat. 3,152,084 (1964), to Monsanto Co.
100. A. M. Leas and E. A. Thompson, U.S. Pat. 3,163,632 (1964), to Ash-
     land Oil and Refining Co.
101. Y. Naoki and I. Shuko, Nurex Process, A New Process for Separation
     of n-Paraffins from Kerosene or Gas Oil, Paper presented at the 156th
     Natl. Meeting of the Am. Chem. Soc., Atlantic City, N.J., Sept. 9,
     1968.
102. O. Minoru, Nurex Process. Manufacture of Normal Paraffins by the
     Nippon Mining Urea Method, Sekiyu to Sckiyu Kagaku, $\underline{12}$ (6), 38-41
     (1968).
103. T. Hutson, Jr., U.S. Pat. 3,402,216 (1968), to Phillips Petroleum Co.
104. H. C. Brown and G. Zweïfel, J. Am. Chem. Soc., $\underline{82}$, 1504 (1960).
105. M. F. Hughes, U.S. Pat. 3,277,205 (1966), to Chevron Research Co.
106. L. L. Ferstandig, U.S. Pat. 3,277,204 (1966), to Chevron Research
     Co.
107. H. S. Bloch and F. J. Reidl, U.S. Pat. 3,360,586 (1967), to Universal
     Oil Products Co.
108. N. A. Fishel, U.S. Pat. 3,383,431 (1968), to Universal Oil Products
     Co.
109. H. S. Bloch, U.S. Pat. 3,391,218 (1968), to Universal Oil Products Co.
110. P. J. Kuchar, U.S. Pat. 3,429,944 (1969), to Universal Oil Products
     Co.
111. H. S. Bloch, U.S. Pat. 3,448,165 (1969), to Universal Oil Products Co.
112. H. S. Bloch, U.S. Pat. 3,459,822 (1969), to Universal Oil Products Co.
113. D. B. Broughton and R. C. Berg, Hydrocarbon Process. Petrol. Re-
     finer, $\underline{48}$ (6), 115-117 (1969).
114. L. J. Hughes, U.S. Pat. 3,248,451 (1966), to Monsanto Co.
115. J. B. Abell et al., U.S. Pat. 3,315,008 (1967), to Monsanto Co.

116. J. F. Roth and A. R. Schaefer, U.S. Pat. 3,356,757 (1967), to Monsanto Co.

117. J. B. Abell and L. W. Fannin, U.S. Pat. 3,435,090 (1969), to Monsanto Co.

118. J. M. McEuen, U.S. Pat. 3,322,849 (1967), to Ethyl Corp.

119. K. A. Keblys, U.S. Pat. 3,433,851 (1969), to Ethyl Corp.

120. K. A. Keblys, U.S. Pat. 3,433,852 (1969), to Ethyl Corp.

121. A. E. Trevillyan and R. A. Sanford, U.S. Pat. 3,429,943 (1969), to Sinclair Res., Inc.

122. R. M. Englebrecht et al., U.S. Pat. 3,409,703 (1968), to Monsanto Co.

123. R. M. Englebrecht et al., U.S. Pat. 3,402,217 (1968), to Monsanto Co.

124. R. M. Englebrecht et al., U.S. Pat. 3,315,009 (1967), to Monsanto Co.

125. J. M. Schuck et al., U.S. Pat. 3,333,017 (1967), to Monsanto Co.

126. J. A. Acciarri et al., Chem. Eng. Progr., 58 (6), 45-90 (1962).

127. Oil, Paint, Drug Reptr. (October 13, 1969).

128. K. Ziegler, Belg. Pat. 546,432 (1956).

129. F. J. Radd and W. W. Woods, U.S. Pat. 3,104,252 (1963), to Continental Oil Co.

130. B. J. Williams and P. A. Lobo, U.S. Pat. 3,382,269 (1968), to Continental Oil Co.

131. H. E. Redman, U.S. Pat. 2,787,626 (1957), to Ethyl Corp.

132. H. E. Redman, U.S. Pat. 2,886,581 (1959), to Ethyl Corp.

133. K. Ziegler, Ger. Pat. 917,006 (1954).

134. P. A. Lobo, U.S. Pat. 2,971,969 (1961), to Continental Oil Co.

135. K. Ziegler and H. G. Gellert, U.S. Pat. 3,207,770 (1965), to K. Ziegler.

136. K. Zosel, U.S. Pat. 3,207,771 (1965), to K. Ziegler.

137. K. Ziegler and K. Zosel, U.S. Pat. 3,207,772 (1965), to K. Ziegler.

138. K. Ziegler et al., U.S. Pat. 3,207,773 (1965), to K. Ziegler.

139. K. Ziegler and H. G. Gellert, U.S. Pat. 3,207,774 (1965), to K. Ziegler.

140. J. A. Acciarri, U.S. Pat. 3,445,494 (1969), to Continental Oil Co.

141. A. F. Meiners and F. V. Morriss, U.S. Pat. 2,962,513 (1960), to Ethyl Corp.

142. W. T. Davis, U.S. Pat. 3,384,651 (1968), to Ethyl Corp.

143. W. T. Davis and C. L. Kingrea, U.S. Pat. 3,415,861 (1968), to Ethyl Corp.

144. M. F. Gautreaux, U.S. Pat. 3,412,126 (1968), to Ethyl Corp.

145. G. C. Feighner, U.S. Pat. 3,293,274 (1966), to Continental Oil Co.

146. D. M. Coyne et al., U.S. Pat. 2,978,523 (1961), to Continental Oil Co.

147. E. F. Kennedy and J. A. Acciarri, U.S. Pat. 3,210,435 (1965), to Continental Oil Co.

148. W. R. Kroll, U.S. Pat. 3,153,075 (1964), to Continental Oil Co.

149. R. L. Poe and H. L. Hackett, U.S. Pat. 3,206,522 (1965), to Continental Oil Co.

150. R. L. Poe and B. J. Williams, U.S. Pat. 3,308,143 (1967), to Continental Oil Co.
151. L. E. Scroggins and T. A. Yokley, Jr., U.S. Pat. 3,367,989 (1968), to Phillips Petroleum Co.
152. W. R. Kroll, U.S. Pat. 3,406,187 (1968).
153. D. L. Crain and F. Kleinschmidt, U.S. Pat. 3,352,894 (1967), to Phillips Petroleum Co.
154. R. L. Poe and E. F. Kennedy, U.S. Pat. 3,309,416 (1967), to Continental Oil Co.
155. C. Roming, Jr., U.S. Pat. 3,227,773 (1966), to Esso Research and Engineering Co.
156. L. D. Boyer, U.S. Pat. 3,458,594 (1969), to Continental Oil Co.
157. W. B. Carter and W. C. Ziegenhain, U.S. Pat. 3,317,625 (1967), to Continental Oil Co.
158. G. B. Kottong and O. A. Ritter, U.S. Pat. 3,389,161 (1968), to Ethyl Corp.
159. H. B. Fernald et al., U.S. Pat. 3,441,631 (1969), to Gulf Research and Development Co.
160. H. B. Fernald et al., U.S. Pat. 3,444,263 (1969), to Gulf Research and Development Co.
161. H. B. Fernald et al., U.S. Pat. 3,444,264 (1969), to Gulf Research and Development Co.
162. R. L. Poe and H. L. Hackett, U.S. Pat. 3,278,262 (1966), to Continental Oil Co.
163. M. T. Atwood, U.S. Pat. 3,423,444 (1969), to Continental Oil Co.
164. S. A. Miller, Chem. Process. Eng., 50 (10), 103-106 (1969).
165. T. H. Pearson and J. K. Presswood, Br. Pat. 990,748 (1965), to Ethyl Corporation.
166. Gulf Research and Development Co., Belg. Pat. 625,002; Br. Pat. Equiv. 1,020,563 (1966).
167. K. Ziegler et al., U.S. Pat. 3,310,600 (1967), to K. Ziegler.
168. P. R. Stapp, U.S. Pat. 3,366,704 (1968), to Phillips Petroleum Co.
169. R. E. Reusser, U.S. Pat. 3,375,296 (1968), to Phillips Petroleum Co.
170. P. R. Stapp, U.S. Pat. 3,384,678 (1968), to Phillips Petroleum Co.
171. H. B. Fernald et al., U.S. Pat. 3,842,000 (1969), to Gulf Research and Development Co.
172. W. T. Davis, U.S. Pat. 3,391,175 (1968), to Ethyl Corp.
173. W. T. Davis and M. F. Gautreaux, U.S. Pat. 3,391,219 (1968), to Ethyl Corp.
174. F. F. Farley, Alpha-Olefin Production, U.S. Pat. 3,647,906 (March 7, 1972), to Shell Oil.
175. Shell International Research, Higher Alkene Mixture Production, French Pat. 2,132,099 (Nov. 17, 1972).
176. A. J. Berger, Olefin Production, U.S. Pat. 3,726,938 (April 10, 1973).
177. P. W. Glockner et al., Ethylene Oligomerization, U.S. Pat. 3,647,914 (March 7, 1972), to Shell Oil.
178. R. S. Bauer et al., Ethylene Oligomerization, U.S. Pat. 3,647,915 (March 7, 1972), to Shell Oil.

179. R. F. Mason et al., Novel 9-Carboxymethyl-9-Phosphabicyclononanes and Alkali Metal Salts Thereof, U.S. Pat. 3,636,091 (Jan. 18, 1972), to Shell International Research.

180. H. Van Zwet et al., Ethylene Oligomerization in the Presence of Complex Nickel-Fluorine-Containing Catalysts, U.S. Pat. 3,644,564 (Feb. 22, 1972), to Shell International Research.

181. R. F. Mason, Alpha-Olefin Production, U.S. Pat. 3,676,523 (July 11, 1972), to Shell Oil.

182. R. F. Mason, Alpha-Olefin Production, U.S. Pat. 3,686,351 (Aug. 22, 1972), to Shell Oil.

183. R. F. Mason, Alpha-Olefin Production, U.S. Pat. 3,737,475 (June 5, 1973), to Shell Oil.

184. Shell International Research, Isomerization of Alkenes, Neth. Pat. 72,09849 (Oct. 25, 1972).

185. Shell International Research, Double-Bond Isomerization of Alkenes, German Offen. 2,259,995 (July 12, 1973).

186. C. P. C. Bradshaw, Production of C_7 to C_{20} Olefins, U.S. Pat. 3,600,456 (Aug. 17, 1971), to British Petroleum.

187. R. L. Banks, Conversion of Olefins, U.S. Pat. 3,658,929 (April 25, 1972), to Phillips Petroleum.

188. R. J. Sampson et al., Disproportionation of Olefins, Br. Pat. 1,205,677 (Sept. 16, 1970), to ICI.

189. Chevron Research Co., Br. Pat. 848,385 (1960).

190. W. A. Pardee, U.S. Pat. 2,945,076 (1960), to Gulf Research and Development Co.

191. C. L. Dulaney and M. L. Owens, Jr., U.S. Pat. 3,221,077 (1965), to Monsanto Co.

192. R. P. Cahn and W. H. Jones, U.S. Pat. 3,256,265 (1966), to Esso Research and Engineering Co.

193. G. C. Feighner and D. D. Krehbiel, U.S. Pat. 3,291,853 (1966), to Continental Oil Co.

194. W. J. Mattox, U.S. Pat. 3,331,882 (1967), to Esso Research and Engineering Co.

195. G. Caprioli and S. Pistoia, U.S. Pat. 3,444,261 (1969).

196. Castrol, Ltd., Br. Pat. 917,047 (1963).

197. A. G. Stauffer and W. L. Kranich, Ind. Eng. Chem. Fundamentals, 1 (2), 107-111 (1962).

198. Institut Français du Pétrole, Br. Pat. 797,989 (1958).

199. Hydrocarbon Process. Petrol. Refiner, 41 (9), 190 (1962).

200. A. M. Henke and R. T. Sebulsky, U.S. Pat. 3,374,285 (1968), to Gulf Research and Development Co.

201. S. M. Kovack, U.S. Pat. 3,296,331 (1967), to Sinclair Research, Inc.

202. S. M. Kovack, U.S. Pat. 3,311,672 (1967), to Sinclair Research, Inc.

203. S. M. Kovack, U.S. Pat. 3,364,279 (1967), to Sinclair Research, Inc.

204. H. H. Eby and G. L. Nield, U.S. Pat. 3,410,925 (1968), to Continental Oil Co.

205. F. D. Possini et al., Selected Values of Physical and Thermodynamic Properties of Hydrocarbons and Related Compounds, Carnegie Press, Pittsburgh, Pa., 1953, p. 34.

206. N. Irving Sax, Dangerous Properties of Industrial Materials, Reinhold Publishing Corp., New York, 1963.
207. C. Beerman, The Outlets for n-Paraffins, Preprints of the Symposium on Normal Paraffins held in Manchester, U.K., Nov. 16, 1966, p. 25. Heywood-Temple Industrial Publications, Ltd., London, 1967.
208. Chevron Alpha Olefins, New Products for New Profits, Tech. Broch. 2158, Chevron Chemical Co., San Francisco.
209. Product Information Sheets Nos. A08-667 st., A010-667 st., A012-667 st., A014-667 st., A016-667 st., A018-667 st., and A020-667 st., Gulf Oil Corp., New York.
210. J. C. Kirk, F. Kennedy, and A. J. Lundeen, Reactions of Olefins, in Advances in Petroleum Chemistry (John J. McKetta, Jr., ed.), Interscience Publishers, Inc., New York, 1962, pp. 323-367.
211. Alpha Olefins for Detergent Applications; also, Chemical Reactions of Normal Alpha Olefins, Tech. Broch. SP12415CD of Gulf Oil Corp., New York, July 1971.
212. Product Information Sheet Nos. PTR-467 st. and PTE-66-8 st., Gulf Oil Corp., New York.
213. OPD Chemical Buyers Directory, 1969-70, Schnell Publishing Co., Inc., New York, 1970.
214. P. A. Lobo et al., Chem. Eng. Progr., 58 (5), 85-88 (1962).
215. W. T. Davis, U.S. Pat. 3,487,097 (1970), to Ethyl Corp.
216. K. Ziegler, U.S. Pat. 2,892,858 (1959).
217. Pat W. K. Flannagan, U.S. Pat. 3,070,616 (1962), to Continental Oil Co.
218. H. J. Cragg and D. A. Nolen, U.S. Pat. 3,475,476 (1969), to Ethyl Corp.
219. N. D. Guzick and J. H. McCarthy, U.S. Pat. 3,475,501 (1969), to Ethyl Corp.
220. L. Regovin and A. O. Wikman (assignors of one-half each to P. and G. Co. and Ethyl Corp.), U.S. Pat. 3,505,414 (1970).
221. A. J. Lundeen and C. M. Starks, U.S. Pat. 3,461,176 (1969), to Continental Oil Co.
222. Deutsche Erdoel A.G., Br. Pat. 1,149,281 (1969).
223. O. Roelen, Ger. Pat. 849,548 (1962).
224. O. Roelen (Vested in Alien Property Custodian), U.S. Pat. 2,327,066 (1943).
225. L. F. Hatch, Higher Oxo Alcohols, Wiley, New York, 1957.
226. D. L. Richards, Oil Paint Drug Reptr., 195 (5) (1970).
227. R. N. Shiras, U.S. Pat. 2,490,283 (1949), to Shell Development Co.
228. R. E. Shexnailder, U.S. Pat. 2,500,913 (1950), to Standard Oil Development Co.
229. J. T. Harlan, Jr., U.S. Pat. 2,504,682 (1950), to Shell Development Co.
230. P. C. Johnson and N. R. Cox, U.S. Pat. 3,014,970 (1961), to Union Carbide Corp.
231. R. A. Heimsch and W. E. Weesner, U.S. Pat. 3,113,974 (1963), to Monsanto Chemical Co.
232. E. Field, U.S. Pat. 2,683,177 (1954), to Standard Oil Co.

233. H. W. B. Reed and O. O. Lenel, U.S. Pat. 2,911,443 (1959), to I.C.I.

234. Br. Pat. 1,067,543 (1967), to Chemische Werke Hüls, A.G.

235. L. H. Slaugh and R. D. Millineaux, U.S. Pat. 3,239,569 (1966), to Shell Oil Co.

236. Br. Pat. 1,109,787 (1968), to Shell Int. Res. Maatchappij N.V.

237. E. R. Tucci, Ind. Eng. Chem. Prod. Res. Develop., 7 (1), 32–38 (1968).

238. L. H. Slaugh and R. D. Millineaux, U.S. Pat. 3,239,566 (1966), to Shell Oil Co.

239. L. H. Slaugh and R. D. Millineaux, U.S. Pat. 3,239,570 (1966), to Shell Oil Co.

240. L. H. Slaugh and R. D. Mullineaux, U.S. Pat. 3,239,571 (1966), to Shell Oil Co.

241. C. R. Greene and R. E. Meeker, U.S. Pat. 3,274,263 (1966), to Shell Oil Co.

242. C. R. Greene, U.S. Pat. 3,278,612 (1966), to Shell Oil Co.

243. L. G. Cannell et al., Ger. Pat. 1,186,455 (1965), to Shell Int. Res. Maatchappij, N.V.

244. J. L. Van Winkle et al., U.S. Pat. 3,420,898 (1969), to Shell Oil Co.

245. J. L. Van Winkle et al., U.S. Pat. 3,440,291 (1969), to Shell Oil Co.

246. C. M. Starks and E. F. Kennedy, U.S. Pat. 3,346,614 (1967), to Continental Oil Co.

247. M. A. McMahon, Jr. and H. Chafetz, U.S. Pat. 3,410,913 (1968), to Texaco, Inc.

248. Chem. Week, 100 (4), 59 (1967).

249. Chem. Eng. News, 42 (23), 31–33 (1964).

250. O. Norman, U. Sonntag, and K. T. Zelch, Synthetic Fatty Acids, in Fatty Acids and Their Industrial Application (E. Scott Patterson, ed.), Marcel Dekker, Inc., New York, 1968, p. 356.

251. B. S. Alayev, N. K. Man'Kovskayo, and A. M. Shiman, Production of Synthetic Fatty Acids, in Fatty Acids and Their Industrial Application (E. Scott Patterson, ed.), Marcel Dekker, Inc., New York, 1968, pp. 356–361.

252. Karl T. Zilch, J. Am. Oil Chemists' Soc., 45, 11 (1968).

253. A. F. Ellis, Ger. Pat. Appl. 2,047,102 (1971).

254. E. F. Jason and E. K. Fields, U.S. 3,076,842 (1963), to Standard Oil Corp.

255. E. Scott Patterson, ed., Fatty Acids and Their Industrial Applications, Marcel Dekker, Inc., New York, 1968.

CHAPTER 3

LIPID AND OTHER NONPETROCHEMICAL
RAW MATERIALS

Frank Scholnick

Eastern Regional Research Center
U.S. Department of Agriculture
Philadelphia, Pennsylvania

I. INTRODUCTION

Anionic surfactants are principally synthesized either from petroleum or
from lipid and nonpetrochemical raw materials. The first source has been
discussed in Chapter 2. Although petroleum derivatives continue to be a
very important source of anionic synthetic detergents, considerable quantities of the latter have also been prepared from many naturally occurring fats
and oils.

In 1973, U.S. production of anionic surfactants approached 3.0 billion
lb. Of this total, as shown in Table 1, almost 1.7 billion lb were obtained

TABLE 1

U.S. Production of Nonpetrochemical Surface-Active Agents, 1973

Anionic surface-active agent	Production, 1000 lb
Fatty, rosin, and tall-oil acids and their salts	939,159
Sulfonic acids and salts	
Lignosulfonates, total Ca, NH_4, Na salts	689,495
Sulfosuccinic acid amides and esters	19,490[a]
Taurine derivatives	3,504[a]
Sulfuric acid esters and their salts	
Sulfated natural fats and oils	31,853
Sulfated fatty and tall-oil acids, amides, and esters	8,251
Sulfated linear alcohols	835[a]
Phosphoric acid esters of linear alcohols	3,955[a]
Total	1,696,542

[a]Probably includes some petroleum-based substrates.

from nonpetrochemical sources such as (1) potassium and sodium salts of fatty, rosin, and tall-oil acids; (2) calcium, ammonium, and sodium salts of lignosulfonates; (3) sulfated fatty and tall-oil acids and their derivatives; (4) sulfated fatty and sperm-oil alcohols; and (5) sulfated natural fats and oils [1]. Each of these raw materials is an important source for anionic surfactants and will be discussed below.

II. SOURCES

A. Carboxylic Acids

1. Fatty Acids

In 1973, almost 690 million pounds of potassium and sodium salts of fatty acids from animal and vegetable sources were produced for use as anionic surface-active agents. Some of the production figures are listed in Table 2 [1].

These fatty acids are obtained from the glycerides present in natural fats and oils (see Sec. C). The fatty acids of natural fats have almost always an even number of carbon atoms ranging from 4 to 24 and may be saturated, monounsaturated, or polyunsaturated. The unsaturation is usually of the cis

TABLE 2

U.S. Production of Salts of Fatty Acids, 1973

Potassium and sodium salts	Production, 1000 lb
Acids	
Coconut oil	149,601
Corn oil	782
Mixed vegetable oil fatty	3,109
Oleic	1,450
Soybean oil	579
Stearic	1,750
Tallow	528,151
Total	685,422

type. Fatty acids are particularly suitable for the production of surface-active agents, especially those acids containing 12 to 18 carbon atoms, such as lauric, myristic, palmitic, stearic, oleic, and ricinoleic acid. Highly unsaturated acids are subject to oxidation and rancidity, thus limiting their usefulness. Table 3 gives the fatty acid composition of selected animal and vegetable fats and oils.

The fatty acids are prepared from the glyceride fat or oil by several processes involving either saponification followed by acidification, or hydrolysis with or without a catalyst. Most of the soap produced throughout the world is prepared by batch saponification or by continuous processes [2]. The details of soap manufacture will not be covered here. In the United States hydrolysis of glycerides to prepare fatty acids is most commonly effected by a continuous, countercurrent, high-temperature, and pressure process. The crude product is purified by fractional crystallization [3], especially where unsaturation is a factor, or by distillation [4]. Some of the physical properties of fatty acids are given in Table 4. Their chemical reactions used in preparation of anionic surfactants will be discussed in Sec. III.

2. Tall-Oil Fatty and Rosin Acids

These acids occur in nature in pine wood as esters of glycerol or sterols. In the sulfate process for making paper pulp from pine wood, most of the esters are saponified to form salts of the fatty and rosin acids. In 1973, over 28 million lb of potassium and sodium salts of tall-oil acids were produced [1]. Acidification of the soaps obtained by saponification furnishes crude tall oil containing an approximately equal mixture of fatty and rosin acids contaminated

TABLE 3

Fatty Acid Composition of Selected Fats and Oils, %

Acid	C atoms; double bonds	Castor	Coconut	Corn	Lard	Mustard seed	Neats-foot	Olive	Palm	Palm kernel	Peanut	Soybean	Tallow beef	Tallow mutton	Rapeseed
Caprylic	8; 0		8							4					
Capric	10; 0		7							4					
Lauric	12; 0		48							50					
Myristic	14; 0		17		1		1		1	16			2	1	
Palmitic	16; 0		9	12	26		17	14	46	8	11	11	35	21	3
Stearic	18; 0		2	2	11		3	3	4	2	3	4	16	30	2
Oleic	18; 1	8	6	27	49	32	67	68	38	12	46	25	44	43	32
Ricinoleic	18; 1	88													
Linoleic	18; 2	4	3	57	12	18		13	10	3	31	50	2	5	19
Linolenic	18; 3			1	1	3					2	8			1
Erucic	22; 1					42									24

TABLE 4

Physical Properties of Fatty Acids

Acid	Titer, °C	Value			Unsaponifiable, %
		Acid	Saponification	Iodine	
Lauric, %					
94/96	39–41	279–284	279–284	0.5[b]	0.5
98/100	43–43.6	279–282	279–282	0.2[b]	0.2
Stripped coconut oil	25–28	250–260	252–262	8–14	0.5
Stripped palm-kernel oil[a]	23–27	250–260	250–260	16–22	1.0
Myristic, %					
94/96	50–53	244–254	245–255	1.0[b]	0.5
98/100	51–54	244–254	245–255	1.0[b]	0.5
Palmitic, %					
94/96	60–61	215–225	215–225	2.0[b]	0.5
98/100	61–62	219–224	219–224	2.0[b]	0.5
Stearic, %					
94/96	64–66	195–198	196–199	1–3	1.5
98/100	66–68	195–199	196–200	2.0[b]	1.5
Stearin, pressed					
single	52–53	207–211	207–211	8–10	0.5
double	53–54	208–211	208–211	5–8	0.5
triple	54–56	207–211	207–211	1.5–5	0.5[b]

TABLE 4 (Continued)

Acid	Titer, °C	Value			Unsaponifiable, %
		Acid	Saponification	Iodine	
Tallow[a]	39–43	204–208	205–209	53–57	1.5[b]
Oleine light colored	7–10	186–204	188–206	85–92	2–6
Palm oil[a]	42–46	207–213	209–215	44–54	1.5[b]
Refined tall oil[a]	--	163–168	165–170	153–167	7–8.0
light colored	--	188–193	190–195	155–165	1.5–2.2

Source: Ref. 5, p. 123. Courtesy, Chemical Rubber Publishing Co., Cleveland, Ohio.
[a]Distilled.
[b]Maximum.

with about 10% neutral materials (polycyclic hydrocarbons, sterols, and high-molecular-weight alcohols). The composition of some crude tall oils is given in Table 5.

Distillation of crude tall oil yields a product having the composition and properties given in Table 6, while tall oil refined by treatment with concentrated sulfuric acid has the composition and properties shown in Table 7.

Detailed descriptions are available of the processing and refining of tall oil in Refs. 6 and 7. Fractional distillation effects a good separation of the rosin and fatty acids. A typical composition of commercial tall-oil fatty acids is given in Table 8. It is evident that oleic and linoleic acids are the principal fatty acids found in low-rosin tall oil. Some of the properties of the tall oil fatty acids are given in Table 9.

TABLE 5

Composition of Some Crude Tall Oils

Pine-tree source	Fatty acids, %	Rosin acids, %	Neutral fraction, %
Swedish	40–58	30–50	6–15
Finnish	37–59	32–49	6.8–11.3
Finnish (whole trees)	29–36	52–57	9–11
Finnish (sawmill waste)	49–62	30–39	5–12
Danish	50	43	6.7
American	18–53	35–65	8–24
Canadian mixed sample	46	28	25
S.E. Virginian	55–56	39–40	4.8–5.8
N.E. North Carolinan	53–54	40–41	5.3–6.1
S.W. North Carolinan	51–53	41–42	6.1–6.9
E. Central S. Carolinan	46–48	45–47	6.2–7.4
E. Central Georgian	45–48	45–48	6.2–7.4
S.E. Georgian	43–46	47–51	6.3–7.4
S.E. Texas	45–48	46–48	5.3–7.1

Source: Ref. 6, p. 617. Courtesy, John Wiley & Sons, Inc., New York.

The composition and properties of rosin obtained by fractionation of crude tall oil are listed in Table 10.

TABLE 6

Distilled Tall Oil, Composition and Properties

Average composition, %		Typical properties	
Fatty acids	60–85	Color, Gardner	4–12
Rosin	14–37	Specific gravity, 25° C (77%F)	0.940–0.950
Neutral materials	1–3	Acid number	180–190
		Saponification number	185–195
		Viscosity (Gardner-Holdt)	B–E
		Flash point (open cut)	
		°C	182–210
		°F	360–410

Source: Ref. 6, p. 617. Courtesy, John Wiley & Sons, Inc., New York.

TABLE 7

Acid-Refined Tall Oil, Composition and Properties

Average composition, %		Typical properties	
Fatty acids	50–70	Color, Gardner	8–12
Rosin	25–42	Specific gravity, 25° C (77%F)	0.900–1.000
Neutral materials	5–7	Acid number	155–170
		Saponification number	160–175
		Viscosity (Gardner-Holdt)	$Z–Z_2$
		Flash point (open cup)	
		°C	204–216
		°F	400–420

Source: Ref. 6, p. 620. Courtesy, John Wiley & Sons, Inc., New York.

The major carboxylic acids found in tall-oil rosin are derivatives of alkyl hydrophenanthrene. Typical formulas and percentages are given in Table
11.

TABLE 8

Typical Composition of Tall Oil Fatty Acids in Low-Rosin Tall Oil

Fatty acid	Percent in low-rosin tall oil
Palmitic	1
Palmitoleic	--
Stearic	2
Oleic	51
Linoleic (nonconjugated)	40
Linoleic (conjugated)	5
Unknown	4

Source: Ref. 7, p. 26. Courtesy, Pulp Chemicals Association, New York.

TABLE 9

Properties of Commercial-Type Tall-Oil Fatty Acids

Property	Type I[a] Min	Type I[a] Max	Type II[a] Min	Type II[a] Max	Type III[a] Min	Type III[a] Max
Acid number	197		192		190	
Rosin acids, %		1.0		2.0		10.0
Neutral materials, %		1.0		2.0		10.0
Fatty acids, %	98		96		90	
Color, Gardner		4		5		10.0
Iodine number	125	135				

Source: Ref. 6, p. 621. Courtesy, John Wiley & Sons, Inc., New York.
[a]Grouped by rosin content.

TABLE 10

Tall-Oil Rosin, Composition and Properties

Average composition, %		Typical properties	
Rosin acids	90–95	Color, USDA rosin scale	X–N
Fatty acids	2–3	Acid number	162–172
Neutral materials	3–7	Saponification number	170–180
		Softening point	
		°C	73–83
		°F	163–181

Source: Ref. 6, p. 619. Courtesy, John Wiley & Sons, Inc., New York.

TABLE 11

Principal Resin Acids Found in Tall Oil Rosin

Component	Structure	Approximate percentage
Abietic acid	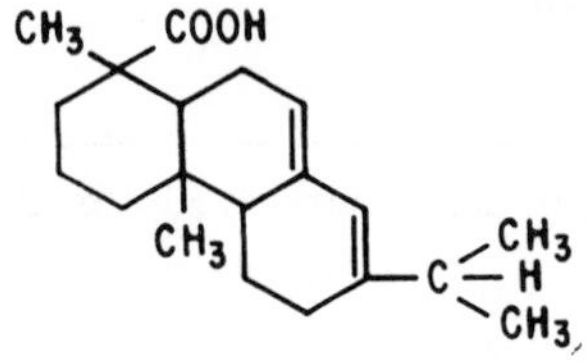	30–40
Dehydroabietic acid	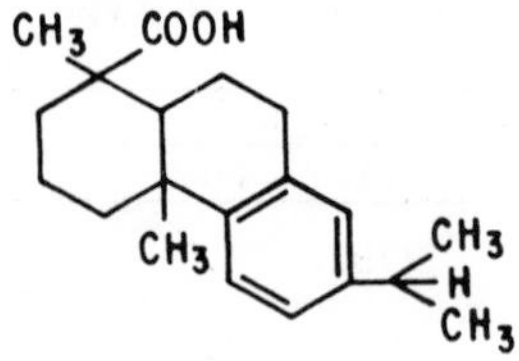	5
Neoabietic acid	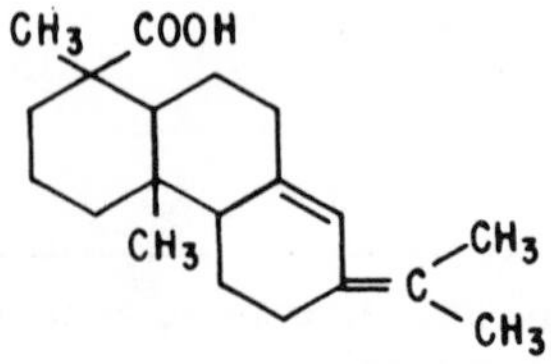	10–20

TABLE 11 (Continued)

Component	Structure	Approximate percentage
Dextropimaric acid		8
Isodextropimaric acid		8
Dihydroabietic acid		14
Tetrahydroabietic acid		14

Source: Ref. 6, p. 626, and Ref. 7, p. 33. Courtesy, John Wiley & Sons, Inc., New York, and Pulp Chemicals Association, New York, respectively.

Reactions of carboxylic acids found in tall oil will be discussed subsequently; however, mention should be made of their neutralization. Water-soluble soaps of tall-oil fatty acids are prepared by neutralization of these acids with sodium, potassium, or ammonium hydroxide. Oil-soluble "metallic" soaps can be prepared by treating water-soluble soaps with an appropriate magnesium, aluminum, or calcium salt. Rosin acids form similar salts or soaps although the carboxyl group is somewhat sluggish to react because of the bulkiness of the groups attached to it. Other chemical reactions are described in Sec. II.

B. Lignosulfonates

Sulfonates prepared from lignin are an important source of anionic surfactants (see Table 1). They are prepared from coniferous (soft) and deciduous (hard) varieties of wood, as well as various other plants. The lignin content of these woods depends upon the source and location of the tree and on the age of the plants. Tables 12 and 13 list the lignin content of various woods and plants. Lignin is believed to be a complex system of polymers derived from 4-hydroxyphenylpropane and related compounds [8]. Sulfonates of lignin are formed when wood chips are heated at 125-145° C under pressure with a bisulfite (usually calcium bisulfite) and sulfur dioxide. Over half of the wood is dissolved forming soluble lignosulfonic acids and salts. Insoluble basic calcium salts are prepared by addition of calcium hydroxide. These salts can be separated and converted to other appropriate derivatives particularly effective as dispersing agents of carbon in aqueous systems.

C. Natural Fats and Oils

In 1973, almost 32 million lb of natural fats and oils were sulfated for use as anionic surfactants. The natural fats were obtained from vegetable sources

TABLE 12

Lignin Content of Some Varieties of North American Wood

Coniferous	Percent[a]	Deciduous	Percent
Douglas fir	27.2	Beech	21.0
Noble fir	29.3	Trembling aspen	19.3
Englemann spruce	26.3	White birch	20.0
Jack pine	26.7	Yellow birch	22.7
Slash pine	28.0	Chestnut oak	24.3
Western hemlock	27.8	Red maple	22.8

Source: Ref. 9, p. 368. Courtesy, John Wiley & Sons, Inc., New York.
[a]Lignin as percentage of dry unextracted wood.

TABLE 13

Lignin in Miscellaneous Plant Materials

Source	Percent	Source	Percent
Spruce wood (normal)	26	Rice hulls	40.0
Spruce wood (compression)	38	Peanut shells	28.0
Eucalyptus (normal)	22	Bune from flax	24.4
Eucalyptus (tension)	16	Barley straw	16–22
Loblolly pine (early sapwood)	28.1	Bagasse	20.3
		Coconut shells	31.9
Loblolly pine (late sapwood)	26.8	Oat straw	14–22
Loblolly pine (early heartwood)	26.8	Clover	4.8
		Hay	7.3
Loblolly pine (late heartwood)	24.2	Alfalfa	23
		Club moss	37
Black spruce bark (outer)	33.9	Peat moss	4.5
Black spruce bark (inner)	6.6	Flax straw	21.8
Black spruce (cambium)	1.8	Pine needles	23.9
Jute	14.2	Wheat straw	13.9
Jute sticks	19.6	Corncobs	13.4
Bamboo	29–35		

Source: Ref. 9, p. 368. Courtesy, John Wiley & Sons, Inc., New York.

such as castor, coconut, and soybean oil, from mixed fish oils, cod and sperm oil, and from animal sources including tallow and neatsfoot oil. Production figures for 1973 [1] are listed in Table 14. Statistics are not available regarding the production of other sulfated natural fats and oils, including mustard seed, peanut, ricebran, herring, and whale oil, and lard and grease although their use has been reported [1].

The technology concerned with separation of fats and oils from their natural sources is extensive and highly specialized [10]. Oils from animal tissues are extracted by means of heat treatment or "rendering." Two general methods of rendering exist. In "dry" rendering the animal material is cooked with agitation at atmospheric, high, or reduced pressures. In "wet" rendering, often used in extraction of lard, tallow and whale oil, the fat is heated in the presence of water or with steam. Vegetable oil seeds are

TABLE 14

U.S. Production of Sulfated Natural Fats and Oils, 1973

Sulfated natural fats and oils, sodium salt	Production, 1000 lb
Castor	5,920
Coconut	871
Cod	1,666
Herring	690
Mixed fish	4,023
Neatsfoot	2,066
Ricebran	9
Soybean	614
Sperm	778
Tallow	5,860
All other	9,356
Total	31,853

cleaned, dehulled, and reduced in size before they are pressed or extracted with solvents.

As previously discussed, fats and oils are usually complex mixtures of glycerides of fatty acids. The esters are randomly distributed between the 1- and 2-positions of glycerol and their location can affect the physical properties such as melting point and crystal structure of the fats. To a minor extent, mono- and diglycerides exist in natural fats and result in the presence of free hydroxyl groups. For detailed discussions and description of the glyceride composition of animal and vegetable fats, see Ref. 11. The composition of some selected fats and oils is given in Table 3.

In contrast with most fats and oils, sperm oil is a mixture of esters of fatty acids bound mainly to higher fatty alcohols rather than glycerol. Industrial sperm oil is obtained from both the blubber and head cavities of the sperm whale. Hilditch [12] reported sperm-head oil to consist of 74% of fatty esters and 26% of triglycerides, while blubber oil contained 66 and 34%, respectively, of these components. The fatty acid and alcohol composition of the oils obtained from sperm whales was determined by the same workers [13] and is given in Table 15 and 16. The preparation of long-chain alcohols from sperm oil and other sources will be discussed in Sec. D.

TABLE 15

Weight Percent of Fatty Acid Components of Sperm Whale
Head and Blubber Oils

Acid	Weight %	
	Head oil	Blubber oil
Saturated		
Decanoic	3.5	--
Lauric	16	1
Myristic	14	5
Palmitic	8	6.5
Stearic	2	--
Unsaturated		
C_{12}	4 (-2H)[a]	--
C_{14}	14 (-2H)	4 (-2H)
C_{16}	15 (-2H)	26.5 (-2H)
C_{18}	17 (-2H)	37 (-2H)
C_{20}	6.5 (-2H)	19 (-2.5H)
C_{22}	--	1 (-4H)

Source: Ref. 12, p. 72. Courtesy, John Wiley & Sons, Inc., New York.
[a]Denotes double bond.

During sulfation of natural fats and oils, sulfuric acid attacks the carbon-carbon double bonds and leads to formation of sulfates, $R-CH_2-CH(OSO_2Na)-R'$, as well as some by-products caused by hydrolysis or formation of lactones, lactides, and polymers. Completely saturated fatty acid chains are for the most part unaffected. Where hydroxyl groups are present, as in castor oil or in monoglycerides, sulfation of these functional groups occurs to yield esters. These types of surfactants are discussed in Chapter 7.

D. Fatty Alcohols

Unlike fatty acids, aliphatic alcohols are not found in large quantities in nature. However, both cetyl and oleyl alcohols make up a significant portion of the component alcohols found in sperm oil (see Table 16). In general, the fatty alcohols are obtained from the oil by saponification and are purified by crystallization procedures similar to those used for fatty acids.

In order to supplement the quantities of fatty alcohols available from natural sources, the Bouveault-Blanc method of reduction of esters using sodium and a lower alcohol was extended to glycerides. Thus C_{12}-C_{16} alcohols
were obtained from coconut and palm kernel oils and C_{16}-C_{18} alcohols from
tallow. The process has been described by Schroeder [14]. The overall reaction is shown in Eq. (1):

$$\underset{\displaystyle \text{R—C—O—R'}}{\overset{\displaystyle \overset{\textstyle O}{\|}}{}} + 4\,\text{Na} + 2\,\text{R'—OH} \longrightarrow \text{R—CH}_2\text{ONa} + 3\,\text{R'ONa} \qquad (1)$$

Isolated or nonconjugated double bonds are unaffected.

High-pressure catalytic hydrogenation has been used with fatty acids,
their lower esters, or their glycerides [15]. The reaction is carried out at
200 to 250 atm at about 300° C, usually using copper chromite as catalyst, as
shown in Eq. (2):

$$\underset{\displaystyle \text{RC—OH}}{\overset{\displaystyle \overset{\textstyle O}{\|}}{}} + 2\,\text{H}_2 \longrightarrow \text{RCH}_2\text{OH} + \text{H}_2\text{O}$$

or

$$
\begin{array}{l}
\text{CH}_2\text{—O—}\overset{\overset{\textstyle O}{\|}}{\text{C}}\text{—R} \\
\quad\big| \\
\text{CH—O—}\overset{\overset{\textstyle O}{\|}}{\text{C}}\text{—R} \\
\quad\big| \\
\text{CH}_2\text{—O—}\underset{}{\overset{\overset{\textstyle O}{\|}}{\text{C}}}\text{—R}
\end{array}
+ 8\,\text{H}_2 \longrightarrow
\begin{array}{l}
\quad\text{CH}_3 \\
\quad\big| \\
\text{CH—OH} \\
\quad\big| \\
\quad\text{CH}_3
\end{array}
+ 3\,\text{RCH}_2\text{OH} + 2\,\text{H}_2\text{O} \qquad (2)
$$

The reaction is nonselective and causes hydrogenation of double bonds and
formation of saturated alcohols. Other methods for catalytic reduction of animal fats to yield unsaturated alcohols such as oleyl alcohol have been studied
[16, 17]. Yields of 60-70% of oleyl alcohol from red-oil grade oleic acid have
been reported recently using a mixed catalyst systems of Cr—Zn—Cd—Al at
350° C and 3000 psi [18].

Some of the properties of commercial fatty alcohols obtained by high-
pressure hydrogenation of fats and oils are given in Tables 17 and 18. Preparation of higher alcohols from petrochemical sources has been discussed in
Chapter 2. The alcohols, natural or synthetic, find wide use in the preparation of anionic detergents such as alcohol sulfates and sulfates of ethylene
oxide adducts of the alcohols. The synthesis of these surfactants will be described in a subsequent chapter.

TABLE 16

Weight Percent of Fatty Alcohol Components of Sperm Whale Head and Blubber Oils

Alcohol	Weight %	
	Head oil	Blubber oil
Saturated		
C_{14} Tetradecyl	8	--
C_{16} Hexadecyl (cetyl)	44	25
C_{18} Octadecyl	6	1
Unsaturated		
C_{16} Hexadecenyl	4	--
C_{18} Octadecenyl (oleyl)	28	66
C_{20} Eicosenyl	10	8

Source: Ref. 12, p. 73. Courtesy, John Wiley & Sons, Inc., New York.

TABLE 17

Properties of Pure Fatty Alcohols

Alcohol	Formula	Molecular weight	Melting point, °C	Boiling point, °C	Hydroxyl number
Decanol	$C_{10}H_{21}OH$	158	7	231^{760}	355
Undecanol	$C_{11}H_{23}OH$	172	14	131^{15}	326
Dodecanol (lauryl)	$C_{12}H_{25}OH$	186	24	$135-137^{10}$	301
Tridecanol	$C_{13}H_{27}OH$	200	30	155^{13}	280
Tetradecanol (myristic)	$C_{14}H_{29}OH$	214	38	$159-161^{10}$	262
Pentadecanol	$C_{15}H_{31}OH$	228	44		246
Hexadecanol (palmitic or cetyl)	$C_{16}H_{33}OH$	242	49	$179-182^{12}$	232
Heptadecanol	$C_{17}H_{35}OH$	256	54		219
Octadecanol (stearyl)	$C_{18}H_{37}OH$	270	58	202^{10}	208
Octadecanol (oleyl)	$C_{18}H_{35}OH$	268	15-16	$177-183^{3}$	209

Source: Ref. 5, p. 73. Courtesy, Chemical Rubber Publishing Co., Cleveland, Ohio.

TABLE 18

Properties of Typical Commercial Fatty Alcohols
Derived from Oils and Fats

Alcohol	Number of carbons	Iodine number	Solidification point, °C	Boiling range, [a] °C	Hydroxyl number
Lauryl					
Wide-range	$C_{10}-C_{18}$	<0.5	17-21	220-320	275-285
80% C_{12}	$C_{12}-C_{14}$	<0.5	17-23	255-85	283-293
Myristyl,					
95% C_{14}	C_{14}	<0.5	36-38	280-95	255-62
Cetyl,					
95% C_{16}	C_{16}	<0.5	46-49	316-30	225-35
Stearyl,					
95% C_{18}	C_{18}	<0.5	55-57	340-55	203-10
Tallow fatty	$C_{14}-C_{18}$	<0.5	48-52	120-90	210-20
Oleyl-cetyl					
mixture	Mainly C_{18} [b]	45-120	4-35	310-65	200-20

Source: Ref. 5, p. 72. Courtesy, Chemical Rubber Publishing Co.,
Cleveland, Ohio.
[a]At 760 mm Hg.
[b]Saturated and unsaturated.

III. CHEMICAL MODIFICATIONS OF NONPETROCHEMICAL RAW MATERIALS

A few of the reactions necessary to convert nonpetrochemical raw materials
into anionic surfactants have been alluded to in the previous part of this chapter. These have included saponification of fats and oils, sulfation of naturally occurring materials, and sulfation of fatty alcohols obtained by reduction or hydrogenolysis of glycerides and fatty acids. Other modifications will
now be presented. Several of these reactions will be discussed in depth in
later chapters.

A. Esters and Ether-Esters

The formation of long-chain esters has been amply documented [19]. Their
sulfated derivatives are important anionic surfactants. Some of the various
types of esters are briefly described below.

1. Sulfated Esters

Esters of low-molecular-weight alcohols and oleic or ricinoleic acid yield anionic surfactants when sulfated. In 1973, over 4.7 million lb of sulfated oleic acid esters were produced [1]. They are itemized in Table 19.

2. Sulfated Monoglycerides

These materials are prepared by sulfation of monoglycerides [20, 21] and ideally have the structure:

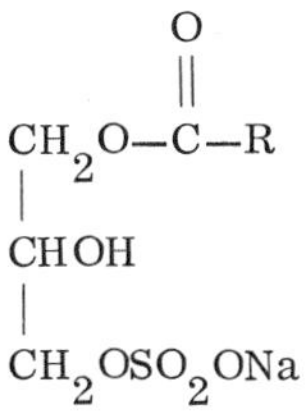

$$
\begin{array}{l}
\overset{\displaystyle O}{\overset{\displaystyle \|}{}} \\[-2pt]
CH_2O-C-R \\
| \\
CHOH \\
| \\
CH_2OSO_2ONa
\end{array}
$$

The monoglycerides are usually synthesized by direct esterification of glycerol with a fatty acid, although other methods have been used [22]. A complete discussion follows in Chaps. 6 and 7.

3. Esters of Isothionic Acid (Igepon A)

These esters are prepared by the reaction of fatty acids with sodium isothionate, using an acidic catalyst [23], as shown in Eq. (3):

$$RCOOH + HOCH_2CH_2SO_3Na \longrightarrow$$

$$RCOOCH_2CH_2SO_3Na + H_2O \qquad (3)$$

TABLE 19

U.S. Production of Sulfated Esters of Oleic Acid, 1973

Sulfated oleate, sodium salt	Production, 1000 lb
Butyl	1,458
Propyl	517
Others	2,784
Total	4,759

More recently, Bistline and co-workers [24] have used isopropenyl stearate as acylating agent for sodium isothionate at 200° C, see Eq. (4). Yields of 95% were reported.

$$RCOOC(CH_3){=}CH_2 + HOCH_2CH_2SO_3Na \longrightarrow$$

$$RCOOCH_2CH_2SO_3Na + CH_3COCH_3\uparrow \quad (4)$$

4. Sulfated Ethylene Oxide Derivatives

The reaction of ethylene oxide with long-chain acids, alcohols, etc., has been extensively studied [25, 26]. The ethoxylated materials, particularly of higher alcohols, give products with properties resembling those of fatty alcohol sulfates [27].

B. Nitrogen Derivatives

Nonpetrochemical raw materials, principally long-chain acids, are used to prepare nitrogen-containing derivatives. Although the latter are of particular importance in synthesis of cationic surfactants [28], they can be useful starting materials for preparation of anionics.

1. Fatty nitriles and amines

Fatty nitriles are obtained industrially by the reaction of fatty acids with ammonia in a counter-current apparatus at 280-330° C under pressures of 100 psi [29, 30], as shown in Eq. (5).

$$RCOOH + NH_3 \rightleftharpoons RCONH_2 + H_2O\uparrow \rightleftharpoons RCN + H_2O\uparrow \quad (5)$$

Reduction of nitriles using a Raney nickel catalyst results in formation of the corresponding amine, see Eq. (6).

$$RCN + 2H_2 \longrightarrow RCH_2NH_2 \quad (6)$$

Special conditions are necessary to avoid reduction of double bonds during the hydrogenation with Raney nickel [31].

2. Fatty Amides

a. Simple Amides

Simple amides of fatty acids are produced in an autoclave by heating the acid with ammonia at about 200° C. Water and ammonia are removed continuously, see Eq. (7).

$$RCOOH + NH_3 \rightleftharpoons RCOONH_4 \rightleftharpoons RCONH_2 + H_2O \qquad (7)$$

Addition of alkylene oxides and subsequent sulfation will be discussed in Chaps. 6 and 7.

b. Taurine Derivatives

Fatty amides of N-methyltaurine have been synthesized from oleoyl chloride and methyltaurine using the Schotten-Baumann technique, see Eq. (8).

$$RCOCl + CH_3-\overset{\overset{\displaystyle H}{|}}{N}-CH_2CH_2SO_3Na \xrightarrow{NaOH}$$
$$RCON(CH_3)CH_2CH_2SO_3Na + NaCl \qquad (8)$$

The use of isopropenyl stearate as acylating agent for N-methyltaurine has been reported and leads to formation of acetone as a volatile byproduct [24], see Eq. (9).

$$RCOCl + CH_3-\overset{\overset{\displaystyle H}{|}}{N}-CH_2CH_2SO_3Na \xrightarrow{NaOH}$$
$$RCON(CH_3)CH_2CH_2SO_3Na + CH_3COCH_3 \uparrow \qquad (9)$$

c. Derivatives of Amino Acids and Proteins

Acylated amino acids and proteins have been shown to possess surface activity and continue to grow in importance. Over 4.5 million lb of salts of N-lauroylsarcosine, $C_{11}H_{23}CON(CH_3)-CH_2-COONa$, were produced in 1973 [1]. Coconut fatty acid and oleic acid or their acid chlorides have been reported to convert low-molecular-weight proteins from casein or leather into surfactants [32]. These types of surfactants are discussed in Chap. 16.

REFERENCES

1. United States Tariff Commission, Synthetic Organic Chemicals, United States Production and Sales, 1973. T. C. Publication 728, U.S. Government Printing Office, 1975.
2. R. H. Potts and V. J. Muckerheide, in Fatty Acids and Their Industrial Applications (E. S. Pattison, ed.), Marcel Dekker, Inc., New York, 1968, pp. 25-29.
3. R. H. Potts and V. J. Muckerheide, in Fatty Acids and Their Industrial Applications (E. S. Pattison, ed.), Marcel Dekker, Inc., New York, 1968, pp. 38-44.

4. R. H. Potts and V. J. Muckerheide, in Fatty Acids and Their Industrial Applications (E. S. Pattison, ed.), Marcel Dekker, Inc., New York, 1968, pp. 29-38.

5. A. Davidsohn and B. M. Milwidsky, Synthetic Detergents, Chemical Rubber Publishing Co., Cleveland, Ohio, 1968.

6. Dan C. Tate, Tall Oil, in Encyclopedia of Chemical Technology, Vol. 19, 2nd ed (A. Standen, ed.), John Wiley & Sons, Inc., New York, 1969, pp. 614-629.

7. L. G. Zachary, H. W. Bajak, and F. J. Eveline, Tall Oil and Its Uses, Tall Oil Products Division of Pulp Chemicals Association, New York, 1965, Chaps. 1 and 2.

8. I. A. Pearl, The Chemistry of Lignin, Marcel Dekker, Inc., New York, 1967, pp. 4-5.

9. D. W. Goheen, D. W. Glennie, and C. H. Hoyt, Lignin, in Encyclopedia of Chemical Technology, Vol. 12, 2nd ed. (A. Standen, ed.), John Wiley & Sons, Inc., New York, 1967, pp. 361-381.

10. F. A. Norris, in Bailey's Industrial Oil and Fat Products (D. Swern, ed.), John Wiley & Sons, Inc., New York, 1964, Chap. 15.

11. D. Swern, in Bailey's Industrial Oil and Fat Products (D. Swern, ed.), John Wiley & Sons, Inc., New York, 1964, pp. 8-11.

12. T. P. Hilditch and P. N. Williams, The Chemical Constitution of Natural Fats, 4th ed., John Wiley & Sons, Inc., New York, 1964, p. 488.

13. T. P. Hilditch and P. N. Williams, The Chemical Constitution of Natural Fats, 4th ed., John Wiley & Sons, Inc., New York, 1964, pp. 72-73.

14. K. P. Schroeder, J. Am. Oil Chemists' Soc., 33, 565-568 (1956).

15. E. F. Hill, G. R. Wilson, and E. C. Steinle Jr., Ind. Eng. Chem., 46, 1917-1921 (1954).

16. H. Bertsch, H. Reinheckel, and K. Haage, Fette, Seifen, Anstrichmittel, 71, 357-362 (1969).

17. J. D. Richter and P. J. Van den Berg, J. Am. Oil Chemists' Soc., 46, 155-166 (1969).

18. R. S. Klonowski, T. W. Findley, C. M. Josefson, and A. J. Stirton, J. Am. Oil Chemists' Soc., 47, 326-328 (1970).

19. K. S. Markley, Fatty Acids, 2nd ed., Interscience Publishers, Inc., New York, 1961, pp. 757-984.

20. A. C. Bell and W. G. Alsop, U.S. Pat. 2,478,354 (1949), to Colgate-Palmolive Co.

21. M. H. Ittner, U.S. Pat. 2,496,328 (1950), to Colgate-Palmolive Co.

22. H. A. Goldsmith, Chem. Rev., 33, 257 (1943).

23. L. M. Schenk, U.S. Pat. 3,004,049 (1961), to General Aniline & Film Corp.

24. R. G. Bistline, E. S. Rothman, S. Serota, A. J. Stirton and A. N. Wrigley, J. Am. Oil Chemists' Soc., 48, 657-660 (1971).

25. A. N. Wrigley, F. D. Smith, and A. J. Stirton, J. Am. Oil Chemists' Soc., 34, 39 (1957).

26. W. B. Satkowski and C. G. Hsu, Ind. Eng. Chem., 49, 1875 (1957).

27. T. P. Matson, Soap Chem. Specialties, 39, 52 (1963).

28. W. M. Linfield, in Cationic Surfactants (E. Jungermann, ed.), Marcel Dekker, Inc., New York, 1970, Chap. 2.
29. R. H. Potts and R. S. Smith, U.S. Pat. 2,808,426 (1957), to Armour & Co.
30. R. H. Potts, U.S. Pat. 3,299,117 (1967), to Armour & Co.
31. R. H. Potts and C. W. Christensen, U.S. Pat. 2,314,894 (1943).
32. H. L. Sanders and M. Nassau, Soap Chem. Specialties, 36 (1), 57 (1960).

CHAPTER 4

THE MECHANISMS OF SULFONATION
AND SULFATION

Ben E. Edwards[*]

Department of Chemistry
University of North Carolina at Greensboro
Greensboro, North Carolina

---

[*]Present affiliation: Old-North Manufacturing Company, Inc., Lenoir, North Carolina.

I. INTRODUCTION

Organic sulfates and sulfonates do not occur in nature in appreciable quantities, and the production of anionic surfactants bearing these groups invariably involves the reactions of sulfonation and sulfation. In the synthesis of
sulfonates a new carbon-sulfur bond is formed, while in sulfation either the
C—O or the O—S bond of the C—O—S sequence may be the new bond formed.
There have been developed both a variety of classes of sulfonates and sulfates suitable for surfactant applications, and a variety of reagents capable
of carrying out the required reactions. In this chapter these reactions will
be identified and the mechanisms by which they are believed to occur will be
discussed.   Since the economics of surfactant production causes continuous
change in what types of products are manufactured and sold, the discussion
will not be limited to those materials currently being marketed.

II. RAW MATERIALS

The classification of raw materials according to sources is a convenient or-
ganizational device, although we shall see that reaction mechanisms are generally insensitive to the origins of the functional groups involved. A commercial product may consist of a mixture of components formed in a single
reaction, but to study mechanisms we consider the reaction of a pure starting
material undergoing conversion to a single product. Nevertheless, conditions
which convert a pure alcohol to its sulfate, also convert a mixture of fatty
alcohols or of petroleum-derived alcohols to their sulfates, and the understanding of reaction mechanisms is important at the production level as well
as from the academic standpoint. The listing of raw materials which follows
is indicative rather than comprehensive.

A. Petroleum

Having successfully changed over from the production of branched-chain alkylbenzenes to linear alkylbenzenes, the petroleum industry still supplies the
bulk of all surfactant raw materials. Linear alkylbenzenes are prepared by
alkylation of benzene with monochloro-n-paraffins derived from purified
petroleum fractions, and with α-olefins derived from Ziegler type polymerization of ethylene or from cracking of n-paraffins. In addition to their use in
the synthesis of alkylbenzenes, the α-olefins provide feed stock for the manufacture of alkyl sulfates and sulfonates and the n-paraffins may be used directly in some of the sulfonation processes to be discussed. Numerous more
complex starting materials are also derived more or less directly from
petroleum.

B. Fats and Oils

Though fats and oils are historically the primary raw material of the surfactant industry, the advent of synthetic detergents, particularly of the alkylbenzene sulfonates, has relegated them to a smaller but still important role. Included in the category of fats and oils are animal fats and tallow, fish oils, and oils from plant sources such as cottonseed, coconut, peanut, soy bean, etc. Besides their direct conversion to soap, a number of starting materials for sulfonates and sulfates can be derived from them. The properties of the fatty acids themselves have been modified by sulfation and by sulfonation. Partial hydrolysis of glycerides, followed by sulfation, yields monoglyceride sulfates having desirable surfactant properties. Reduction of fatty acids or esters yields long-chain alcohols which may be converted to sulfates or, via dehydration, to α-olefins.

C. Others

Cellulose and sugar offer a potentially large source of raw material, but to date no practical surfactant product involving sulfonation or sulfation of these materials has been developed. The sulfation of lignin is a highly developed technology, but does not lead to products which we here consider as surfactants.

III. REAGENTS

Gilbert's excellent monograph on sulfonation [1] provides coverage in depth of the many reagents which may be employed. For our purposes we shall list some of the more important and discuss the characteristics of each which may contribute to an understanding of the mechanisms by which they react.

A. Sulfur Trioxide and Its Complexes

Sulfur trioxide is the simplest chemical entity involved in sulfonation and sulfation. The proper representation of the bonding in sulfur trioxide and its derivatives has been the subject of considerable debate. In the vapor state, it is monomeric and has the configuration of an equilateral triangle with sulfur at the center. The S—O bond length is 1.43 Å, indicating considerable π bonding [2]. The molecule is best represented by the set of Lewis structures [1a], [1b], and [1c], having two single bonds and one double, although structures having two or no double bonds may be drawn. Some additional π bonding between sulfur d orbitals and oxygen p is invoked to explain the extremely short S—O distance.

$$\left[\underline{1a}\right] \qquad \left[\underline{1b}\right] \qquad \left[\underline{1c}\right]$$

The electrophilicity of the sulfur atom implied by its tendency toward π bonding is even more apparent in its chemical behavior. In the uncombined form, sulfur trioxide boils at 44.5° C and melts at 16.8° C. However, it combines with itself to yield a cyclic trimer [2] and various liquid and solid polymers [3]. In these polymers and in sulfates, sulfonates, and other derivatives, the sulfur atom becomes the center of a tetrahedral σ bonding array, with the oxygens participating in some pπ-dπ bonding to sulfur [3].

$$[2] \qquad\qquad [3]$$

The commercial use of sulfur trioxide depends on the addition of inorganic or organic stabilizers, such as borates and sulfonic acids, which inhibit polymerization and allow handling of the material as a liquid [4]. Unlike sulfuric acid and oleum, liquid sulfur trioxide (which may consist largely of the trimer) is completely miscible with liquid sulfur dioxide and many halogenated organic solvents. It is so reactive that it must be diluted in some fashion, even in gas-phase reactions, to prevent charring of the reactants.

An alternate to dilution is to use sulfur trioxide in the form of its complexes. Combination of sulfur trioxide with various Lewis bases yields complexes whose reactivity depends primarily on the strength of the base, and to a lesser degree on steric factors. Suitable bases are ethers, tertiary amines, tertiary amides, thioethers, and trialkylphosphates. Typically they are somewhat soluble in organic solvents and deliver sulfur trioxide to a reaction site in a controllable fashion. The complexes with trimethylamine, pyridine, dimethyl formamide, dioxane, and triethyl phosphate are typical of those which have received considerable laboratory attention, but none of them is widely used in commercial production.

B. Sulfuric Acid and Oleum

Sulfuric acid (100%) may be regarded as the 1:1 complex of sulfur trioxide and water. Oleum results when excess sulfur trioxide is present. Compounds having 1:2, 1:3, and 1:5 ratios of sulfur trioxide to water have been detected by freezing-point studies of sulfuric acid [5], and spectral studies of oleum indicate formation of compounds with ratios 2:1 through 4:1 [6].

Sulfonation is carried out with sulfuric acid and with oleum at many different concentrations. The main feature distinguishing these reactions from those using sulfur trioxide is the requirement of additional thermal energy for the former. Since sulfur trioxide is probably still the active electrophilic species in these reactions, this difference in thermal requirements is interpreted as reflecting the considerable stability of the sulfur trioxide-water complex. Other differences in the process are mainly a function of differences in physical properties, e.g., volatility, viscosity, freezing point, and solubility.

C. Chlorosulfonic Acid

Although chlorosulfonic acid is usually represented as the monacid chloride of sulfuric acid, $ClSO_3H$, its mode of preparation and some of its reactions suggest that it may equally well be considered to be the complex $SO_3 \cdot HCl$. It freezes at $-80°C$, boils at $152°C$, is soluble in partially halogenated solvents, but only slightly soluble in perhalogenated, and dissociates to SO_3 and HCl at the boiling point [7].

Chlorosulfonic acid has been largely displaced by sulfur trioxide in industrial applications because of the difficulties inherent in the disposal of by-product HCl, but it remains an important laboratory reagent for sulfonation and sulfation, or when a sulfonyl chloride is the desired product.

D. Others

Other reagents which may be involved in sulfonation and sulfation are sulfuryl chloride (SO_2Cl_2), sulfamic acid (H_2NSO_3H), sulfur dioxide, and sulfite ion. This listing is by no means exhaustive, but it covers all important types.

Sulfuryl chloride is formed by the combination of sulfur dioxide and chlorine, and may act either as a chlorinating or sulfonating reagent according to the nature of the substrate and the reaction conditions. It may also react with alcohols to give alkylchlorosulfonates. In a reaction catalyzed by pyridine, alkenes yield α-chlorosulfonyl chlorides, and hydrocarbons yield sulfonyl chlorides and HCl [8].

Sulfamic acid is prepared by the reaction of sulfur trioxide and sulfuric acid with urea [Eq. (1)].

$$H_2NCONH_2 + H_2SO_4 + SO_3 \longrightarrow 2\,H_2NSO_3H + CO_2 \uparrow \qquad (1)$$

It is a stable, nonhygroscopic solid, m.p. 205°C, with an acidity about equal to sulfuric acid. It resembles the sulfur trioxide tertiary amine complexes in reactions involving sulfonation or sulfation, but differs from them in that it is used at elevated temperatures in anhydrous media, while the tertiary amine complexes are employed at low temperatures in aqueous alkali. Sulfamic acid is used primarily in the sulfation of alcohols, though even here it is not often the reagent of choice [9].

The chemistry of sulfur dioxide has been reviewed by Schroeter [10]. It is used directly in the reactions of sulfochlorination and sulfoxidation and indirectly as sulfite ion in a number of possible syntheses of sulfonates. Like sulfur trioxide, sulfur dioxide is electrophilic, but reacts most frequently via free-radical processes. The sulfite ion is a surprisingly strong nucleophile and reacts both by nucleophilic substitution processes and by free-radical processes. Even when the reagent is nominally a bisulfite, the more nucleophilic sulfite ion may be the active nucleophile despite the fact that it is present only in small amounts. The high nucleophilicity of the sulfite ion is explained in terms of structure [4] in which sulfur is in an sp^3 hybrid state with considerable pd-π bonding between oxygen and sulfur. This provides a readily available and highly polarizable orbital for bond formation. The easy oxidation of sulfite ion to the ion radical accounts for its activity in free-radical reactions.

$$\left[\begin{array}{c} \overset{\displaystyle \cap}{O} \\ O \cdots S \cdots O \\ \downarrow \\ O \end{array} \right]^{=}$$

[4]

Though these reactions are not currently significant commercially, sulfur dioxide is an important chemical raw material for other reasons.

IV. MECHANISMS OF SULFONATION

Reaction mechanisms are generally classified according to the nature of the bond-breaking and bond-making processes and the character of the reagent species. Through an understanding of reaction mechanisms it may be possible to alter product compositions, or even to obtain entirely different products as a result of modifying the character of the actual reactive intermediates. As will be shown, this control is often exercised merely by changes in reaction temperatures or reagent concentrations.

A. Aromatic Substitution

The generally accepted mechanisms for electrophilic aromatic substitution is set forth in every modern elementary organic textbook. In simplest form it is a two-step process, involving the combination of an electrophile with the aromatic system to give an intermediate σ-complex which subsequently eliminates a proton to give the observed product [Eq. (2)].

$$Ar{-}H + X^+ \longrightarrow \left[Ar \overset{H}{\underset{X}{\diagup\!\!\!\diagdown}} \right]^+ \longrightarrow Ar{-}X + H^+ \tag{2}$$

The mechanisms of aromatic sulfonation have been reviewed recently by Cerfontain [11] in a monograph and by Cerfontain and Kort [12] in a short summary. Another excellent short review, including 383 references, was prepared by Nelson [13]. The main unanswered question concerning sulfonation by electrophilic aromatic substitution is the exact nature of the electrophile. Among other possibilities, it may be sulfur trioxide itself, $H_2S_2O_7$ or SO_3H^+, depending on the reaction conditions. Kinetic studies of sulfonation must consider the equilibria in the reagents which produce attacking electrophile, as well as those leading to products from the initial σ-complex elimination products.

Temperature and acid strength have a strong influence on orientation in sulfonation. After a short reaction time with toluene in 85% H_2SO_4 at 120°C the toluene sulfonate isolated was 20% ortho, 5% meta, and 75% para, but at 160°C the percentages were 13, 4, and 83, respectively. At 25°C in 80.6% acid the product was 25% ortho, but in 98.5% acid it was 50% ortho [14]. These data indicate that the species $H_3SO_4^+$ is more selective and more subject to steric hindrance than $H_2S_2O_7$ [15]. For the long-chain alkylbenzenes important in detergent manufacture, steric effects lead to a product nearly exclusively para substituted, whatever the attacking species.

In solution in anhydrous aprotic solvents such as carbon tetrachloride and trichlorofluoromethane, sulfur trioxide is present mainly as the monomer, whereas in the pure liquid state it may be mostly trimer in equilibrium with some monomer [16]. The sulfonating activity of sulfur trioxide is lower in nitromethane or nitrobenzene than in trichlorofluoromethane, indicating a greater degree of complexing with the solvent in the former cases.

The sulfonation of p-dichlorobenzene with sulfur trioxide in trichlorofluoro methane is first order in both substrate and reagent,

$$\text{rate} = k(ArH)(SO_3)$$

and shows no isotope effect. The rate-determining step in the reaction may therefore follow Eq. (3):

$$ArH + SO_3 \rightleftharpoons \left[Ar \overset{SO_3^-}{\underset{H}{\diagup\!\!\!\diagdown}} \right]^+ \tag{3}$$

Sulfonation in nitrobenzene or nitromethane is first order in substrate, but second order in sulfur trioxide:

$$\text{rate} = k(\text{ArH})(\text{SO}_3)^2$$

Wadsworth and Hinshelwood [17] proposed that the attacking species might be S_2O_6, but no evidence for the existence of such a dimer has been reported, even in liquid sulfur trioxide. Cerfontain [12] postulates that the initial complex formation [Eq. (2)] is followed by a rate-determining reaction with another sulfur trioxide molecule [Eq. (4)], and finally by removal of the aromatic proton [Eq. (5)].

$$\left[\text{Ar} \begin{array}{c} \diagup \text{SO}_3^- \\ \diagdown \text{H} \end{array} \right]^+ + \text{SO}_3 \rightleftharpoons \left[\text{Ar} \begin{array}{c} \diagup \text{S}_2\text{O}_6^- \\ \diagdown \text{H} \end{array} \right]^+ \qquad (4)$$

$$\left[\text{Ar} \begin{array}{c} \diagup \text{S}_2\text{O}_6^- \\ \diagdown \text{H} \end{array} \right]^+ \longrightarrow \text{Ar}-\text{S}_2\text{O}_6\text{H} \qquad (5)$$

He further suggests that this mechanism applies also to reactions in trichlorofluoromethane, but in this medium the greater reactivity of sulfur trioxide makes the reaction in Eq. (4) fast relative to that in Eq. (5), which becomes the rate-determining step. Other pathways involving solvent-S_2O_6, solvent-SO_3^+H, and ArS_2O_6H as the active sulfonating species may also be important. Reaction in anhydrous sulfur dioxide probably proceeds in much the same way as in nitromethane.

Sulfonation in oleum is more complex because of the presence of species such as $H_2S_2O_7$ and $H_2S_4O_{13}$. Rate studies have been carried out only on substrates of very low reactivity such as phenyltrimethylammonium ion and sulfonic acids. The observed rates are approximately

$$\text{rate} = k(\text{ArH})(\text{SO}_3)(\text{H}^+)$$

indicating involvement of a protonated species in the rate-determining step. The mechanism assumed in weak oleum (up to 105% sulfuric acid) is shown in Eqs. (6) and (7),

$$\text{ArH} + \text{H}_3\text{S}_2\text{O}_7^+ \rightleftharpoons \left[\text{Ar} \begin{array}{c} \diagup \text{SO}_3\text{H} \\ \diagdown \text{H} \end{array} \right]^+ + \text{H}_2\text{SO}_4 \qquad (6)$$

$$\left[\text{Ar} \begin{array}{c} \diagup \text{SO}_3\text{H} \\ \diagdown \text{H} \end{array} \right]^+ + \text{H}_2\text{SO}_4 \longrightarrow \text{ArSO}_3\text{H} + \text{H}_3\text{SO}_4^+ \qquad (7)$$

with sulfuric acid being the strongest base available to aid in proton removal.

The mechanism most probably at higher oleum concentrations is shown in Eqs. (8) and (9),

$$ArH + H_2S_4O_{13} \rightleftharpoons \left[Ar \begin{smallmatrix} S_2O_6H \\ \\ H \end{smallmatrix} \right]^+ + HS_2O_7^- \qquad (8)$$

$$\left[Ar \begin{smallmatrix} S_2O_6H \\ \\ H \end{smallmatrix} \right]^+ + HS_2O_7^- \longrightarrow ArS_2O_6H + H_2S_2O_7 \qquad (9)$$

reflecting the higher concentration of polymeric sulfur trioxide species in the medium.

In aqueous sulfuric acid (80-95%) the rate equation, found by correlating the first-order rate constants with the concentrations of the species present in sulfuric acid, is the following:

rate $= k(ArH)(H_2S_2O_7)$

It was concluded that the most probable rate-determining step is formation of the σ-complex from the substrate and $H_2S_2O_7$ [19]. A mechanistic scheme, shown in Fig. 1, involving four reactions is proposed to account for the observed results. At sulfuric acid concentrations from 80-95%, step A is rate-limiting and the product is formed mainly by steps C and D. As sulfuric acid concentration increases, step D becomes partly rate-limiting because of decreased HSO_4^- concentration and step B becomes the main route to the product [20].

With more active substrates in sulfuric acid $> 80\%$, the kinetic expression becomes

rate $= k(ArH)(H_3SO_4^+)$

with the rate-limiting step shown in Eq. (10):

$$ArH + H_3SO_4^+ \longrightarrow \left[Ar \begin{smallmatrix} SO_3H \\ \\ H \end{smallmatrix} \right]^+ + H_2O \qquad (10)$$

$$ArH + H_2S_2O_7 \overset{A}{\rightleftharpoons} \left[Ar \begin{smallmatrix} SO_3H \\ \\ H \end{smallmatrix} \right]^+ + HSO_4^- \overset{C}{\rightleftharpoons} \left[Ar \begin{smallmatrix} SO_3^- \\ \\ H \end{smallmatrix} \right] + H_2SO_4$$

$$\downarrow B \qquad\qquad + HSO_4^- \downarrow D$$

$$ArSO_3H + H_2SO_4 \qquad\qquad ArSO_3^- + H_2SO_4$$

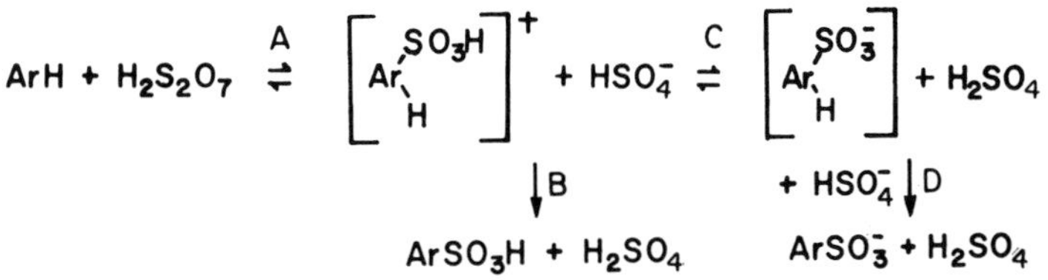

FIG. 1. Mechanism of sulfonation by aqueous sulfuric acid.

Sulfonation in sulfuric acid, unlike the majority of electrophilic aromatic substitution reactions, is an easily reversible process. At elevated temperatures in diluted acid, sulfonate groups may be effectively removed. This explains the considerable differences observed between kinetic and thermodynamic product distributions in sulfonation. It also underlines the necessity of removing or otherwise dealing with the water generated in the process of sulfonation by sulfuric acid.

The presence of long alkyl chains in the arenes being sulfonated for surfactant use causes several other deviations from the behavior observed in simpler systems. Although the alkyl groups do activate the ring towards sulfonation, solubility may become a problem in aqueous sulfuric acid systems. A more serious problem is the dealkylation of the alkyl benzenes in strong acid to give olefins by a reversal of the Friedel-Crafts alkylation. This problem is most serious in sulfuric acid, less so in oleum and most easily controlled when SO_3 is the reagent. The presence of small quantities of the olefins impart an unpleasant smell to the product sulfonate and hence is especially undesirable [21]. Presumably dealkylation occurs when the strong acid protonates the aromatic ring at the carbon bearing the alkyl group and a carbonium ion is ejected from the σ-complex.

Formation of sulfones can be an important side reaction in sulfonation of simple aromatic compounds with sulfur trioxide or chlorosulfuric acid [22] but with long-chain alkylbenzenes it does not seem to occur [23]. Sulfonic anhydrides may also be formed in sulfonations with sulfur trioxide [24] but are readily hydrolyzed to the acids [Eq. (11)].

$$2\,RH \ + \ 3\,SO_3 \ \longrightarrow \ (RSO_2)_2O \ + \ H_2SO_4 \tag{11}$$

$$\big\downarrow H_2O \quad 2\,RSO_2OH$$

B. Addition to Alkenes

Sulfonation by addition to alkenes may involve ionic or free-radical intermediates and may yield products which represent overall addition to the double bond, or apparent substitution. The latter invariably is the result of a combination of addition and elimination steps. Alkene sulfonation requires the use of sulfur trioxide or one of its complexes since sulfuric acid reacts with alkenes to give sulfates (see Sec. V,A).

1. Ionic Mechanisms

As in electrophilic aromatic substitution, the initial step in electrophilic addition to alkenes is the formation of a bond between the electrophile and the π-electron system of the alkene. However, unlike the aromatic σ-complex, pathways leading to several types of products are available. Because the C—S bond of the product sulfonate is extremely stable; no isomerization by

migration of the sulfonate group of the initial product can occur, in contrast
to the observed isomerization of aromatic sulfonates and alkyl sulfates.
Puschel [25] has presented a well-reasoned mechanistic scheme for the sulfonation of α-olefins. Variations in type and quantity of products are observed with changes in temperature, solvent, and complexing agents, and
molar ratio of sulfur trioxide to olefin.

In accord with the predictions of Markovnikov's rule, the products of
α-olefin sulfonation are generally terminal sulfonates. In the gas phase or in
inert solvents, the scheme shown in Fig. 2 applies. The initially formed di-
polar ion [5] may react further by elimination of a proton to give the 2-alkene-
1-sulfonic acid [8]. With liquid sulfur dioxide as solvent, up to 10% of the
1-alkene-1-sulfonic acid [7] was formed, but in other solvents none was detected [26]. Apparently, abstraction of the proton at C-3 is favored over that
at C-1, perhaps because of the assistance of a cyclic intermediate involving

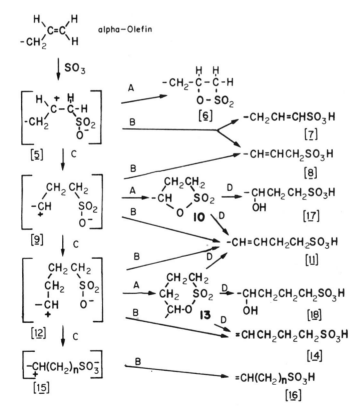

FIG. 2. Sulfonation of α-olefins with SO_3. A = Ring closure; B = Proton abstraction; C = Hydride shift; D = Hydrolysis.

the sulfonate group. Sulfonation of 1-hexadecane with sulfur trioxide diluted with air in a continuous falling-film reactor [27] gave a mixture of alkene-sulfonic acids in which the double bond was found at all positions from 1 to 10, indicating the rapid occurrence of isomerization of the initially formed carbonium ion by hydride shifts. Additional experiments involving rapid quenching with aqueous sodium hydroxide of material from a continuous reactor indicate very rapid disappearance of [5]. The yields of 2-hydroxysulfonate formed in < 0.1, 3, 9, and 180 sec were 30, 18, 10, and 0.1%, respectively. These points make a nearly linear semilog plot [28] in agreement with rapid, first-order isomerization of [5].

In addition to proton elimination to give alkenesulfonic acid, the carbonium ion may also undergo cyclization with the negative oxygen of the sulfonate group to give sultones. Cyclization of the initial adduct would give the highly strained 4-membered ring 1,2-sultone [6]. 1,2-Sultones from styrene [29] and from perfluoroalkenes [30] have been isolated and characterized, but none have been identified in α-olefin sulfonation products. The more stable 5-membered 1,3-sultones [10] have been identified [26, 31] in the sulfonation products of α-olefins and the 6-membered 1,4-sultones [13] are probably also present. Larger rings are considered unlikely to form. Hydrolysis [32] of the water-insoluble sultones gives mixtures of alkene sulfonates [8], [11], and [14] and hydroxyalkanesulfonates [17] and [18]. If the sulfonation is carried out without excess sulfur trioxide, saponification gives a mixture of alkene sulfonates and 3- and 4-hydroxyalkanesulfonates, all of which are water soluble, and possess desirable detergent characteristics. 2-Hydroxyalkanesulfonates are exceptionally insoluble in water, and their absence in the above product mixtures may be taken as evidence that the 1,2-sultones are not intermediates in these reactions [25]. Other possible intermediates include the dimeric 8-membered ring disultone [19] and a 1,4-sultone [20], formed via alkylation of an olefin by the intermediate carbonium ion [5] followed by ring closure. Both of these products yield alkenesulfonates or hydroxy alkane sulfonates on saponification.

$$
\begin{array}{cc}
\underset{\underset{\text{SO}_2\text{O}}{\overset{\overset{\text{O}-\text{SO}_2}{\diagup\ \diagdown}}{\underset{\text{H}_2\text{C}}{\overset{\text{RCH}}{\underset{\diagdown}{\ }}\quad\underset{\diagup}{\overset{\text{CH}_2}{\ }}}}}{[19]} &
\underset{\underset{\underset{\text{O}}{\text{RCH}}}{\overset{\text{RCH}}{\underset{\diagdown}{\ }}}{[20]}
\end{array}
$$

Some secondary sulfonates are also formed [28], presumably by sulfonation of previously isomerized olefin. Small amounts of disulfonates of undetermined structure can also be detected among the products [25]. These presumably arise by further sulfonation of the product alkenesulfonates.

The mechanism of α-olefin sulfonation with complexes of sulfur trioxide is somewhat obscured by some unsupported structural assignments in the literature discussed below. The general scheme is outlined in Fig. 3. The primary difference from sulfonation with free sulfur trioxide hinges on the ability of Lewis bases used as complexing agents to stabilize the initially formed carbonium ion.

Weil, Stirton, and Smith [33] studied the reaction of 1-hexadecene and 1-octadecene with the sulfur trioxide-dioxane complex at below 10° C and at 50° C. They report that at 0-10° C the sulfonated product from 1-octadecene after hydrolysis consists mainly of water-insoluble vicinal hydroxysulfonic acids; 64% of the 2-hydroxy-1-sulfonic acid [23], 33% of the 3-2-isomer, and 3% of the 4-3-isomer. Since the starting alkene was shown to be of >97% purity, it must undergo acid-catalyzed isomerization before sulfonation. For the reaction at 50° C, they deduce a product composition of 52% octadecene-sulfonates, 28.5% hydroxyoctadecanesulfonates, 15.5% sulfated hydroxyocta-decanesulfonates, and 4% 1,2-sultone [6], on the basis of elemental analysis and iodine number of the mixture. No specific evidence for the presence of the 1,2-sultone is presented, though reference is made to a reported isola-tion of such a compound [34]. Migration of the double bond before sulfonation is more pronounced at the higher temperature; oxidation of the octadecene-sulfonates yielded 30% heptadecanoic, 45% hexadecanoic, 15% pentadecanoic, 7% tetradecanoic, 2% tridecanoic, and 1% dodecanoic acids, corresponding to migration of the double bond up to five carbons along the chain. These au-thors assume that there is no isomerization of carbonium ion [5] before eli-mination, and that all of the alkene sulfonates are of the vinyl rather than the alkyl type. No evidence bearing on these assumptions is given. The effect

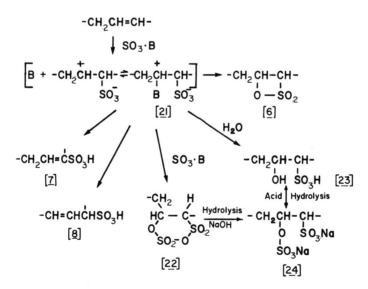

FIG. 3. Sulfonation of olefins in the presence of a Lewis base.

of the dioxane in this reaction is threefold. First, it inhibits isomerization
of the intermediate carbonium ion [5] by complexation [21]. Second, the stabilized carbonium ion may react with another mole of sulfur trioxide to give
the relatively stable carbyl sulfate [22]. Finally, the reduced reactivity of
the complexed sulfur trioxide allows acid-catalyzed alkene isomerization to
become an important side reaction. The observed differences in products
obtained at 10 and at 50° C reflect shifts in the stability of the dioxane-
carbonium ion complex, and changes in the relative rates of formation of
carbyl sulfate [22] and of alkenesulfonates from [5], [21], and [22].

Turbak and Livingston [35] have used butyl phosphate as the complexing
agent and were able to obtain complete conversion of α-olefins to alkene sulfonates, provided that a 2:1 mole ratio of sulfur trioxide to olefin was used.
According to these authors, butyl phosphate provides greater carbonium-ion
stabilization and also accelerates the reaction of the second mole of sulfur
trioxide to give the carbyl sulfate. At room temperature and above, the butyl
phosphate also accelerates the elimination reaction via the carbyl sulfate to
give 1-alkenesulfonic anhydride [25], and ultimately 1-alkenesulfonic acid as
shown in Fig. 4. The assignment of the terminal position to the double bond
is based on unpublished nmr and ir specta.

Püschel [25] concludes that the manner in which the sulfur trioxide and
olefin are combined determines the composition of the product. Addition of
the olefin to excess sulfur trioxide promotes formation of the carbyl sulfate,
leading at least in part, to the undesirable 2-hydroxyalkanesulfonates. The
presence of Lewis bases in the reaction mixture also promotes carbyl sulfate
formation, but at higher temperatures the carbyl sulfate may be converted
almost exclusively to the alkenesulfonic anhydride which may subsequently be
hydrolyzed to alkenesulfonic acid. The obvious disadvantages of this route
are the consumption of two moles of sulfur trioxide per mole of product and
the consequent formation of a mole of sulfuric acid (or sodium sulfate) in the
hydrolysis. If, however, sulfur trioxide is added to the α-olefin in a 1:1 mole

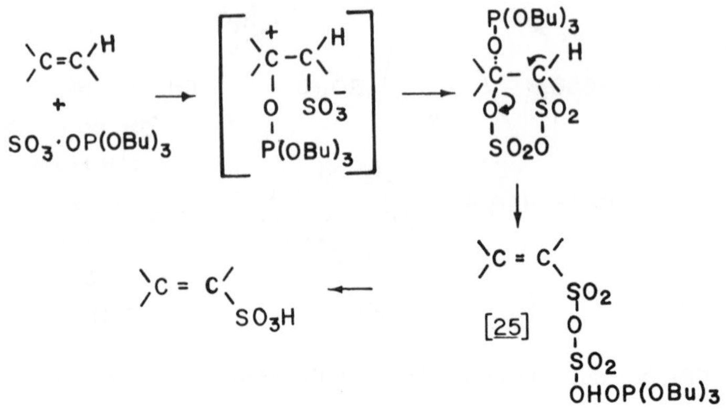

FIG. 4.   Participation of tributylphosphate in olefin sulfonation.

ratio, the products consist of alkenesulfonic acids and 1,3- and 1,4-sultones, from which the useful 3- and 4-hydroxysulfonic acids may be formed. The disadvantages of direct sulfonation with sulfur trioxide include the greater formation of dark colored by-products in the reaction mixture which necessitate a bleaching step in a production process.

Alkenes may also be sulfonated by chlorosulfonic acid [36] or by acetic anhydride-sulfuric acid (acetyl sulfate) [36] to give products similar to these discussed above. No mechanistic studies of these reactions have been reported.

Unsaturated acids such as oleic have also been sulfonated [37]. Reaction of sulfur trioxide in sulfur dioxide at $-10°$ C with oleic acid gave a sulfonated product which was formulated as an allylsulfonate, formed by allylic substitution via an acylsulfonic anhydride [Eq. (12)].

$$C_8H_{17}CH{=}CH(CH_2)_7CO_2H + SO_3 \longrightarrow C_8H_{17}CH{=}CH(CH_2)_7\overset{\overset{\displaystyle O}{\|}}{C}OSO_3H$$

$$C_8H_{17}CH{=}CHCH(CH_2)_6CO_2H \quad (12)$$
$$\underset{SO_3H}{|}$$

No evidence bearing on the actual position of the double bond is available. This material accounted for 54.30% of the total product, the balance consisting of hydroxysulfonate and sulfate-sulfonate. The absence of sulfonation α to the carboxyl group is remarkable (cf. Sec. IV, C). When n-propyl oleate was substituted for the oleic acid, the product was essentially saturated, presumably composed of hydroxysulfonate and sulfate-sulfonates [Eq. (13)]. This reaction seems to proceed by addition via a carbyl sulfate intermediate.

$$C_8H_{17}CH{=}CH(CH_2)_7CO_2C_3H_7 + 2SO_3 \longrightarrow C_8H_{17}\underset{\substack{\diagup\\O\\\diagdown\\SO_2{-}O}}{CH}{-}\underset{\substack{\diagdown\\SO_2\\\end{}}}{CH}(CH_2)_7CO_2C_3H_7 \;\downarrow H_2O$$

$$C_8H_{17}\underset{\substack{|\\O\\|\\H}}{C}H\underset{|}{C}H(CH_2)_7CO_2C_3H_7 + C_8H_{17}\underset{\substack{|\\O\\|\\SO_3H}}{C}H\underset{|}{C}H(CH_2)_7CO_2C_3H_7 \quad (13)$$
$$\qquad SO_3H \qquad\qquad\qquad SO_3H$$

Since the structure of the products were not firmly established other possible reaction paths cannot be ruled out, though the markedly different products 'from the acid and the ester make the proposed pathways attractive.

Long-chain esters of maleic acid and other activated alkenes add bisulfite by a Michael addition mechanism [Eq. (14)] to give the Aerosol type of sulfosuccinate esters [38].

$$\underset{CH{=}CH}{\overset{ROOC}{\diagdown}\overset{COOR}{\diagup}} + NaHSO_3 \longrightarrow \underset{CH_2CH}{\overset{ROOC}{\diagdown}\overset{COOR}{\diagup}} \quad (14)$$
$$\qquad\qquad\qquad\qquad\qquad\qquad\qquad \underset{SO_3Na}{\diagdown}$$

2. Free Radical

Sulfonation of alkenes may also occur by a free-radical addition process [39].
Addition of ammonium bisulfite in an aqueous ammoniacal solution to an α-
olefin in aqueous alcohol with a peracid initiator gives alkane sulfonates.
The rate of the reaction is markedly dependent on the chain length of the α-
olefin; almost a thousandfold decrease in the reaction rate was observed in
going from 1-hexene to 1-hexadecene under the same reaction conditions.
This effect is attributed to the decreasing solubility of the α-olefins in the
aqueous reaction medium. Use of alcohol as a cosolvent, maintenance of a
pH of 7-9, and thorough mixing with slow addition of the bisulfite to avoid
precipitation leads to yields of 95% and better in as little as 2 hours.

It has not been established whether the chain-carrying radical in this
reaction is SO_3^- or $HSO_3\cdot$ [40]. Presumably it could be either, depending on
the pH of the reaction mixture. The chain-carrying steps may be depicted as
in Eqs. (15) and (16):

$$SO_3^{\cdot-} + CH_2{=}CH{-}R \longrightarrow {}^-SO_3{-}CH_2{-}\overset{\cdot}{C}H{-}R \tag{15}$$

$$^-SO_3{-}CH_2{-}\overset{\cdot}{C}H{-}R + HSO_3^- \longrightarrow {}^-SO_3{-}CH_2{-}CH_2{-}R + SO_3^{\cdot-} \tag{16}$$

Oxygen may also serve as an initiator, as shown in Eq. (17).

$$HSO_3^- + O_2 \longrightarrow SO_3^- \cdot + HO_2\cdot \tag{17}$$

Various other olefinic compounds such as 10-undecenoic acid, oleic ani-
lide, and some unsaturated fatty oils and esters are sulfonated by bisulfite
ion and air, but nothing is specifically known of the mechanisms of these reactions [41].

C. Substitution at Saturated Carbon

1. Sulfonation

Direct sulfonation of hydrocarbons with sulfur trioxide alone or in one of its
combined forms has been attempted, but the products consist of such a complex mixture that the reaction is of no practical use. In addition to sulfonates and sulfates, carbonyl, hydroxyl, and carboxylic acid groups may be
introduced [42].

Saturated long-chain fatty acids react with sulfur trioxide, chlorosulfonic acid, or dioxane/sulfur trioxide to give monosulfonation in the α-position
[43]. The initial step appears to be the formation of a mixed anhydride.
Upon further heating, the anhydride is converted to the α-sulfo acid, probably via an intermediate with some enolic character [Eq. (18)].

$$RCH_2CO_2H + SO_3 \longrightarrow RCH_2CO_2SO_3H \longrightarrow$$

$$\longrightarrow RCHCO_2H \quad (18)$$
$$\underset{SO_3H}{|}$$

Only monosubstitution occurs, and only in the α-position. Studies to determine specific mechanistic details have not been undertaken.

2. Sulfochlorination and Sulfoxidation

Although direct sulfonation of hydrocarbons is impractical, there is an indirect method for the introduction of the elements of a sulfonic acid group to a hydrocarbon in one operation. Sulfochlorination [44] and sulfoxidation [45] are reactions in which sulfur dioxide combines with a paraffin under the influence of an oxidizing agent such as chlorine or oxygen. They are both free-radical chain-reactions.

Sulfochlorination is formulated as occurring via the following steps [Eqs. (19)-(23)]:

$$Cl_2 \xrightarrow{h\nu} 2\ Cl\cdot \qquad \text{Chain initiating} \qquad (19)$$

$$RH + Cl\cdot \longrightarrow R\cdot + HCl \qquad\qquad\qquad\qquad (20)$$

$$R\cdot + SO_2 \longrightarrow RSO_2\cdot \qquad \text{Chain propagating} \qquad (21)$$

$$RSO_2\cdot + Cl_2 \longrightarrow RSO_2Cl + Cl\cdot \qquad\qquad\qquad (22)$$

$$Cl\cdot + Cl\cdot \longrightarrow Cl_2$$

$$RSO_2\cdot + Cl\cdot \longrightarrow RSO_2Cl \qquad \text{Chain terminating} \qquad (23)$$

Wall reactions, etc.

Irradiation of the reaction mixture generates chlorine radicals [Eq. (19)] which may abstract hydrogen from the alkane to give alkyl radicals and hydrogen chloride [Eq. (20)]. The alkyl radicals combine immediately with sulfur dioxide to form the alkylsulfonyl radical which, in turn, reacts with chlorine to give the alkanesulfonyl chloride and a new chain-initiating chlorine radical [Eq. (22)].

The reaction of the alkyl radical with sulfur dioxide is at least 100 times faster than its reaction with chlorine, making it relatively easy to avoid simple halogenation of the alkane. However, oxygen reacts with alkyl radicals 10^4 times faster than sulfur dioxide, and the reaction must therefore be carried out at a minimum oxygen content to promote maximum conversion. In the gas phase, "wall" reactions are the main chain-terminating reactions since they most effectively disperse the energy of the reactants, but in liquid phase, recombination with the aid of collisions with other molecules becomes important. The quantum yield in the reaction may be 30,000 - 40,000 under laboratory conditions and 2000 - 3000 under industrial production conditions. The rate of formation of heptanesulfonyl chloride in the laboratory was found to be expressed by

$$\text{rate} = k(I_{absol.})^{1/2}(C_7H_8) \qquad (I_{absol.} = \text{light density})$$

when the sulfur dioxide concentration is above a certain small minimum. The fact that the quantum yield for the formation of sulfuryl chloride from chlorine and sulfur dioxide ($SO_2 + Cl_2 \longrightarrow SO_2Cl_2$) is less than one also contributes to the success of the sulfochlorination reaction.

As with other free-radical processes, compounds which form free radicals by decomposition in the absence of light may also serve as initiators. Substances such as diazomethane, tetraethyl lead, α,α'-azobis-(isobutyronitrile) and numerous peroxides, hydroperoxides, and even ozonides provide chain initiators. Some control of reaction rates and induction times has been obtained by careful selection of chemical initiators. Nitrogen compounds, particularly aromatics such as pyridine, isoquinoline, and aniline which may be found in paraffin feed stocks, act as inhibitors of the sulfochlorination reaction, and hence must be rigorously removed in commercial processing. When production of surfactants is the aim of the process, the alkane sulfonyl chlorides are hydrolyzed with aqueous NaOH to give a mixture of alkanesulfonate and NaCl.

The process of sulfoxidation follows logically from that of sulfochlorination. Alkyl radicals produced by any means may be made to combine with SO_2 and then with oxygen to give alkanesulfonic acids. For C_7 and smaller hydrocarbons the reaction, once initiated, will proceed autocatalytically, but for higher hydrocarbons, initiators such as ozone, peracids, or irradiation must be continuously provided. This may be the consequence of higher relative rates of chain-terminating processes for the higher-molecular-weight hydrocarbons. Both γ-irradiation and photoradiation have proven useful as initiators.

Whatever the source of the initial alkyl radical, it may react with sulfur dioxide to give the intermediate sulfonyl radical which, in turn, gives rise to an alkanepersulfonic radical by combination with oxygen [Eq. (24)].

$$R\cdot + SO_2 \longrightarrow RSO_2\cdot \xrightarrow{O_2} RSO_2OO\cdot \qquad (24)$$

The chain process is extended by abstraction of hydrogen from an alkane by the alkanepersulfonyl radical [Eq. (25)].

$$RSO_2OO\cdot + RH \longrightarrow RSO_2OOH + R\cdot \qquad (25)$$

The potential autocatalytic nature of the reaction arises from the fact that the alkanepersulfonic acid may now decompose to two new radicals capable of acting as initiators [Eq. (26)]. In the presence of water, however, the persulfonic acid is immediately reduced by the sulfur dioxide, removing this potential source of initiators [Eq. (27)].

$$RSO_2OOH \longrightarrow RSO_2-O\cdot \ + \ \cdot OH$$

$$\downarrow RH \qquad\qquad \downarrow RH \qquad\qquad (26)$$

$$(RSO_2OH + R\cdot)(R\cdot + HOH)$$

$$RSO_2OOH + H_2O + SO_2 \longrightarrow RSO_2OH + H_2SO_4 \qquad (27)$$

The products are heavy oils which separate from the reaction mixture, making the isolation relatively simple.

Long-chain alkyl hydroperoxides, prepared by air oxidation of paraffins yield sulfonates [Eq. (28)] by reaction with excess aqueous bisulfite [46]. The raw materials and product in this process are the same as for sulfoxidation, but the process operates stepwise.

$$RH \xrightarrow{\ O_2\ } ROOH \xrightarrow{\ 2\,NaHSO_3\ } RSO_3Na + NaHSO_4 + H_2O \qquad (28)$$

3. Nucleophilic Substitution

The reaction of alkyl halides with sulfite ion, the Strecker reaction, provides yet another means of introducing a sulfonic acid group at a saturated carbon atom. An aqueous solution of sodium, potassium, or ammonium sulfite is heated with the halide to give the alkanesulfonate salt by displacement of halogen in an S_N2 process. The fact that a C–S bond rather than a C–O is formed, even though O is more electronegative, reflects the greater nucleophilicity of sulfur in the sulfite ion resulting from the greater polarizability of sulfur and the availability of d-orbitals. The Triton-type surfactants made by Rohm and Haas Co. [Eq. (29)] and dodecylsulfoacetate [Eq. (30)] are representative of compounds manufactured by this process.

$$C_8H_{17}-C_6H_4-OC_2H_4OC_2H_4Cl + Na_2SO_3 \longrightarrow$$

$$C_8H_{17}-C_6H_4-OC_2H_4OC_2H_4SO_3Na \qquad (29)$$

$$C_{12}H_{25}O\overset{\overset{\displaystyle O}{\|}}{C}CH_2Cl + Na_2SO_3 \longrightarrow$$

$$C_{12}H_{25}O\overset{\overset{\displaystyle O}{\|}}{C}CH_2SO_3Na + NaCl \qquad (30)$$

V. MECHANISMS OF SULFATION

Sulfation mechanisms may be divided into two classes, addition to alkenes, and esterification of alcohols. This division also reflects to some extent the sources of the raw materials, alkenes coming principally from petroleum, and alcohols often being prepared by hydrogenation of fatty esters. The reagents used for sulfonation are also employed, under appropriate conditions for sulfation.

A. Addition to Alkenes

The addition of sulfuric acid to ethylene is the most widely used process for the manufacture of ethanol and this reaction has therefore been thoroughly studied from the practical standpoint. The detailed general mechanism is, however, still unclear [47].

Addition definitely follows the Markovnikov rule, and migration of the functionality is observed in long-chain alkenes under some conditions. These data, as well as kinetic evidence, leave no doubt that there is a carbonium-ion intermediate. The point in question is what are the steps leading to this carbonium ion. Taft [48] proposed an equilibrium formation of a protonated π-complex [Eq. (31)] followed by a rate-determining transformation to the carbonium ion [Eq. (32)].

$$\text{C=C} \;+\; H^+ \;\rightleftharpoons\; \overset{H^+}{\text{C=C}} \quad \text{Equilibrium} \tag{31}$$

$$\overset{H^+}{\text{C=C}} \;\rightarrow\; \overset{H}{-\text{C}-\overset{+}{\text{C}}-} \quad \text{Slow} \tag{32}$$

The observation that the ratio of rate constants for hydration in D_2O vs. H_2O is close to unity is not consistent with the above interpretation and direct protonation to form the carbonium ion [Eq. (33)] has also been suggested as the rate-determining step [47].

$$\text{C} = \text{C} \;+\; H^+ \;\longrightarrow\; -\overset{H}{\underset{\wedge}{\text{C}}} - \overset{+}{\text{C}} \tag{33}$$

In the formation of sulfates, the intermediate carbonium ion combines with HSO_4^- to give the product. By-products are those which would be predicted for a carbonium-ion process and may include dialkyl sulfates, which are readily hydrolyzed to alkyl sulfates, isomerized alkenes, and tars formed by alkene polymerization. Efficient mixing is critical in the sulfation of of long-chain alkenes for surfactant use since the acid and hydrocarbon are immiscible.

Addition of 96% acid to 1-dodecene at 0°C leads at first almost exclusively to dialkyl sulfate, but this is converted to alkyl sulfate as more acid is added. Under these conditions, the product is mostly the 2-isomer, but if the alkene is added to excess acid the sulfate consists of all the possible secondary isomers [34].

B. Esterification

The sulfation of alcohols may be regarded formally as esterification of sulfuric acid, although reagents other than sulfuric acid are most commonly used. Primary and secondary alcohols may be sulfated readily, though for the secondary alcohols, dehydration may become important in the presence of sulfuric acid or sulfur trioxide vapor [49].

The reaction of alcohols with sulfuric acid proceeds by a bimolecular displacement mechanisms like that of acid-catalyzed esterification. The observed rate expression is:

$$\text{rate} = k[\text{ROH}][\text{H}_2\text{SO}_4] \; (\text{H}^+ \text{ activity})$$

Primary alcohols react an order of magnitude faster than secondary. Since the reaction is an equilibrium process, optimization of yields depends on driving the equilibrium to the right by use of excess reagent or by water removal.

The reaction of alcohols with sulfur trioxide may be considered as solvolysis of an anhydride [Eq. (34)].

$$\text{ROH} + \text{SO}_3 \longrightarrow \text{ROSO}_3\text{H} \tag{34}$$

It may give rise to dialkyl sulfates as well. Sulfur trioxide gives good yields of sulfates from primary alcohols, though the products are generally darker than those obtained by other processes. Excessive dehydration occurs with long-chain secondary alcohols in the presence of sulfur trioxide.

An interesting combination of sulfation and sulfonation occurs in the formation of ethionic acid from ethanol and two moles of sulfur trioxide [Eq. (35)]. The first mole of sulfur trioxide adds at 0°C to give ethyl sulfate

$$\text{EtOH} + 2\,\text{SO}_3 \longrightarrow \text{HO}_3\text{SCH}_2\text{CH}_2\text{OSO}_3\text{H} \tag{35}$$

and the second adds at 50°C. It has been suggested that this sulfonation reaction occurs via electrophilic displacement of hydrogen by sulfur trioxide with assistance in removal of the hydrogen by a cyclic intermediate [26] involving the sulfate group [50].

[26]

Many sulfations using sulfur trioxide complexes have been reported but none are used in commercial production. As is the case for aromatic sulfonation, complexing the sulfur trioxide moderates its reactivity and thereby promotes cleaner reactions, but the necessity of removing or recycling the complexing agent limits the commercial application of these reactions.

Sulfation of alcohols by chlorosulfonic acid is probably the most generally useful laboratory method [Eq. (36)].

$$ROH + ClSO_3H \longrightarrow ROSO_3H + HCl \tag{56}$$

In this case too, the mechanism parallels that of ester formation from alcohols and the carboxylic acid chloride. Because of the efficiency and high yields of the reaction it has also been used in batchwise production of sulfates as well as in experimental continuous processes, but the corrosiveness of the by-product HCl calls for special equipment and handling techniques.

In addition to fatty alcohols, mono- and diglycerides and hydroxy stearates such as are found in castor oil may be sulfated to give surfactants [51]. For these, and the many other more complex sulfated or sulfonated structures which have been elaborated for use as surfactants, specific mechanistic studies have not been made, and it can only be assumed that the general mechanistic principles which operate in the cases discussed above are generally applicable.

REFERENCES

1. E. E. Gilbert, Sulfonation and Related Reactions, John Wiley & Sons, Inc., New York, 1965.
2. F. A. Cotton and G. W. Wilkinson, Advanced Inorganic Chemistry, 2nd ed., John Wiley & Sons, Inc., New York, 1966, p. 542.
3. D. J. Rogers, Bonding Theory, McGraw-Hill, New York, 1968, pp. 121-122.
4. C. F. P. Bevington and J. L. Pegler, Chem. Soc. (London) Spec. Publ., 12, 283 (1958).
5. T. S. Harrer, in Kirk-Othmer Encyclopedia of Chemical Technology, Vol. 19, 2nd ed., John Wiley & Sons, Inc., New York, 1969, pp. 441-482.
6. Reference 1, p. 5.
7. J. R. Donovan, in Kirk-Othmer Encyclopedia of Chemical Technology, Vol. 5, 2nd ed., John Wiley & Sons, Inc., New York, 1964, pp. 357-363.

8. P. Macaluso, in Kirk-Othmer Encyclopedia of Chemical Technology, Vol. 19, 2nd ed., John Wiley & Sons, Inc., New York, 1969, pp. 401-403.
9. Reference 1, p. 20.
10. L. C. Schroeter, Sulfur Dioxide, Pergamon, New York, 1966.
11. H. Cerfontain, Mechanistic Aspects in Aromatic Sulfonation and Desulfonation, John Wiley & Sons, Inc., New York, 1968.
12. H. Cerfontain and C. W. F. Kort, in Mechanisms of Reactions of Sulfur Compounds (N. Kharasch, ed.), Vol. 3, Intrascience Research Foundation, Santa Monica, Cal., 1968, pp. 23-28.
13. K. L. Nelson in Friedel-Crafts and Related Reactions (G. Olah, ed.), Vol. 3, Pt. 2, John Wiley & Sons, Inc., New York, 1964, Chap. 40.
14. E. E. Gilbert, in Kirk-Othmer Encyclopedia of Chemical Technology, Vol. 19, 2nd ed., John Wiley & Sons, Inc., New York, 1969, p. 291.
15. C. W. F. Kort and H. Cerfontain, Rec. Trav. Chim., 67, 24 (1968).
16. R. J. Gillespie and E. A. Robinson, Can. J. Chem., 39, 2189 (1961).
17. K. D. Wadsworth and C. N. Hinshelwood, J. Chem. Soc., 1944, 469.
18. C. W. F. Kort and H. Cerfontain, Rec. Trav. Chim., 88, 1298 (1969).
19. A. W. Kaandorp, H. Cerfontain, and F. L. J. Sixma, Rec. Trav. Chim., 81, 969 (1962).
20. C. W. F. Kort and H. Cerfontain, Rec. Trav. Chim., 88, 860 (1969).
21. Reference 1, p. 74.
22. C. M. Suter and A. W. Weston, Organic Reactions (R. Adams, ed.), Vol. III, John Wiley & Sons, Inc., New York, 1946, pp. 141-197.
23. Reference 14, p. 294.
24. N. H. Christensen, Acta Chem. Scand., 15, 1507 (1961).
25. F. Püschel, Tenside, 4 (9), 186 (1967).
26. F. Püschel and C. Kaiser, Chem. Ber., 98, 735 (1965).
27. D. M. Marquis, Hydrocarbon Process. Petrol. Refiner, 47, 109 (1968).
28. D. M. Marquis, Private communication, 1970.
29. F. G. Bordwell, M. L. Peterson, and C. S. Rondestvedt, Jr., J. Am. Chem. Soc., 76, 3945 (1954).
30. D. C. England, M. A. Dietrich, and R. V. Lindsay, Jr., J. Am. Chem. Soc., 82, 6181 (1960).
31. D. M. Marquis, S. H. Sharman, R. House, and W. A. Sweeney, J. Am. Oil Chemists' Soc., 43, 607 (1966).
32. C. Kaiser and F. Püschel, Chem. Ber., 97, 2926 (1964).
33. J. K. Weil, A. J. Stirton, and F. D. Smith, J. Am. Oil Chemists' Soc., 42, 873 (1965).
34. W. D. Nielsen, Paper presented before the 148th National Meeting of the Am. Chem. Soc., Chicago, Ill., Aug. 30-Sept. 4, 1964.
35. A. F. Turbak and J. R. Livingston, Jr., Ind. Eng. Chem., Prod. Res. Develop., 2 (3), 229 (1963).
36. S. Miron and G. H. Richter, J. Am. Chem. Soc., 71, 453 (1949).
36a. R. Stern and P. Baumgartner, Compt. Rend., 257 (10), 1713 (1963).
37. T. W. Sauls and W. H. C. Rueggeberg, J. Am. Oil Chemists' Soc., 33, 383 (1956).
38. M. Morton and H. Landfield, J. Am. Chem. Soc., 74, 3523 (1952).

39. E. Clippinger, Ind. Eng. Chem. Prod. Res. Develop., $\underline{3}$ (1), 3 (1964).
40. C. Walling, Free Radicals in Solution, John Wiley & Sons, Inc., New York, 1957, p. 327.
41. Reference 1, p. 150.
42. Reference 1, p. 32.
43. A. J. Stirton, J. Am. Oil Chemists' Soc., $\underline{39}$, 490 (1962).
44. F. Assinger, Paraffins, Chemistry and Technology (B. J. Hazzard, transl.), Pergamon, New York, 1968, Chap. 5.
45. Reference 44, Chap. 7.
46. Reference 14, p. 290.
47. F. A. Long and M. A. Paul, Chem. Rev., $\underline{57}$, 935 (1957).
48. R. W. Taft, Jr., J. Am. Chem. Soc., $\underline{74}$, 5372 (1962).
49. Reference 1, pp. 345-354.
50. D. S. Breslow, R. R. Hough, and J. T. Fairclough, J. Am. Chem. Soc., $\underline{76}$, 5361 (1954).
51. W. M. Linfield, in Fatty Acids and Their Industrial Applications (E. S. Pattison, ed.), Marcel Dekker, Inc., New York, 1968, pp. 155-186.

CHAPTER 5

ALCOHOL AND ETHER ALCOHOL SULFATES

Samuel Shore*

Mazer Chemicals, Inc.
Gurnee, Illinois

Daniel R. Berger

The Richardson Company
Melrose Park, Illinois

---

*Deceased.

135

I. INTRODUCTION

As a group, salts of primary alkyl sulfates rank as the oldest anionic surfactants after soap. The first alkyl sulfate was prepared by Dumas in 1836.
Fatty alcohols were obtained from the hydrogenolysis of oils in a major technical advance by Schrauth in 1928 [1], followed by the commercial production
of alkyl sulfates in 1930 and their incorporation into retail detergents in the
United States in 1932 [2].

Sulfated surfactant products are among the best wetting, dispersing,
emulsifying, and cleaning agents. They are used in heavy-duty cotton detergents, light-duty liquid detergents, liquid dish-washing preparations, shampoos, textile auxiliaries, rug and upholstery cleaners, toothpastes, emulsifiers, hard surface cleaners, plating bath additives, etc.

Attachment of the hydrophilic group, the half sulfuric acid ester $ROSO_3$,
to the hydrophobe is through an oxygen, i.e., the C—O—S linkage. The additional oxygen makes the sulfate a stronger solubilizing group than the sulfonate but it is more easily hydrolyzed than the C—S linkage of the sulfonates,
$R—SO_3$. Susceptibility to hydrolysis in acid media places some restrictions
on the utility of sulfates. They would not be suitable for use in acidic preparations if the product is required to have a long shelf life. However, they
can and are being used where exposure to hydrolytic conditions is short and
does not result in the total depletion of surfactant during the time its effectiveness is needed.

The C_{12} and C_{14} alcohols are usually the most desirable for sulfation
and are the most expensive of the C_8 to C_{18} alcohols obtainable from the hydrogenation of coconut oil. Their sulfates possess the best balance of solubility, foaming, and detergency properties. Increased solubilization of
hydrophobes through a combination of oxyethylation and sulfation is used to
modify and improve the properties of sulfates obtained from cheaper raw
materials that cannot be adequately solubilized by sulfation alone. The ether
sulfates derived from tallow alcohols are an example.

Synthetic long-chain alcohols which have been produced from olefins and
hydrocarbons by a variety of processes are now available. These are blended,
either as the alcohols or the ethoxy alcohols, and sulfated to produce a finished product with the properties desired for a particular end use.

The sodium and ammonium salts of oxyethylated alkylphenol sulfates
were widely used in retail and industrial detergent products. Usage grew
rapidly from their introduction in 1950 to large volume by 1966 when sales
began to decline. The shift of the detergent industry to more biodegradable
products in 1965 started a trend away from alkylphenol-based surfactants towards oxyethylated and sulfated aliphatic alcohols [3].

Secondary alcohol sulfates produced by the addition of sulfuric acid to
an olefin have never been widely used in the United States. In Britain, however, their usage once exceeded that of the alkylbenzenesulfonates. A lively

interest in the research and development of these products remains for several reasons. This synthetic route, especially since the availability of linear α-olefins, is potentially capable of producing high-quality alkyl sulfates at low cost. Lower-grade olefins, such as those produced from the pyrolysis of shale oil and cracked wax, can be converted to acceptable commercial surfactant products. Research activity is greatest in Eastern European countries where vegetable and animal oils are at a premium.

II. PREPARATION OF SULFATES

A survey of the literature pertaining to the preparation of alkyl and ethoxyalkyl sulfates reveals the fact, not too surprising, that the traditional sulfating reagents have continued to receive most attention. This is reasonable since the commercial importance of these sulfates requires that they be made as inexpensively as possible, so that reagents such as sulfur trioxide and sulfuric acid become the preferred sulfating reagents, followed closely by chlorosulfonic acid. Some interest has been shown in the use of more exotic sulfating agents, including some mention of them in the patent literature, but little commercial interest is evident in these on a practical level.

The sulfation of an alcohol, or any molecule containing a reactive hydroxyl group, involves replacement of the O—H bond with an O—S bond [Eq. (1)]:

$$RO-H + SO_3 \longrightarrow RO-SO_3H \tag{1}$$

Reagents useful for this purpose are sulfur trioxide, sulfuric acid, oleum, chlorosulfonic acid, sulfamic acid, numerous complexes of these, and some exotic reagents. With the exception of sulfur trioxide itself, all may in effect be considered as its derivatives, although the mechanism of the sulfation reaction with these may not be the same. Their structures are:

Sulfur trioxide	SO_3
Sulfuric acid	$H_2SO_4 = H_2O \cdot SO_3$
Oleum	$H_2SO_4 \cdot nSO_3 = H_2O \cdot (n+1)SO_3$
Chlorosulfonic acid	$ClSO_3H = HCl \cdot SO_3$
Sulfamic acid	$H_2NSO_3H = NH_3 \cdot SO_3$

A. Sulfur Trioxide as a Sulfating Agent

The use of this most basic sulfating agent has been made possible by the commercial availability of a stable liquid form of sulfur trioxide. It exists in three forms, known as the α, β, and γ forms. The former, a solid at room temperature, is the more stable form, and in the absence of inhibitors, the liquid β and γ forms revert to the α. By using small amounts (approx. 0.5%) of inhibitors, which may include boric acid derivatives, and by rigorously excluding moisture, sulfur trioxide may be manufactured, stored, transported, and used as a liquid. The α and β forms are asbestos-like solid polymers, but the β is lower melting (32.5 vs 62.3° C) and has a considerably lower heat of fusion. The usual commercial product consists of a liquid mixture of β and γ (mp 16.8° C) forms.

The stoichiometry involved in sulfation with sulfur trioxide is quite simple [Eqs. (2) and (3)]; there are no by-products other than whatever neutralized inorganic sulfate results from the small amount of excess sulfur trioxide trapped in the alkylsulfuric acid.

$$ROH + SO_3 \longrightarrow ROSO_3H \tag{2}$$

$$ROSO_3H + NaOH \longrightarrow ROSO_3Na + H_2O$$

$$SO_3(xs) + 2\,NaOH \longrightarrow Na_2SO_4 + 2\,H_2O \tag{3}$$

As is usually the case whenever there is a choice of reagents, sulfur trioxide has both advantages and disadvantages compared to other potential sulfation reagents. Specific studies comparing reagents will be described later, but the general advantages and disadvantages are as follow:

Sulfur trioxide has the advantage of being inexpensive, highly reactive, reacting quantitatively without requiring excess reagent, and giving sulfates with minimal inorganic salts (unlike sulfuric acid, oleum, or chlorosulfonic acids, where varying amounts of sulfates or chlorides are always present). Since it must be vaporized for proper reaction, its reactivity can be controlled rather easily by varying its concentration in an inert carrier gas. Furthermore, its reactivity can be modified by forming Lewis-base complexes (such as with pyridine or dioxane) although this is not generally practical on a commercial level.

Although sulfur trioxide must be vaporized and diluted with an inert carrier gas to prevent burning of the sulfate, some color is usually formed. In order to vaporize it, complex equipment is required. It is a dangerous material, quite difficult to handle, and must be kept reasonably warm and quite dry. Secondary fatty alcohols cannot be sulfated with it successfully. Some ring sulfonation is expected in the sulfation of oxyethylated alkylphenols.

1. Primary Alcohol Sulfates

Sulfur trioxide, in its stabilized form, continues to receive primary attention as a sulfating reagent. Several laboratory reactors intended to simulate commercial systems have been developed for studying sulfur trioxide reactions. Two examples are a batch system described by Sheely and Rose [4] and the so-called "Shell continuous bench sulfator" [5, 6]. In each case, data have been gathered concerning process variables in the sulfation of linear alcohols as well as of the corresponding oxyethylated alcohols, and suggestions were made for improving process conditions and product quality. Numerous studies are available concerning both batch [7-11] and continuous [12-24] sulfation. An interesting two-step process has been reported [25], wherein either sulfates or ether alcohol sulfates may be prepared, by treating the alcohol with 70-90% of the required amount of sulfur trioxide in a conventional system. In the second (continuous) step, the reaction is finished by treatment of the product from step 1 with additional sulfur trioxide. The first step requires one-half to two hours at 40-70° C, while the finishing step takes only a matter of minutes, giving yields of 95-100% and requiring only moderate cooling. The claimed advantage of this two-step process is that it avoids the fast reaction and consequently drastic cooling necessary to minimize decomposition of the esters formed in standard one-step batch or continuous processes.

Sources of sulfur trioxide in these sulfations vary. In general, stabilized sulfur trioxide was used, but some attention was also paid to the use of contact or converter gas [13-16] and oleum [26, 27]. Considerable data have been generated on a commercial scale for the semicontinuous sulfation of primary fatty alcohols using sulfur trioxide from converter gas [16].

In general, the alcohols were sulfated neat while the sulfur trioxide was diluted with dry air or nitrogen [28]. A methylene chloride addition at a level of 10-12% by weight to the alcohol before sulfation was suggested [238]. The advantage of the solvent was said to be in the absorption of the heat of reaction and the lowering of the viscosity of the sulfate half esters. Similarly, solvents such as ethyl ether and petroleum ether were used [8] to eliminate the formation of colored products and resins. This was also applicable to sulfation with chlorosulfonic acid. A patent [10] described the use of undiluted sulfur trioxide by sulfation under vacuum.

2. Aliphatic Ether Alcohol Sulfates

References 4-6, 12, 20, 21, 25, and 29-31 describe continuous and batch processes for the sulfation of oxyethylated alcohols with sulfur trioxide. The merits of sulfur trioxide and chlorosulfonic acid are discussed by Gilbert and Veldhuis [30] and process and product variables are described. On balance, sulfur trioxide was found to be the preferred reagent, yielding better cost, process time, and engineering factors. No hydrogen chloride was generated by the use of sulfur trioxide, and thus it did not need to be scavenged; furthermore, no sodium chloride was introduced into the product. Factors favoring chlorosulfonic acid were (1) less heat was evolved, (2) the acid did not need to be vaporized, and (3) lighter colored products were obtained, although acceptable colors were obtained with sulfur trioxide, especially if the product was bleached. Performance factors of either type sulfate were comparable.

3. Aromatic Ether Alcohol Sulfates

Batch [7] and continuous [12] processes for the sulfur trioxide sulfation of
ethylene oxide condensates of alkyl phenols have been described. Gilbert and
Veldhuis studied the sulfation of such compounds, particularly the nonylphenol
4-mole ethoxylate with both sulfur trioxide and sulfamic acid [32]. As in the
comparison between sulfur trioxide and chlorosulfonic acid above [30], there
were advantages and disadvantages to each, but on balance sulfur trioxide
was preferred. In its favor were much lower cost and reaction-time factors,
formation of lighter colored products, and the ability to form salts other than
those of ammonia with ease. Disadvantages for sulfur trioxide were foaming
during sulfation, some ring sulfonation, and the fact that sulfur trioxide
needed to be vaporized. As mentioned previously [30] performance factors
were comparable.

4. Sulfation with Sulfur Trioxide Complexes

Several patents issued in recent years describing the use of complexed sulfur trioxide as the sulfating reagent. In general, it provided a more manageable form of sulfur trioxide since vaporization was no longer necessary, and
milder reaction conditions could be maintained. Specific complexes described
were those with nitriles, as cyanomethane [33], phosphates [34-37], and N-
alkylethylene carbamates [38]. The phosphates involved could be either organic or inorganic. Both alkyl sulfates and alkyl ether sulfates could be prepared by the use of complexed forms of sulfur trioxide.

B. Sulfuric Acid and Oleum as Sulfating Agents

Sulfuric acid and oleum continue to play a role in the commercial preparation of sulfates. Whether oleum properly belongs in the same category as
sulfuric acid, rather than with sulfur trioxide, is problematical; oleum
probably acts as sulfur trioxide in sulfation, but is handled like sulfuric acid,
and therefore is discussed here for convenience. Sulfuric acid is generally
defined as up to 100% sulfuric acid, whereas oleum is sulfuric acid of greater
strength, where the excess is available as free sulfur trioxide. Sulfur trioxide is sometimes generated from oleum by passing a stream of dry, inert
gas, such as air or nitrogen, through the oleum and carrying the sulfur trioxide to the reaction system as a vapor [26, 27].

     In general, the use of sulfuric acid is much more convenient than that of
sulfur trioxide, since it can be metered as a liquid into the reaction system,
either slowly or as a single charge, depending on the chemistry involved.
Other advantages include its relative mildness compared to sulfur trioxide or
chlorosulfonic acid, and the fact that its reactivity can be controlled to a considerable extent by addition of water or sulfur trioxide. Disadvantages in
the use of sulfuric acid or oleum include the fact that the sulfation is an
equilibrium reaction, requiring an excess of acid.   Thus, a great deal of

spent acid remains at the end of the reaction, which must be disposed of or recycled. Furthermore, excess acid remaining in the alkylsulfuric acid is neutralized, giving a high inorganic sulfate content to the neutralized fatty sulfate. Side reactions leading to the formation of ethers and dialkylsulfates are also to be expected as is some dehydration to form olefins, which may or may not be further sulfated or sulfonated to give useful side products. Yields of sulfates with the aid of sulfuric acid are generally lower than those obtained with other reagents. With H_2SO_4 sulfations short contact times with immediate neutralization are necessary to avoid color formation and product degradation.

1. Primary Alcohol Sulfates

Major attention had been given to the sulfation of linear alcohols with sulfuric acid, as with other reagents, because of the commercial importance of the products. Many studies of the products and of process conditions have been carried out. Most laboratory studies were on batch processes, but interest was also shown in the continuous method [13, 39-43]. In one study laboratory batch and commercial continuous sulfations were carried out in a comparison between natural and synthetic fatty alcohols [44]. The synthetic alcohols were obtained by hydrogenation of fatty acids from paraffin oxidation, and contained both odd and even carbon chains, as well as some branched alcohols and 3-4% unsaponifiable hydrocarbons, whereas the natural alcohols contained only even carbon chains. In general, the natural alcohols were easier to process by either technique and gave a higher degree of sulfation.

The use of adducts with sulfuric acid was investigated. Boric acid was used [45, 46], as was urea [47] dissolved in sulfuric acid to give the ammonium sulfates, with equimolar quantities of carbon dioxide and ammonia generated from the urea. Approximately equimolar ratios of urea and 90% sulfuric acid were used, and conversions ran over 90%.

Combinations of sulfuric acid and acetic anhydride (acetyl sulfate or acetylsulfuric acid) were of interest. In one study [46], the effects of time, temperature, and concentration on yield from several reagents were given. Maximum conversions obtained from these reagents are shown in Table 1. The high rate of conversion in the case of the sulfuric acid–acetic anhydride sulfating reagent was apparent rather than real. Side reactions produce much acetylated alcohol.

The milder sulfating reagents converted a maximum of 60% of the fatty alcohols and could cause dehydration, while severe reagents could cause resinification of the products as well.

Thiourea or its derivatives were added in small amouts to fatty alcohols prior to sulfation with sulfuric acid, as well as chlorosulfonic acid and sulfamic acid [50]. The resulting sulfuric acid esters had less color buildup (greater light transmittance) than controls lacking the thiourea.

TABLE 1

Comparison of Sulfation Reagents

Reagent	Conversion, %
Oleum–acetic anhydride	44
Sulfuric acid (99%) + phosphorous pentoxide	nil
Acetyl sulfate	18
Boric ester interchange	66
Acetic ester interchange	17
Sulfuric acid	46
Sulfuric acid–acetic anhydride	77

Vacuum techniques were used to sulfate alcohols with sulfuric acid, followed by addition of acetic anhydride, all at low temperatures of 6 to 8° C, to give yields of 83% [48]. Sodium acetylsulfate in dimethylformamide [49] was an interesting system for sulfating polyols and sugar derivatives. Lower-molecular-weight alcohol sulfates were also described.

The effect of a large excess of sulfating reagent was evaluated [51] and comparisons were made between sulfuric and chlorosulfonic acid. With the use of a 250% excess of sulfuric acid at 37° C and neutralization with caustic at 40-47° C, the degree of sulfation reached 74%. There was evidence of possible dehydration; some oxidation to fatty acids also occurred. Side reactions occurred to a greater extent with chlorosulfonic than with sulfuric acid. As the molecular weight of the alcohol increased, the tendency of unsaturated by-products to polymerize also increased. The effect was more pronounced for chlorosulfonic acid which produced 33-35% hydrocarbons in the unsulfated portion, in contrast to sulfuric acid which gave only 20-22% hydrocarbon.

2. Secondary Alcohol Sulfates

Little has been published on the use of sulfuric acid to sulfate the secondary alcohols, the most likely reason being that the competitive dehydration to olefin predominates. A Russian patent [52] suggests the use of a solvent such as benzene although the extent of side reactions was not discussed. A study of the sulfation [53] of a series of secondary alcohols with 98.5% sulfuric acid to obtain optimum yields was reported. It was found that the conversion decreased as the chain length increased from 2-undecanol to 2-tetradecanol and 2-octadecanol. Conversion to the sulfate also decreased as the hydroxyl group was moved toward the center of the hydrocarbon chain. Conversions for 2-, 4-, 5-, and 7-tetradecanol were 80, 60, 50, and 43%, respectively.

Attention was also given to the problem of removing excess sulfuric acid from secondary alcohol sulfations [53]. A mixture of one part of gasoline and one part of the acid sulfation paste was countercurrent extracted with 0.25 parts of 20% sulfuric acid. The spent acid was absorbed into the dilute sulfuric acid. The alkylsulfuric acid in the gasoline phase was neutralized.

3. Ether Alcohol Sulfates

Sulfation of oxyethylated alcohols could be accomplished with the use of sulfuric acid without cission of the ether chain if carried out at low temperature. Oxyethylated branched tridecyl alcohol was treated with a large excess of sulfuric acid at 15° C, followed by neutralization at moderate temperature, 35-50° C. The sulfated ether alcohol acid was added to the base in order to reduce the amount of hydrolysis [29] during neutralization. Alkylphenol oxyethylates gave the ammonium salt directly on sulfation with urea in sulfuric acid [47].

C. Sulfation with Chlorosulfonic Acid

Chlorosulfonic acid is an important agent for the commercial preparation of sulfates; a study comparing it to sulfur trioxide has been discussed earlier [30]. Advantages of chlorosulfonic acid lie in its ease of handling as a liquid without dilution, although its reactivity may be modified by use of solvents or complexing agents. Lighter colored products are obtained with it than with sulfur trioxide or sulfuric acid, especially in batch processes. The reaction is stoichiometric and irreversible being driven to completion by the loss of hydrogen chloride.

Disadvantages involve, for the most part, the evolution of hydrogen chloride from the reaction. First of all, chloride salts are inevitably present in the final product. Secondly, evolved hydrogen chloride must be trapped and either recovered as concentrated acid or neutralized and disposed of. Thirdly, the hydrogen chloride evolved is quite corrosive to equipment, requiring glass reactors. Finally, heat and gas evolution are unbalanced. Approximately 60% of the heat is evolved when only 20% of the acid has been added, since the hydrogen chloride is trapped exothermically as the alkoxonium chloride; see Eqs. (4) and (5):

$$ROH + ClSO_3H \longrightarrow ROSO_3H + HCl \tag{4}$$

$$ROH + HCl \longrightarrow ROH_2^+Cl^- \tag{5}$$

When the gas is evolved toward the end of the sulfation, much foaming occurs. These last two factors, corrosiveness and unbalanced heat evolution, have prevented continuous sulfations with chlorosulfonic acid from becoming as important as batch processes [54, 55], although work in this direction has progressed, as discussed below.

1. Primary Alcohol Sulfates

A number of important patents have issued covering the continuous sulfation of fatty alcohols with chlorosulfonic acid. An apparatus was described in patents issued to Chemithon Corp. in which sulfating agent and reactant were continuously fed into a reaction zone consisting of a series of concentric tubes, where the reaction mass was mixed with recirculated reaction product and then fed into a degassing reactor. Part of the degassed product was finally fed into a neutralization system where it was continually neutralized with alkali. This process, which is of commercial importance, was also applicable to the sulfonation of alkylbenzenes with sulfur trioxide [22].

In Italy, Bozetto has developed commercial processes for sulfation with chlorosulfonic acid [56]. In this process, the materials were reacted under vacuum on a cone or dish which rotated rapidly close to the wall of a cylindrical vessel. Hydrogen chloride was removed almost instantaneously as the product was thrown to the cylinder wall by centrifugal force. Since heat was removed quite efficiently by the expansion and degassing involved, little cooling of the cylinder wall was necessary. The product was then continuously neutralized.

Several additional patents issued describing continuous chlorosulfation [41, 43, 57, 58]. Some showed remarkable similarity to others. For example, an East German patent [42] described a process quite similar to the Bozetto patent [56], although an additional claim was made for using sulfuric acid or oleum. Yields were 27–36%. Another patent [59] involved saturation of the alcohol with hydrogen chloride before contact with chlorosulfonic acid. This was said to help provide lighter colored products by minimizing local heat and the resulting side reactions and decomposition of the intermediate acid esters before neutralization. By the saturation of the alcohol with hydrogen chloride, cooling to just above the freezing point, and then allowing the system to find its own temperature during reaction, side effects were minimized.

A study was made [13] of the costs of continuous sulfation of fatty alcohols with chlorosulfonic acid and with sulfur trioxide from contact gas. For a given production of active material, costs for chlorosulfonic acid sulfation were 29% greater than with sulfur trioxide, while production involving sulfuric acid was 19% greater than with sulfur trioxide. In batch chlorosulfation processes, a solvent was sometimes necessary. When solvents were used, temperature and local heating were more easily controlled [8, 58] and lighter colored products could be expected. Solvent recovery became a consideration, as well as effects of residual solvent on the properties of the sulfate.

In general, light colors were obtained even in the absence of solvent. An interesting process [60] for neutralization of chlorosulfated fatty alcohols involved the use of powdered carbon dioxide ("dry ice") mixed with the base (sodium bicarbonate or sodium carbonate). The carbon dioxide gas evolved during neutralization helped maintain a low neutralization temperature. Solvents were not used, and powdered surfactants were obtained.

Studies were carried out on the degree of sulfation as a function of reaction conditions. Much of this information is unpublished, but some is available. For example [58], when primary C_{10}-C_{18} synthetic fatty alcohols were sulfated with 5% excess chlorosulfonic acid over theoretical, a 84-90% sulfation was obtained, the exact amount depending upon the purity of the starting material. With 10% excess chlorosulfonic acid, the degree of sulfation increased by 2-2.5%, but additional acid did not further increase the degree of sulfation. Reaction temperature (20-35°C preferred) seemed to affect the yield more than molecular weight. By use of solvents, such as petroleum ether, the reaction temperature was lowered by some 20°C, giving an increase in yields of 3-5%. The reason for this appeared to be that at higher temperatures side reactions such as decomposition of sulfate esters occurred.

Complexes of chlorosulfonic acid have been studied. Phosphate complexes [35] and sulfation in the presence of boric acid [45] were used to moderate the reaction and to provide for improved product quality.

2. Secondary Alcohol Sulfates

The chemistry of secondary alcohol sulfates is similar to that of the primary products, but much less work has been done because of the lesser commercial importance of the secondary derivatives and because of stability problems with the intermediate acid esters. Specifically, they are prone to split out water easily, giving olefins or derivatives of olefins.

In general, complexes were utilized to moderate the reactions. The chlorosulfonic acid was complexed with ether [61] or with acetic acid to give acetyl sulfate [62] prior to the reaction with the secondary alcohol. Alternately an ester was prepared from the alcohol and phosphoric acid or phosphorus pentoxide and the ester then treated with chlorosulfonic acid [63, 64]. Side products formed in the latter case were alkyl phosphates which possessed valuable cleaning properties, and thus did not have to be removed from the product. In a direct comparison between sulfation of the alcohol and the phosphated alcohol, 7.7% unsulfated product ("free oil") was obtained in the former case, compared to only 1.8% using the phosphate ester which would suggest the absence of extensive side reactions.

3. Ether Alcohol Sulfates

Linear oxyethylated alcohol sulfates have assumed a commercial importance comparable to that of the alcohol sulfates. The chemistry of the two is

similar. In general, processes for sulfation of the alcohols with chlorosulfonic acid apply also to the ether sulfates. Several process patents specific
for the oxyethylated products were developed [34, 63, 64-66]. Several involved the use of complexes as in the case of the alcohols [34, 63, 64].
Aromatic ethoxysulfates were also covered in a few references [67, 68].
Reference 30, previously mentioned, discussed the relative merits of sulfur
trioxide and chlorosulfonic acid in batch sulfation of oxyethylated alcohols.

Several ethoxy derivatives of fatty acids, such as oleic and stearic,
were also sulfated with chlorosulfonic acid as well as sulfuric acid [67].
Larger molar amounts of sulfuric acid than chlorosulfonic acid were required
for the sulfation of unsaturated acid ethoxy derivatives because sulfuric acid
added to the double bonds present.

D. Sulfation with Sulfamic Acid and Derivatives

Sulfamic acid found some use as an agent for the commercial preparation of
sulfates. Reference 30 discusses the relative advantages and disadvantages
of sulfamic acid vis-a-vis sulfur trioxide. In general, advantages of sulfamic acid as a sulfating agent are the preservation of double bonds and other
groups reactive toward stronger reagents, absence of ring sulfonation in
alkoxyphenols, low inorganic salt content in products, simultaneous sulfation
and neutralization, absence of corrosive substances, and safety in the handling of the reagent [60]. Disadvantages include the need to handle a solid,
the use of a solvent for the reaction, the difficulty in preparing salts other
than ammonium salts, and the frequent need for a catalyst for the reaction.
Sulfamic acid reacted well [Eq. (6)] with primary alcohols to form the ammonium alkyl sulfates [70]:

$$ROH + H_2NSO_3H \longrightarrow ROSO_3NH_4 \tag{6}$$

Tertiary alcohols did not react, while a catalyst was usually required for
secondary alcohols [69, 70]. Amides were frequently used, although secondary alcohols have been sulfated in the absence of catalysts [71].

Several papers describe the use of sulfamic acid as a sulfating agent
in the presence of some modifying agent or catalyst. Thus, urea was used
[72-76], as were amine oxides [76], phosphates [77], and sulfuric acid [78].
N-Substituted sulfamic acids could be used as catalysts [79] or as sulfating
reagents [80]. In the latter case, reaction rates were obtained for the sulfation of various alcohols with N-cyclohexylsulfamic acid. The order is primary alcohols > secondary alcohols > tertiary alcohols > phenols. Salts
of imidodisulfonic acid have been investigated as sulfating reagents. Typical
of these was ammonium imidodisulfonate which was prepared as shown in
Eq. (7):

$$H_2NSO_3H + H_2NSO_3NH_4 \longrightarrow HN(SO_3NH_4)_2 \tag{7}$$

Yields of 60-80% of the higher alcohol sulfates were obtained with the aid of ammonium imidodisulfonate in the presence of trimethylamine-sulfur trioxide complex, dimethylformamide, or urea [81]. Another example suggested the use of the imidodisulfonate treated with sulfuric acid [82]. In the presence of dimethylformamide, the imidodisulfonate was used to sulfate a wide range of higher fatty derivatives [83], including preparation of the ammonium salts of octadecylamine polyethylene glycol ether sulfate (85% yield), and nonylphenol polyethylene oxide ether sulfate [92%). Pyridinium imidodisulfonate has been used to sulfate many lower and higher alcohols or other molecules containing available hydroxyl groups [84].

The sulfation of oxyethylated alkylphenols was generally carried out with the same methods as used for fatty alcohols; several specific references are available [71, 75, 77]. Oxyethylated aliphatic alcohols were also sulfated in the same manner [75].

E. Miscellaneous Sulfation Agents

Thionyl chloride was used to prepare sulfates by reaction with active methylene groups [240] as shown in Eq. (8).

$$(CH_3)_2CHCH_2C(CH_3)_2CH_2OR + SO_2Cl \longrightarrow$$

$$(CH_3)_2CHCHC(CH_3)_2CH_2OR + (CH_3)_2CHCH_2C(CH_3)_2CHOR \qquad (8)$$
$$\underset{OSO_3H}{|} \qquad\qquad\qquad\qquad\qquad \underset{OSO_3H}{|}$$

The sodium salts were obtained after neutralization of the corresponding acids. The identity of the products as sodium 1- or 3-alkoxysulfates was not established. Compounds sulfated included those where R was ethyl, isobutyl, 2-methylpentyl, 2-ethylhexyl, and decyl. Yields were 42, 38, 22, and 69% for the first four and not specified for the decyl derivative. Ascorbic acid sulfate gave almost quantitative yields of sulfates in the presence of an oxidizing agent, such as bromine [85]. At room temperature, n-octanol did not sulfate in the absence of oxidizing conditions but reacted under non-oxidizing conditions at 100°C. A combination of one mole of phenylsulfuryl chloride and one mole of pyridine sulfated two moles of an alcohol on heating the reaction mixture to 70°C for two hours [86]. Alcohols ranging from methyl to octadecyl and including both linear and branched alcohols were sulfated in yields of 45-73%. The highly hindered tertiary butyl alcohol gave only an 11% conversion. The sulfates, prepared as the potassium salt, were converted to the S-benzylthiuronium salt for characterization by melting point. Ammonium bisulfate, in the presence of 1% hydrazine or phosphoric acid, was used to prepare the sulfates of a variety of fatty alcohols and their ethenoxy derivatives in high yield [87]. Carbyl sulfate, ($\overline{OCH_2CH_2SO_2OSO_2}$), and its alkyl derivatives gave sulfates of dodecyl and behenyl alcohols as well as of lower alcohols [88].

F. Mixed Sulfation-Sulfonation

Most practical uses of sulfates involve "synergistic" combinations with other surfactants. Frequently used mixtures are the sulfates and ether sulfates, or sulfates and alkylarylsulfonates. It seemed fruitful to explore mixed sulfations and sulfonations of the appropriate raw materials to prepare the final mixes in one step, which indeed has been done to some extent.

Continuous treatment with chlorosulfonic acid of alcohols and dodecylbenzene in a rotating film reactor was studied [89]. When the ratio of alcohol to aromatic was 2:1, a 90% conversion was obtained with a 30% molar excess of chlorosulfonic acid. The addition of aliphatic amines lowered the yield but led to a better color. Oleum was also studied in a continuous sulfation-sulfonation scheme [90]. In another study, aromatics were alkylated with primary alcohols. The resulting mixture of alkylate and alcohol was sulfated-sulfonated in the same vessel with sulfuric acid [91]. A stepwise process [92] utilized excess sulfuric acid from the alkylation of benzene to sulfate added alcohol. The overall yields were low. A continuous stepwise process was described [93] where an alkylbenzene was sulfonated with oleum, followed by a second step where the sulfonic acid was mixed with additional oleum and a fatty alcohol. Large amounts of sodium sulfate were formed, but only small amounts of unreacted alkylbenzene or alcohol remained. Mixed primary and secondary alcohols have been sulfated with sulfuric acid. Partial hydrolysis to liberate some secondary alcohol, followed by extraction and neutralization, gave an 85:15 mixture of primary to secondary sodium sulfates [94]. Chlorosulfonic acid was used to sulfonate mixtures of oxyethylated alcohols and alcohols [95] to give 90% conversion to the sulfate.

In a continuous process [96], aromatics were alkylated with olefins using oleum and the resulting alkylates were sulfonated while the excess olefins were sulfated. A continuous process for the sulfation-sulfonation of alcohols, olefins, alkylbenzenes, epoxides or mixtures of these is described in a French patent [15].

G. Sulfation of Unsaturated Molecules

This section is limited to a discussion of the reaction of olefins and unsaturated alcohols to form saturated and unsaturated sulfates. Treatment of olefins with sulfur trioxide to form the so-called α-olefin sulfonates is discussed in another chapter. The reaction of olefins with sulfuric acid to give secondary alkyl sulfates was described in several patents and papers [97-107]. In general, α-olefins or olefins from cracked paraffins were used. In some cases, dialkyl sulfates were formed which tended to lower the yields of monoalkyl sulfates. Means have been developed to minimize this problem [98, 99, 102]. Excess sulfuric acid in these reactions could be removed by dilution of the acid to 30% with water and extraction of the sulfates with isopropyl ether [108].

A recent patent [241] describes the preparation of secondary alkyl sulfates in two steps directly from hydrocarbons. Detergent-range hydrocarbons are dehydrogenated catalytically, the olefin sulfated with sulfuric acid without separation of the unreacted hydrocarbon, and, after neutralization, the hydrocarbon is recycled to the dehydrogenation step. The method is applicable with either broad-range hydrocarbons or relatively pure fractions.

Unsaturated fatty alcohols were converted to unsaturated sulfates by the use of mild sulfating agents. Sulfation of oleyl alcohol with sulfamic acid with [109] or without [110] pyridine or with chlorosulfonic acid and pyridine [109] or with sulfuric acid [110] are described; in all but the case of sulfuric acid, good yields of unsaturated primary sulfates were obtained. Sulfation with sulfuric acid led to a decrease in iodine value, but not as a result of addition to the double bond. Apparently, polymerization of the olefin is catalyzed by the sulfuric acid, and unsaturated polymeric sulfates were derived. Lower-molecular-weight olefins, such as 1,3-butadiene, cyclopentadiene and piperylene, when treated with suspensions of sulfamic acid, benzamide, or oxamide with sulfuric acid in methylene chloride, gave polymeric unsaturated sulfates. For example, the sulfation of 1,3-butadiene with oxamide and sulfuric acid in methylene chloride ultimately led to a polymer containing 39 butadiene moieties, 4.2 hydroxyl groups, and 15.8 double bonds [111]. Sulfation of unsaturates derived from rapeseed oil with chlorosulfonic acid and pyridine, or sulfamic acid, led to unsaturated sulfates [110, 112]. Whale-oil alcohols were reported [9] to be sulfated with sulfur trioxide to give an 80% retention of the double bond.

A study of the use of chlorosulfonic acid or sulfuric acid showed that a 60% retention of the double bond was obtained when preparing oleyl ether sulfates, compared to only 25% when sulfating oleyl alcohol under comparable conditions [113].

III. PHYSICAL PROPERTIES

A compilation of data from the literature on various properties of linear primary alkyl sulfates and their solutions was prepared by J. E. Gotte [114] and some of the data follow. The alkyl sulfate salts of monovalent metals such as sodium, potassium, and silver did not form hydrates on crystallization from aqueous solutions. The salts of the bivalent alkaline earth and heavy metals, however, contained water of crystallization. The degree of hydration was found to be one mole of water less than that of the comparable inorganic metal sulfate. Thus the magnesium, zinc, and ferrous alkyl sulfates contained six moles of water whereas the copper salt contained four [114].

Lottermoser and Stoll [115] described a technique for preparing the salts whose original crystal morphology was determined. More recently, Bone and O'Day modified this procedure. They precipitated the barium alkyl sulfate from a solution of the sodium salt and barium chloride. The barium

compound was collected and refluxed in 95% ethanol in combination with the
desired metal sulfate [116]. Maurer and Stirton described a technique for
preparing the mono-, di-, and trivalent salts by the direct reaction of the
alkylsulfuric acids with the metal, the metal hydroxide, or salt of the appropriate metal [117].

The potassium salts of the alkyl sulfates were less soluble than the sodium salts, which, in turn, were less soluble than the ammonium salts.
Fig. 1 illustrates the solubility of the sodium and potassium salts of several
alkylsulfates. The calcium C_8 and C_{10} alcohol sulfates were quite water
soluble. The homologs with a longer chain were quite insoluble, as can be
seen in Table 2. It is important to note at this point that a typical commercial product which is a mixture of homologs of various chain lengths is substantially more water soluble than the pure components. The more soluble
short-chain species have a solubilizing effect on the insoluble longer-chain
homologs [114].

The pure alkylsulfuric acids were obtained from the chlorosulfation of
octadecanol, hexadecanol, tetradecanol, and dodecanol. They were white

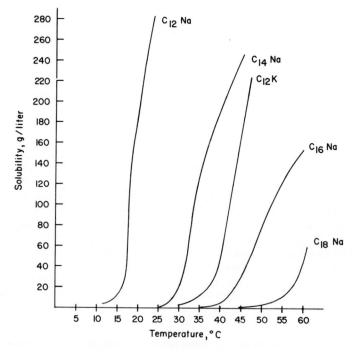

FIG. 1. Water solubility as a function of temperature of the sodium
salts of C_{12}, C_{14}, C_{16}, and C_{18} alkylsulfates compared to that of potassium
dodecylsulfate.

crystalline solids with definite melting points. Their hygroscopic nature increased with increasing chain length. Table 3 [118] lists their melting points in comparison with those of the starting alcohols.

Hydrolysis of a 0.05 molar solution of octadecylsulfuric acid in distilled water at 100° C, or the sodium salt acidified with an equivalent amount of mineral acid, amounted to 50% in less than half an hour. When measured at 60° C, the temperature most often encountered in household laundering, the degree of hydrolysis was only 10% after 3 hr and 17% after 7 hr [118].

The surface-active properties of octadecylsulfuric acid were similar to those of its sodium salt. Interfacial tension, foaming ability, wetting, and detergency were about equivalent for the two as seen in Table 4. The acid was more soluble in water and organic solvents than the salt. The critical micelle concentration (CMC) of the acid of 0.0387 millimoles per liter was one-third of that of the sodium salt [118].

TABLE 2

Water Solubility of Calcium Salts of
Alkyl Sulfates [131]

Number of carbon atoms	Solubility at 25° C, g/liter
8	400
10	250–300
12	0.3–0.4
14	0.03–0.04

TABLE 3

Melting Points of Long-Chain Alcohols and Their Sulfuric Acids, $ROSO_3H$

Alkyl group, R	Melting point, °C		Purity of acid[a], %
	Acid	Alcohol	
Dodecyl	25–27	24.1[b]	97
Tetradecyl	37–39	37.2–38.0	98
Hexadecyl	40–42	49.3–49.6	99
Octadecyl	51–52	58.1–58.6	99

[a]Purity by conversion to the sodium salt, $ROSO_3Na$, and analysis for sodium.
[b]Freezing point.

TABLE 4

Properties of Octadecylsulfuric Acid and Its Amine and Amino Acid Salts

Acid or salt	pH	Solubility at 25°C (%)			Aqueous solutions, 0.1%							
		Water	Butanol	Chloroform	Surface tension, dyn/cm at 25°C	Interfacial tension, dyn/cm at 25°C	Detergency ΔR at 60C[a], cloth		Emulsion stability, sec, at 25°C	Foam height, mm at 60°C[b]	Wetting time, sec, at 60°C
							A	B			
Octadecylsulfuric acid	3.13	1	5	>10	41.6	10.4	40.2	23.4	1190	195	11
Amine salts											
Triethylamine	5.15	1	>10	>10	38.4	7.0	13.8	12.4	690	190	21
Triethanolamine	5.15	10	1	0.1	40.9	7.0	19.0	19.8	730	190	20
THAM salt[c]	4.90	1	0.1	0.1	40.1	9.1	29.7	21.9	730	205	23
Amino acid salts											
Glycine	3.40	0.1	0.1	0.1	41.1	6.5	39.7	22.9	1050	210	19
DL-Leucine	3.30	0.5	5	5	36.1	4.3	9.9	16.6	1170	180	26
L-Methionine	3.30	1	10	5	37.4	5.9	13.4	18.4	840	200	20
Sodium octadecylsulfate	--	--	--	--	--	14.2[d]	41.3	--	760[e]	210	18

[a]Measured as increase in reflectance after washing in the Terg-O-Tometer. Cloths A and B represent different soil-removal problems in washing cotton.

[b]Ross-Miles pour foam test [Oil & Soap, 18, 99-102 (1941)].

[c]THAM = Tris-(hydroxymethyl)-aminomethane, $(HOCH_2)_3CNH_2$.

[d]Not completely in solution at 25°C.

[e]Measured at 50°C because of limited solubility.

Maurer and Stirton and his co-workers published the physical properties of the salts prepared directly from the acid and metals, metallic bases, and nitrogenous bases [117, 119-121, 123, 129]. The data are summarized in Tables 4-15.

Table 4 shows that the salts prepared from amino acids had the lowest interfacial tensions with mineral oil. This trend is reinforced by the fact that the data show them to be good emulsifiers. Under the conditions of the test, it was found that detergency decreased with increasing degree of substitution at the nitrogen atom but increased with a greater content of hydrophilic hydroxyl or carboxylic acid groups. Glycine and iminodiacetic acid salts were about equal to octadecylsulfuric acid (ΔR, 40.2) and sodium octadecyl sulfate in detergency as measured by the change in reflectance (ΔR) of a standard soiled cloth [117].

The melting points of the nitrogen base salts are given in Table 5. Solubilities and surface-active properties appear in Table 4. Additional emulsifying properties are listed in Table 6 [120].

The metal alkyl sulfates not only have utility as detergents but many are suitable also in lubricant greases and as additives to lubricating oils as well [119]. Melting points of the pure salts are given in Table 7, which also gives solubilities in plasticizers and lubricants, as well as in water and some organic solvents.

TABLE 5

Melting Points of Amine and Amino Acid Salts of
Octadecylsulfuric Acid

Salt	Melting point, °C
Amine salt	
Triethylamine	70-72.5
Triethanolamine	86.0-86.8
THAM	124-127
Urea	113-114
Guanidine	145-146.4
2-Benzyl-2-thiopseudourea	95.8-97.2
Aniline	124.8-125.8
Pyridine	103-106.5
Amino acid salt	
Betaine	108-109

TABLE 6

Emulsifying Properties of Octadecylsulfuric Acid and Its Salts

Acid or salt	Relative stability of emulsion with immiscible organic solvent[a]		Relative stability of emulsion with paraffin oil, sec[b]
	Solvent	Time	
Octadecylsulfuric acid	--	--	1190
Salts			
Triethylamine	Butanol	5 hr	690
Triethanolamine	Butanol	5 hr	730
THAM	Butanol	5.5 hr	730
Glycine	--		1050
DL-Leucine	Chloroform	72 hr	1170
	Carbon tetrachloride	36 hr	
	Tetrachloroethylene	24 hr	
	o-Dichlorobenzene	48 hr	
	Turpentine	2 hr	
L-Methionine	--	--	840
Sodium oleate	Chloroform	84 sec	350
Commercial alkyl-phenol type nonionic surface-active agents	Butanol	240 sec	--

[a]Method of Atlab emulsion testing apparatus [246]. Emulsions prepared by mechanically shaking 25 ml organic solvent with 25 ml 0.2% solution of emulsifying agent in water, noting the time required for 10% separation from the emulsion.

[b]Method of manual, violent, intermittent shaking [247]. Time required for 10 ml to break from an emulsion of 40 ml paraffin oil with 40 ml 0.1% solution of emulsifying agent in distilled water.

The oxyalkalation of a fatty alcohol [1] with propylene oxide produced an ether-secondary-alcohol [2], see Eq. (9).

$$ROH + CH_2CHCH_3 \longrightarrow ROCH_2CHCH_3 \tag{9}$$

$$[1] \qquad\qquad OH$$

$$[2]$$

TABLE 7

Metal Salts of Octadecylsulfuric Acid

Metal ion	Melting point, °C	Solubility[a], % in			
		Water	Butanol	Aniline	Plasticizers and lubricants[b]
NH_4^+		0.1	5	1	i
Li^+	184.5-185 dec	1	1	0.1	DOP, DOS, TOF
Na^+	189.5-191	--	--	--	--
K^+	182-3	0.1	0.1	i	DBS, SAE-10, TOF
Ag^+	153-8 dec	i	1	10	TOF
Be^{++}	c	i	10	0.1	DBS, DOP, DOS, SAE-10, TOF
Mg^{++}	c	0.1	0.1	i	TOF
Ca^{++}	c	i	i	i	i
Sr^{++}	165-166 dec	i	i	i	DBS, DOS, TOF
Ba^{++}	172.8-173 dec	i	i	i	i
Co^{++}	180 dec	i	10	1	i
Cu^{++}	135-140 dec	i	10	10	i
Zn^{++}	c	0.1	0.1	1	TOF
Cd^{++}	193-196 dec	i	1	1	TOF
Pb^{++}	151.8-152 dec	i	i	1	DOP, DOS, TOF
Al^{++}	162 dec	i	0.1	5	--

[a]The symbol "i" indicates a solubility of less than 0.1%.

[b]Solubility of 1% or greater. DBS = dibutyl sebacate, DOP = dioctyl phthalate, DOS = dioctyl sebacate, SAE-10 = petroleum lubricating oil, TOF = trioctyl phosphate.

[c]Does not have a sharp, definite melting point.

Ethylene oxide, on the other hand, gave an ether-primary-alcohol. Since the secondary alcohol was less susceptible to subsequent oxyalkylation than a primary alcohol, it was possible to obtain a high yield of monooxyalkylation product [2] with propylene oxide, [1] is thus more reactive than [2]. Taking advantage of this property, Weil and co-workers produced pure mono- and dioxypropylated alcohols which were converted to their sodium sulfates. Their surface-active properties are given in Table 8 [123].

A comparison of Krafft points, i.e., the temperature at which a 1% turbid dispersion changed sharply to a clear solution on gradual heating, showed
that a single oxypropyl group increased the solubility of the ether sulfate over
that of the parent alcohol sulfate to a greater extent than a single oxyethylene
group.

Consistent with the fact that the oxypropyl group was more hydrophobic
than the oxyethyl group, insertion of the former lowered the CMC to a greater
extent than did the latter. A second oxypropyl group had less effect, presumably due to coiling in the ether chain as occurred with oxyethylation [123].

A series of ether alcohols was prepared from the alkyl bromide, sodium, and the appropriate glycol. They were sulfated to give eight individual
pure compounds of the general formula $R(OC_2H_4)_nOSO_3Na$, where $R = C_{16}H_{33}$
or $C_{18}H_{37}$ and $n = 1$, 2, 3, or 4. Since they were pure compounds rather
than mixtures of homologs, they could more clearly illustrate the relation between chemical composition and properties as shown in Tables 9 and 10. The
sodium salts of these ether sulfates did not have sharp melting points but
passed through a glassy state before melting with some decomposition.
Where $n = 1$, 2, 3, and 4, the melting points for the hexadecyl series were
about 184, 177, 161, and 93° C, and for the octadecyl series 193, 187, 191,
and 96° C. The melting points of the triethanolammonium salts of the 12, 14,
16, and 18 carbon alkylsulfuric acids were 121-122.5, 106-107, 82.4-83.0,
and 86.0-86.8° C, respectively. The above compounds with four oxyethyl
groups were most soluble in water as well as in organic solvents [129].

The preparation of a pure sample of sodium dodecyl sulfate which would
show no minimum in its surface-tension vs log-concentration curve was difficult. The presence of even a trace of water led to hydrolysis and gave a
minimum in the curve. A sample was prepared by a technique using a combination of liquid-liquid extraction and crystallization. Aqueous solutions of
this material began to show evidence of a minimum within 24 hr and it became quite distinct at the end of four days [125].

The C_{16}- and C_{18}-alkyl sulfates were less soluble in water than the
C_{12}- and C_{14}-alkyl sulfates. Higher temperatures were thus needed before
they exhibited their maximum detergency and wetting properties. Oxyalkylation of the starting fatty alcohol before sulfation increased their water solubility. Hexadecyl and octadecyl alcohols are saturated alcohols obtained
from tallow. The sulfates prepared from them are known to be excellent detergent but their general usefulness is limited by their low solubility in water at room temperature. Oxyethylation of the base alcohol before sultation
in order to improve solubility was studied by Bistline and co-workers [121].
They found that incorporation of two oxyethyl groups improved solubility
without loss in detergency as compared with the parent sodium hexadecyl and
octadecyl sulfates. Sodium polyoxyethylene octadecyl sulfate (oxyethylene,

10 moles) had greater solubility than the two-mole ethoxamer, but was a less effective detergent as shown in Tables 11 and 12.

Ether sulfates were also more soluble in organic solvents than the corresponding alcohol sulfates and then ten-mole ethylene oxide adduct had a greater solubility in organic solvents than the two-mole adducts [121].

In contrast to the tallow alcohol ether sulfates, the two- and three-mole oxyethylene adducts of lauryl alcohol after sulfation gave sodium, potassium, ammonium, calcium, magnesium, and triethanolamine salts that were capable of forming clear, free-flowing solutions at 30% concentration at room temperature [122]. Because of their more hydrophilic nature these materials could also tolerate a high concentration of electrolyte before their solutions separate. Up to 8.5% of sodium chloride could be dissolved in a 20% active solution of sodium lauryl ether sulfate [124] and the solution remained clear and free-flowing.

Among the tallow derivatives, solubility was increased by the presence of sodium oleyl sulfate [126] or sodium 9,10-dichlorooctadecyl sulfate [127, 130] as well as by the addition of oxyethyl groups. Improved solubility in built solutions and in hard water made them useful in detergent formulations [128]. Another way in which solubility and related properties of the tallow alcohol sulfates could be improved was to prepare the triethanolammonium salt rather than the sodium salt; see Tables 9 and 10 [129].

Normal alcohols ranging from C_{11} to C_{18} and five isomeric C_{13} aliphatic alcohols were sulfated or oxyethylated and then sulfated for a study of the effect of structure on surfactant properties [132]. The cotton-washing performance of the sodium sulfates of the n-primary alcohols (PA I) and a mixture of 75% n-primary and 25% 2-methyl alcohols (PA II) was determined; the results are shown in Table 13. The C_{11} and C_{12} members in both series of sulfated alcohols were much less effective detergents than their higher homologs. The C_{13} sulfates were intermediate and highest cotton-washing performance was attained at a chain length of C_{14} through C_{18}. There was a significant difference in performance between the PA I and the PA II series. The clear points for the above series of alcohol sulfates versus their carbon number are given in Table 14. The PA II sulfates (those containing 25% of the 2-alkyl isomers) were shown to be more soluble than PA I derivatives. The Krafft point for the n-primary alcohol sulfates increased linearly for the even-carbon-number compounds. The foam performance of the two series of primary alcohol sulfates in a heavy-duty high-foaming solid (HDHFS) detergent composition in which 10% of the active material was replaced by 10% of lauric isopropanolamide (LIPA) was greatest at a carbon number of C_{15} for both series. While the peak in foaming occurred at C_{15}, there was a sharp rise between C_{13} and C_{14} and a sharp drop-off between C_{16} and C_{17}. The same two series of alcohols which were

TABLE 8

Surface Active Properties of Pure Ether Alcohol Sulfates

Alcohol sulfate	CMC[a] mmoles per liter	Krafft point, 1% solution, °C	Ca^{++} stability, ppm CaCO$_3$	Lime soap dispersing power, %	Detergency, 60°C, ΔR[b]		Foam height, 60°C, mm	
					0.25%, distilled water	0.05% + 0.2% builder,[c] 300 ppm	0.25%, distilled water	0.05% + 0.2% builder,[c] 300 ppm
C$_{12}$H$_{25}$OCH$_2$CHCH$_3$ OSO$_3$Na	2.69	Clear at zero	>1800	14	21	22	200	210
C$_{14}$H$_{29}$OCH$_2$CHCH3 OSO$_3$Na	0.58	14	>1800	8	23	26	215	220
C$_{14}$H$_{29}$[OCH$_2$CH(CH$_3$)]$_2$OSO$_3$Na	0.36	Clear at zero	>1800	9	20	23	210	200
C$_{16}$H$_{33}$OCH$_2$CHCH$_3$ OSO$_3$Na	0.16	27	780	8	25	27	200	185

$C_{18}H_{37}OCH_2CHCH_3$ $\phantom{C_{18}H_{37}OCH_2CH}OSO_3Na$	0.07	43	n.s.[d]	8	30	28	160	95
$C_{18}H_{37}OCH_2CH_2OSO_3Na$	0.11	46	n.s.[d]	9	29	29	160	100
$C_{12}H_{25}OSO_3Na$	6.8[e]	16	650	30	23	18	220	185
$C_{16}H_{33}OSO_3Na$	0.42[e]	45	n.s.[d]	40	29	28	245	240
$C_{18}H_{37}OSO_3Na$	0.11[e]	56	n.s.[d]	n.s.[d]	34	28	225	190

[a] Pinacyanole chloride method.
[b] ΔR is the increase in reflectance after washing standard soiled cotton in the Terg-O-Tometer.
[c] Builder: 55% $Na_5P_3O_{10}$, 24% Na_2SO_4, 10% $Na_4P_2O_7$, 10% Na metasilicate, 1% CMC.
[d] Not soluble enough for the test conditions.
[e] Measured at 50°C.

159

TABLE 9

Solution and Surface-Active Properties of Alcohol Sulfates

Alcohol sulfate	Abbreviation	Krafft point, °C [a]	CMC, mmoles/liter [b]	Surface and interfacial tension, 0.1% solution, 25°C, dyn/cm		Wetting time 0.1% solution, at 60°C, sec
				S.T.	I.T.	
$C_{12}H_{25}OSO_3Na$	12	16°	6.8	49.0	20.3	19.1
$C_{16}H_{33}OSO_3Na$	16	45°	0.42	35.0[c]	7.5[c]	11.6
$C_{16}H_{33}OSO_3NH(C_2H_4OH)_3$	16-T	Clear at zero	0.34	41.0	10.0	14.9
$C_{16}H_{33}OC_2H_4OSO_3Na$	16-1	36°	0.24	36.2	7.2	12.1
$C_{16}H_{33}(OC_2H_4)_2OSO_3Na$	16-2	24°	0.14	39.4	8.7	16.6
$C_{16}H_{33}(OC_2H_4)_3OSO_3Na$	16-3	19°	0.12	41.6	10.2	21.1
$C_{16}H_{33}(OC_2H_4)_4OSO_3Na$	16-4	1°	0.12	43.5	11.7	22.9
$C_8H_{17}CH{=}CH(CH_2)_8OSO_3Na$	18-cis	Clear at zero	0.29	35.8	7.4	10.8
$C_8H_{17}CH{=}CH(CH_2)_8OSO_3Na$	18-trans	29°	0.18	36.1	6.5	10.3

		Clear at zero				
$C_8H_{17}CHClCHCl(CH_2)_8OSO_3Na$	18-Cl$_2$		0.26	35.8	5.8	15.2
$C_{18}H_{37}OSO_3Na$	18	56	0.11	40.6[c]	14.2[c]	18.4
$C_{18}H_{37}OSO_3NH(C_2H_4OH)_3$	18-T	26	0.07	40.9	9.0	19.6
$C_{18}H_{37}OC_2H_4OSO_3Na$	18-1	46	0.09	39.0[c]	11.0[c]	21.8
$C_{18}H_{37}(OC_2H_4)_2OSO_3Na$	18-2	40	0.07	39.5	8.5	24.1
$C_{18}H_{37}(OC_2H_4)_3OSO_3Na$	18-3	32	0.07	41.1	8.9	30.5
$C_{18}H_{37}(OC_2H_4)_4OSO_3Na$	18-4	18	0.07	43.1	10.3	32.8

[a]Temperature at which a 1% dispersion became a clear solution on gradual heating.
[b]Pinacyanole chloride method at 50° C.
[c]Turbid dispersion at 25° C.

TABLE 10

Detergency and Related Properties

Alcohol sulfate[a]	Foam height, 0.25% built[b] solutions, 300 ppm, 60°C, mm	Detergency,[c] 0.25% built[b] solutions, 300 ppm, 60°C, ΔR	Dishwashing test,[d] 0.2%, 100 ppm, 50°C, ΔR	Emulsion stability, 0.1%, 25°C, sec	Calcium stability, 0.5%, 25°C, ppm CaCO$_3$	Lime soap dispersing power, 0.25%, 25°C, %
12	185	6.6	13.8	130	650	30
16	240	31.3	20.4	n.s.[e]	n.s.[e]	n.s.[e]
16-T	240	27.8	14.7	470	420	65
16-1	210	20.7	16.2	240	1060	3
16-2	200	11.4	12.7	290	1600	3
16-3	170	8.3	10.6	260	>1800	4
16-4	170	7.6	9.3	240	>1800	4
18-cis	230	33.5	14.8	200	920	10
18-trans	220	32.9	--	--	870	10
18-Cl$_2$	210	31.3	19.6	300	920	7
18	190	30.4	16.6	n.s.[e]	n.s.[e]	n.s.[e]

18-T	190	28.9	15.1	280	320	n.s.^{e}
18-1	140	18.9	10.2	n.s.^{e}	n.s.^{e}	14
18-2	120	12.4	10.1	440	>1800	16
18-3	105	8.1	--	390	>1800	10
18-4	100	7.2	5.2	340	>1800	10

^{a}Abbreviations of Table 9.

^{b}Built solutions: 0.05% with respect to alcohol sulfate, 0.20% with respect to a mixture of $Na_5P_3O_{10}$, Na_2SO_4, $Na_4P_2O_7$, Na_2SiO_7, CMC as in Table 8.

^{c}Launder-Ometer, 1 swatch of standard soiled cotton/100 ml/jar, 30 steel balls, 5 replicates; ΔR = increase in reflectance after washing.

^{d}Terg-O-Tometer, method of Leenerts, glass slides soiled with greasy soil, 6 replicates; ΔR = increase in reflectance.

^{e}Compounds not adequately soluble at 25° C.

TABLE 11

Detergency and Foam Height of Sodium Salts of Ether Alcohol Sulfates

Sodium salt	Detergency,[a] ΔR at 60° C				Initial foam height mm, 60° C			
	90 ppm		300 ppm		90 ppm		300 ppm	
	0.25%	0.05% + 0.2% B[b]	0.25%	0.05% + 0.2% B[b]	0.25%	0.05% + 0.2% B[b]	0.25%	0.05% + 0.2% B[b]
$C_{16}H_{33}(OC_2H_4)_2OSO_3Na$	34.4	35.9	21.1	34.4	210	200	200	200
$C_{18}H_{37}(OC_2H_4)_2OSO_3Na$	37.2	36.8	18.9	36.0	170	130	95	115
$C_{18}H_{37}(OC_2H_4)_{10}OSO_3Na$	26.7	27.0	17.4	25.6	115	105	115	125
Sodium alkylsulfates								
Sodium dodecyl sulfate	30.7	29.9	22.9	24.5	205	195	225	195
Sodium hexadecyl sulfate	37.5	36.9	36.2	35.7	235	235	120	240
Sodium octadecyl sulfate	38.0	37.3	33.7	36.5	205	185	20[c]	190

[a]Measured as increase in reflectance, ΔR, after washing 10 swatches of standard soiled cotton in one liter of detergent solution for 20 min at 60° C and 110 cycles per min.

[b]B = Builder: 55% $Na_5P_3O_{10}$, 24% Na_2SO_4, 10% $Na_4P_2O_7$, 10% sodium metasilicate, 1% CMC.

[c]Very turbid solution.

TABLE 12

Surface–Active and Related Properties of Sodium Salts of Ether Alcohol Sulfates

Sodium salt	Surface and interfacial tension[a] dyn/cm		Sinking time[a] sec	Emulsion stability[a] sec	Ca stability, ppm $CaCO_3$	Lime soap dispersing power, %
	S.T.	I.T.				
$C_{16}H_{33}(OC_2H_4)_2OSO_3Na$	30.4	8.5	42	810	1340	5
$C_{18}H_{37}(OC_2H_4)_2OSO_3Na$	36.2	10.9	280	1540	n.s.[b]	5
$C_{18}H_{37}(OC_2H_4)_{10}OSO_3Na$	38.4	11.7	160	930	1800	5
Sodium alkyl sulfates						
Sodium dodecyl sulfate	49.0	20.3	14	160	720	30
Sodium hexadecyl sulfate	35.0	7.5	180	380	n.s.[b]	n.s.[b]
Sodium octadecyl sulfate	40.6[c]	14.2[c]	n.s.[b]	n.s.[b]	n.s.[b]	n.s.[b]

[a] 0.1% solutions in distilled water at 25° C.
[b] Not soluble enough for test conditions.
[c] Sodium octadecyl sulfate was not completely in solution at 0.1% concentration.

TABLE 13

Effect of Alcohol Carbon Number on Cotton Detergency of
PA I[a] and PA II[b] Sulfates in Heavy Duty
High Foaming Solid Detergents

Alcohol carbon number	Detergency rating of sulfates[c]	
	PA I	PA II
11	59	56
12	61	65
13	96	94
14	113	115
15	117	117
16	119	114
17	121	110
18	114	116

[a] n-Primary alcohols.
[b] 75% n-Primary alcohols + 25% 2-alkyl isomers.
[c] Terg-O-Tometer test, 150 ppm hardness, 0.04% active matter.
LAS $(C_{12.6})$ = 100.

oxyethylated and sulfated to produce the primary alcohol-polyoxyethylene (3 moles) sulfates were studied. In this case, maximum foam performance was reached between C_{13} and C_{14}. A comparison of the two series shows that oxyethylation made the foam performance less dependent on chain length over the range C_{12} to C_{15}.

The amount of n-primary alcohol in blends of C_{12}-C_{15} alcohols with 2-alkyl isomers did not influence the cotton detergency of ethoxylates (62% ethylene oxide) or the foam performance of alkyl ethoxy sulfates; however, at least 65% of the n-primary alcohol was required for maximum detergency performance of the alkyl sulfates.

As can be seen from Table 15, a shift of the hydroxyl group from the end of the alkyl chain toward the center had a large deleterious effect on the performance of the alkyl sulfates. The shift of a methyl group exhibited a smaller effect. The effects noted are more pronounced for the C_{13} alcohol than the C_{15}. The placement of a $-CH_2OH$ group near the center of the chain (Gurbet alcohol) also has an unfavorable effect on performance [132].

TABLE 14

Effect of Alcohol Carbon Number on Clear Point of
Unbuilt Alcohol Sulfate[a] Solutions

Alcohol carbon number	Clear point, °C					
	PA I[b]			PA II[c]		
	1[d]	5	10	1	5	10
11	11	13	14	8	11	13
12	16	18	18	14	16	17
13	27	31		25	27	29
14	31	34	35	28	31	32
15	40	43	44	38	40	42
16	45	46	47	41	44	45
17	51	52	52	49	49	50
18[e]	52		57	54		50

[a]Extracted with petroleum ether to remove unsulfated alcohol. Less
than 0.5% inorganic salts (Na_2SO_4, traces of Na_2CO_3), based on alcohol sulfate (except C_{18} PA II sulfate which has 0.9% inorganic salts).
[b]n-Primary alcohol.
[c]75% n-Primary alcohol + 25% 2-alkyl isomers.
[d]Percentage solution.
[e]Approximate values.

It has been established that sulfation at the center of the alkyl chain in
long-chain alkyl sulfates gives rise to superior wetting and emulsifying properties. Terminal sulfation results in enhanced surface-tension reduction,
increased detersive efficiency and dispersing ability, and a high degree of
foaming and foam stability [133]. Symmetrical secondary alcohol sulfates of
high purity, prepared by the reduction of ketones to the alcohols followed by
sulfation, showed an improvement in wetting properties with increase in the
length of the carbon chain. Thus 10-nonadecyl sulfate has better wetting
properties than 8-pentadecyl sulfate [134]. The eight possible positional isomers of sodium and potassium n-hexadecyl sulfates were prepared and sodium and potassium sulfates of $RCH(OH)C_{11}H_{23}$ (where the R-group was methyl,
ethyl, or n-propyl) were also prepared. These secondary alcohols were obtained by the organozinc reduction of the corresponding ketones [135]. The
wetting properties of the two homologous series increased as the hydrophilic
sulfate group was moved to the center of the hydrophobe chain. Foaming
properties went in the opposite direction; the highest foamers were those
compounds with the sulfate group near the end of the chain. Similarly, as

TABLE 15

Effect of Hydroxyl and Methyl Group Positions on the Performance and Properties of C_{13} and C_{15} Isomeric Alcohol Sulfates

Position	C_{13}				C_{15}			
	Surface tension dyn/cm at 50°C	Clear point[a] °C	Decrease[b]		Surface tension dyn/cm at 50°C	Clear point, °C	Decrease	
			Cotton detergency[c]	Foam performance[d]			Cotton detergency	Foam performance
Hydroxyl group								
1	33.1	27	0	0	33.4	40	0	0
2	35.0	21	34	72	34.3	34	5	34
7	28.2	<0	51	88	24.5		33	98
Methyl group								
n-Primary alcohol	33.1	27	0	0	33.4	40	0	0
1[e]	35.0	21	34	72	34.3	34	5	34
2	35.7	17	31	63	34.0	30	6	34
6	34.2	<0	29	77				

[a] 1% solution in distilled water, using alcohol sulfates that had been extracted with petroleum ether to remove unsulfated alcohol. The sulfates contain less than 0.45% inorganic salts (Na_2SO_4; traces of Na_2CO_3), based on alcohol sulfate.

[b] Percentage decrease in performance.

[c] Concentration, 0.040% alcohol sulfate (HDHFS).

[d] Concentration, 0.025% alcohol sulfate (HDHFS).

[e] Tridecanol-2 and pentadecanol-2.

expected, the detergency of the terminally sulfated materials was better than that of the others [135].

Data compiled by Rosen [245] were used to define in quantitative fashion the efficiency of surfactants in lowering surface tension. In discussing the relationship between surfactant structure and surface tension reduction, a distinction must be made between the efficiency of a surfactant, indicated by the concentration required to produce some significant reduction in the surface tension of the solvent, and its effectiveness, measured by the maximum value to which it can depress the tension, as these can run counter to one another. Rosen demonstrated that by measuring the concentration (C) required to lower the tension by 20 dyn/cm, the log of the reciprocal of C,

$\log (1/C)$

gives a suitable measure of efficiency. This quantity can be related to the free-energy change involved in the transfer of a surfactant from the interior of a bulk phase to the interface and to the various structural groups present in the surface-active molecule. Table 16 lists the concentration of several sulfates required to lower the surface tension in certain interface systems by 20 dyn/cm.

Sodium alkyl sulfates were prepared by the sulfation of randomly isomeric linear secondary alcohols and their properties were compared to compounds prepared by the direct sulfation of α-olefins [105]. Foaming ability of both materials was equivalent and the optimum chain length for foaming was C_{15}-C_{17}; a range of C_{16} to C_{18} was optimum for the cotton detergency of the alcohol-based secondary sulfates. The wetting ability of these materials reaches a maximum at C_{15} and was superior to that of the olefin-based sulfates. The secondary alcohol sulfates were sufficiently thermally stable to be spray-dried and their hydrolytic stability was adequate for a long shelf life in liquid products [105].

The effect of the cation on the surface-active properties of alkyl sulfates was found to be minimal compared to the alkyl chain length which determined their CMC value. Thus foaming power and surface tension were not greatly affected by the nature of the cation. On the other hand, bulk properties such as viscosity and solubility (Krafft point) are greatly influenced by the cationic group [136]. The tolerance of solutions of sodium dodecyl sulfate (SDS) for calcium ion could be greatly improved by the presence of polar organic compounds such as hexanol or hexylamine. Precipitation of calcium dodecyl sulfate occurred with 0.006 moles of calcium chloride and 0.069 moles of a 2% solution of SDS. When hexanol was present 0.018 moles of calcium chloride could be tolerated without precipitation [137].

A comparison of the foaming properties of lauryl ether sulfate with lauryl sulfate showed substantially better foaming for the ether sulfate. Op-

TABLE 16

Efficiency of Interfacial Tension Reduction as a Function of
Hydrophobe Chain Length[a]

Sodium alkyl sulfate	Interface	Temp., °C	Moles/liter[b]	
Decyl	Aqueous/air	27	1.29	$\times 10^{-2}$
Dodecyl	Aqueous/air	25	4.1	$\times 10^{-3}$
Dodecyl	Aqueous/air	25	2.7	$\times 10^{-3}$
Dodecyl	Aqueous/air	60	5.75	$\times 10^{-3}$
Hexadecyl	Aqueous/air	25	2.0	$\times 10^{-4}$
Octyl	Aqueous/heptane	50	2.45	$\times 10^{-2}$
Decyl	Aqueous/heptane	50	7.7	$\times 10^{-3}$
Dodecyl	Aqueous/heptane	50	1.9	$\times 10^{-3}$
Tetradecyl	Aqueous/heptane	50	4.9	$\times 10^{-4}$
Hexadecyl	Aqueous/heptane	50	1.3	$\times 10^{-4}$
Octadecyl	Aqueous/heptane	50	3.8	$\times 10^{-5}$
Tetradecyl (1 E.O.[c])	Aqueous/air	25	1.5	$\times 10^{-4}$
Hexadecyl (1 E.O.)	Aqueous/air	25	3.1	$\times 10^{-5}$
Hexadecyl (2 E.O.)	Aqueous/air	25	2.1	$\times 10^{-5}$
Hexadecyl (3 E.O.)	Aqueous/air	25	2.1	$\times 10^{-5}$
Octadecyl (1 E.O.)	Aqueous/air	25	1.1	$\times 10^{-5}$
Octadecyl (2 E.O.)	Aqueous/air	25	8.3	$\times 10^{-6}$
Octadecyl (3 E.O.)	Aqueous/air	25	1.15	$\times 10^{-5}$
Dodecyl	0.1 M NaCl/air	25	1.6	$\times 10^{-4}$
Hexadecyl	0.1 M NaCl/air	25	5.8	$\times 10^{-6}$

[a]See Reference [245] and sources cited therein.
[b]Concentration required to reduce interfacial tension by 20 dyn/cm.
[c]Moles ethylene oxide.

timum foaming performance was reached at 2 to 4 moles of ethylene oxide
per mole of lauryl alcohol. The same was found to be true of a comparison
of the wetting properties of the two types of materials. The foaming and wet-

ting performance of the alkyl sulfate was reduced in hard water, whereas that of the ether sulfate appeared to be slightly improved over its soft water performance [122].

Viscosities of aqueous ether sulfate solutions decreased with increasing ether chain length. However, the viscosities were very dependent on the sodium chloride and unsulfated ether alcohol content of the product [122].

A study of the surface-active properties of a series of sodium polyoxy-propylated lauryl sulfates (PPLS) containing from 1 to 20 moles of propylene oxide in comparison with sodium lauryl sulfate (SLS) showed them to be generally superior in wetting, dispersing, and emulsifying powers. The surface tension and CMC values of PPLS were lower than those of SLS. The foaming power and foam stability of SLS was superior to that of PPLS, the 1-mole adduct being an exception. The sodium polyoxypropylated lauryl sulfates were, in general, excellent surfactants [139]. A study of sodium polyoxy-butylated lauryl sulfates showed similar results, except that in this case the oxybutylated sulfates had better foaming power and foam stability than did SLS [243]. The wetting power of the products with fewer than 3 moles of butylene oxide was excellent and the products containing more than 10 moles had good emulsifying properties.

Fatty alcohols from C_{12} to C_{18} were oxyalkylated with ethylene oxide, propylene oxide, and 1,2-butylene oxide, and subsequently converted to the ether sulfates (Table 17). Oxybutylation gave slightly lower Krafft points than oxyethylation. Krafft points for the dioxyalkylated products were lower than for the monooxyalkylated products as shown in Table 17. Oxybutylation reduced the foaming properties of the longer-chain alcohol sulfates and dioxybutylated products had low and unstable foams [140].

At a 100° F, C_{14} and C_{16} alcohol sulfates showed significantly better detergency in a liquid dish-wash formulation than C_{12} or C_{18} alcohol sulfates, the C_{16} and C_{18} sulfates were marginally better than the C_{14} sulfate which, in turn, was significantly better than the C_{12} sulfate. Therefore, the C_{16} sulfate appeared to offer the optimum detergency properties [141].

The general properties of ether sulfates mentioned above were again found to hold true here. Solubility decreased proportionately as the ethylene oxide content decreased. However, the foam stability increased as the ethylene oxide content was lowered [138].

A comparison of the cotton detergency of the primary alkyl sulfates, 2-methylalkanoates, and linear alkylbenzenesulfonates (sodium salts) was made using a statistical analysis of the effects of molecular weight, active concentration, phosphate concentration, temperature, water hardness, and calcium-to-magnesium ratio [242]. In general, the alkyl sulfates ranked between the carboxylic acid salts (best) and the alkylbenzenesulfonates.

TABLE 17

Surface-Active Properties of Ether Alcohol Sulfates

$R[OCH_2CH]_n OSO_3Na$			Krafft point, 1% solution	CMC,[a] millimoles, per liter	Calcium stability, ppm $CaCO_3$	Lime-soap dispersing power, %	Detergency, 60°C ΔR^c 0.05% + 0.2% B,[d] 300 ppm	Foam height at 60°C, mm 0.05% + 0.2% B,[d] 300 ppm
R	R'	n						
$C_{12}H_{25}$		0	16°	--	650	30	18	185
	CH_3	1	Clear at 0	2.69	>1800	14	22	210
		2	Clear at 0	1.54	>1800	6	21	205
	C_2H_5	1	Clear at 0	1.8	>1800	10	22	225
		2	Clear at 0	0.8	>1800	10	21	208
$C_{14}H_{29}$		0	30	--	465	19	29	230
	CH_3	1	14	0.58	>1800	8	26	220
		2	Clear at 0	0.36	>1800	9	23	200
	C_2H_5	1	13	0.43	>1800	9	25	200
		2	Clear at 0	0.20	>1800	19	22	175
$C_{16}H_{33}$		0	45	n.s.[b]	n.s.[b]	n.s.[b]	28	240
	H	1	36	0.22	1060	3	--	--
		2	24	0.14	1600	3	--	--

CH$_3$	1	27	0.16	780	8	27	185
	2	19	0.076	>1800	7	26	175
C$_2$H$_5$	1	23	0.12	940	9	26	155
	2	21	0.061	1100	8	24	50
C$_{18}$H$_{37}$	0	56	n.s.[b]	n.s.[b]	n.s.[b]	28	190
H	1	46	0.11	n.s.[b]	9	29	100
	2	40	0.07	>1800	16	--	--
CH$_3$	1	43	n.s.[b]	n.s.[b]	8	28	100
	2	31	0.041	1285	7	25	90
C$_2$H$_5$	1	38	--	560	10	25	75
	2	24	0.051	775	11	26	35

[a] Critical micelle concentration by the pinacyanole chloride method.
[b] Not soluble enough for test conditions.
[c] Increase in reflectance after washing standard soiled cotton in the Terg-O-Tometer.
[d] Builder: 55% Na$_5$P$_3$O$_{10}$; 24% Na$_2$SO$_4$; 10% Na$_4$P$_2$O$_7$; 10% Na metasilicate, 1% CMC.

IV. APPLICATIONS OF ALCOHOL AND
ETHER ALCOHOL SULFATES

A. Uses in Cosmetics and Medicinals

1. Shampoos

Shampoos are products designed to remove dirt, sebum, and sweat from the
hair and scalp. They also should remove the oil from oily hair and replenish
the oil in dry hair, leaving it glossy, manageable, lustrous, etc. These are
properties that are often mutually inconsistent. The functional properties of
a shampoo include the following major categories: (1) odor, color, appearance, viscosity, etc., which are the esthetic qualities of the shampoo;
(2) Working characteristics, including foaming ability, ease of application,
rinseability, effect on the skin or eyes, etc., and (3) Conditioning of hair
and skin after shampooing with respect to luster, manageability, and clean-
liness [141].

Many commercial shampoos were designed to remove only part of the
oily soil from the hair [142]. The rationale was that a thorough cleaning of
the hair resulted in undesirable side effects such as poor gloss, harsh feel,
and electrostatic "fly." These shampoos were advertised as allowing the
"natural oils" of the hair (sebum) to remain as though it were desirable.
There is, in fact, little correlation between the amount of sebum removed
from the hair by shampooing and the esthetic quality and condition of the hair.
The aftereffects of shampoos appear to be the result of an interaction between the detergent and material in the outer layers of the hair cuticle. A
suitable shampoo surfactant gives maximum cleansing power combined with
"mildness" to the hair. Auxiliary agents in the shampoo can be used to provide a treatment to the hair reducing electrostatic "fly" and enhancing luster
and softness [143].

Alkyl sulfates and alkyl ether sulfates are the most widely used surfactants in shampoos. Those derived from lauryl alcohol possess the optimum
combination of foaming, detergency, and wetting properties and possess a
low skin-irritation potential. Secondary alcohol sulfates from olefins are
used in Europe in spray-dried powder shampoos [144]. While not as thermally stable as the primary sulfates, they are sufficiently stable to be spray
dried [145]. Linear alkyl sulfates prepared from the odd-chain-length alcohols, such as Shell Oil Company's Neodol alcohols, are a recent innovation.
Both these and their ethoxamers were sulfated for use in shampoos and were
claimed to perform somewhat better than the equivalent naturally derived
even-number materials [146]. The ethoxamers of the natural and synthetic
linear fatty alcohols and secondary alcohols as well have been sulfated for
use in shampoos. These materials produced higher foam in hard water and
were milder to skin than their alkyl sulfate counterparts.

The use of sulfates of oxyethylated alkylphenols has declined steadily to
where they now represent a small fraction of the shampoo market. This is

due to their higher raw-materials cost, greater eye-irritation potential and processing difficulties. Direct sulfation with sulfuric acid, sulfur trioxide, or chlorosulfonic acid resulted in some sulfonation of the aromatic ring as well as sulfation of the hydroxyl group on the polyoxyethylene chain. This produced an undesirable by-product which reduced the foaming ability of the product. Sulfamic acid had to be used for sulfation in order to avoid this by-product. However, sulfamic acid was an expensive reagent and it yielded only the ammonium salt. The product must be digested with caustic soda or caustic potash whereby ammonia is driven off in order to obtain the sodium or potassium salt.

Brief descriptions of a number of shampoo formulations are given below. They are intended to illustrate a variety of surfactant formula combinations and are not intended to be either comprehensive or restrictive. Shampoo preparations usually contain alkyl sulfates with various cations as the principal surfactant ingredient. Other formulations are combinations of the above with amphoteric surfactants, nonionic surfactants, oxyethylated amines, phosphates, amine oxides, etc. An aerosol shampoo composition based on an alkyl sulfate contained a combination of sodium dodecyl sulfate, lauric monoethanolamide, and other ingredients. Lanolin was present to provide superfatting and the propellant was chlorodifluoroethane [147]. A shampoo which improved the manageability of washed hair contained a mixture of C_8-C_{18} alkyl sulfate blended with C_{10}-C_{18} acyl sarcosinate and fatty alkyl phosphonates as well as other ingredients. The formulation appeared to bear some resemblance to a rug shampoo. An apparatus for the measurement of friction forces showed that inclusion of the phosphonate greatly reduced the frictional forces of the formulation as compared with an analogous formulation from which it was omitted.

In a novel approach to a hair shampoo a light mineral oil emulsified in water by a combination of an alkyl sulfate, alkyl aryl sulfonate, and an amphoteric surfactant was used. Monoethanolammonium lauryl sulfate, sodium dodecylbenzenesulfonate, N-lauryl-N-(carboxymethyl)-N'-(2-hydroxy-ethyl)ethylenediamine, isopropyl alcohol, light mineral oil, and perfume were combined with water. The combination formed an oil-in-water emulsion which had good foaming and lathering properties. The amphoteric surfactant was sufficiently substantive to the hair to reduce fly-away and the small amount of residual mineral oil that remained imparted a sheen to the hair and also contributed to manageability. This same combination of ingredients could be compounded as a bath oil by varying their proportions [149].

A mixture of the ethanolamine and diethanolamine salts of lauryl sulfuric acid were combined with a dialkanolamide of a fatty acid as a foam booster and stabilizer, and a nonionic surfactant, polyethylene glycol 400 dilaurate, as an emulsifier and thickening agent. The composition was used for impregnating cloth to produce shampoo pads [150]. Triethanolammonium lauryl sulfate was used as the principal surfactant in a Russian hair-shampoo patent. It was employed in combination with a monoalkanolamide of a fatty acid and an amphoteric surfactant and sulfonpone (the product of the condensation of peptides with sulfonic acid chlorides) [151].

An example of the use of an alkyl sulfate in combination with an oxy-ethylated amine was a formulation consisting of the diethanolamide of lauric acid, the 3-mole ethylene oxide adduct of hexadecylamine and isopropanol-ammonium fatty alcohol sulfate. The clear viscous liquid shampoo base left the hair soft and manageable. The same composition could be milled into a combination soap bar to which it imparted excellent lime-soap dispersing properties [152]. Similar water-clear shampoo compositions were obtained by the solubilization of the alkyl sulfate with a combination of fatty acid monoalkanolamide and a diol such as 3,6-dimethyl-4-octyne-3,6-diol. A synergistic action was claimed for the latter two ingredients [153]. Alkyl sulfates were frequently used in combination with amphoteric surfactants as in the following examples: Combinations of alkyl sulfates, such as diethanol-ammonium or sodium lauryl sulfate, with amides, such as the diethanol-amide of palmitic acid or the isopropanolamide of lauric acid, were formu-lated with amphoteric surfactants, such as lauryldimethylcarboxymethyl betaine or lauryl bis-(2-hydroxyethyl)-carboxymethylbetaine. These prepa-rations reduced static electricity in dry hair and were less irritating to the eyes than a shampoo containing triethanolammonium lauryl sulfate as the principal surfactant [154]. An amphoteric surfactant was prepared from the condensation of aminoethylethanolamine with lauric acid and subsequent oxy-ethylation. The product was then sulfated and finally neutralized with lauric acid. This material was said to cause little eye irritation in shampoos [155].

Amine oxides and amide betaines could be used to profitably replace the alkanolamides conventionally added to alkyl sulfates in shampoos and bubble baths to increase viscosity and stabilize foams. The amine oxides were neutral nonionic surfactants, which dissolved readily in the alkyl sul-fates and were more efficient than alkanolamides. The amphoteric amide betaines increased viscosity, stabilized foam and had some antiseptic proper-ties [156]. A compatible mixture of an alkyl sulfate and an oxyalkylated quaternary ammonium compound formed the basis for a unique shampoo that combined normally incompatible anionic and cationic surfactants. Triethanol-ammonium lauryl sulfate and the quaternary salt prepared by adding methyl chloride to the reaction product of diethanolamine and 25 equivalents of propylene oxide were combined with other ingredients to form a shampoo [157]. Fatty alcohol ether sulfates were used in combination with fatty alco-hol sulfate in liquid hair shampoo [158] and in solid detergents for use as shampoos and toilet soaps [159]. An alkylarylpolyoxyethylene sulfate in com-bination with a fatty acid imidazoline-based amphoteric surfactant made up the principal ingredients for a shampoo with low eye-irritation properties [161]. Fatty alcohol sulfates were used in many hair and personal-care cos-metic products [160, 162], such as aerosol hair sprays [162], hair straight-eners [164], hair-waving compositions [165], and cleaning products for wigs and hair pieces [166]. A water-containing detergent composition that did not contain soap and was stable to phase separation, even on heating, was

prepared from sodium lauryl sulfate. It was used in combination with hy-
drogenated castor oil and various di- or monoalkanolamides [167] of fatty
acids.

A series of clear, moderately viscous shampoos was prepared from a
combination of monoethanolammonium lauryl sulfate and various oxyalkylated nitrogenous derivatives of sulfosuccinic acid. Esters [5] of oxyalkylated mono- or isopropanolamides of fatty acids [3] were prepared from
maleic acid or anhydride [4]. The esters, in turn, were treated [168] with
sodium bisulfite to produce the sulfosuccinate derivative [6] according to
Eq. (10):

$$RCONHCH_2CH_2OH + nCH_2CH_2 \overset{O}{\overbrace{}} \longrightarrow RCONH(CH_2CH_2O)_{n+1}H$$

[3]

$$[3] + \begin{matrix} HC-C=O \\ \parallel \quad \searrow O \\ HC-C=O \end{matrix} \longrightarrow RCONH(CH_2CH_2O)_{n+1}COCH=CHCOOH$$

[4]   [5]   (10)

$$[5] + NaHSO_3 + NaOH \longrightarrow RCOHN(CH_2CH_2O)_{n+1}COCHCH_2COONa$$
$$\underset{SO_3Na}{|}$$

[6]

In general, n was from 1-6. Two parts of [6], where n = 3, and one part of
monoethanolammonium lauryl sulfate gave a good shampoo composition. For
more detail on similar compounds see Chapter 12.

2. Detergent Toilet Bars

Completely synthetic detergent bars with desirable physical properties were
prepared from a combination of sodium dodecyl sulfate, urea, and N-methyl-
N-sorbityllauramide [169]. A detergent bar prepared from higher fatty alcohol sulfates in combination with glycerol monostearate also had good physical
properties [170]. Higher fatty alcohol sulfates, in combination with amphoteric surfactants, produced bars with good milling and finished physical
properties which also had a skin-conditioning effect [171]. Palmitone and
other fatty ketones were found to improve the washing characteristics of detergent bars when they were used in combination with sulfated fatty alcohols
and other ingredients [172].

Soap was used together with synthetic surfactants in "combo bars" in
order to overcome some of the shortcomings of an all-synthetic bar. Such
bars were hard and did not readily become mushy in the soap dish. The
lather had a quality that gave a better feel. The milling and plodding operations

in manufacturing the bars were also improved [173]. Toilet-soap bars that generate copious rich, creamy foam were prepared from a combination of tallow soap and alkali or alkaline earth salts of alkyl sulfates. The alkyl sulfates were prepared from primary or secondary branched C_7-C_{14} alcohols [174]. In another instance, a sodium soap made from tallow and coconut oil was combined with mixed sodium and potassium alkyl sulfates to prepare a super-fatted toilet soap. The mixed sodium and potassium salt resulted in a creamier lather than could be attained with the sodium salt alone [175]. The presence of the alkyl sulfates in combination soap-and-detergent bars enhanced their foam- and lather-producing ability and prevented the formation of a lime-soap curd in hard water. A series of toilet-soap bars was made from combinations of soap prepared from mixtures of fatty acids of tallow and palm kernel oil and the sodium salts of mixed even- and odd-chain synthetic alcohol sulfates [176]. Soap alone gave a foam volume of 166 ml in the Ross Miles foam test, while a combination of 40% soap and 60% alkyl sulfate gave a foam volume of 425 ml. With alkyl sulfate alone, the foam volume was 541 ml.

3. Bath Preparations

Alkyl and alkyl ether sulfates have been used frequently as ingredients in bubble-bath preparations. They helped contribute to the production of copious amounts of foam at very low concentrations, to foam stability over a wide range of temperatures both in hard and soft water, to easy rinseability of the foam even though it is stable, and to low skin and eye irritation. Such formulation also eliminated the lime-soap bathtub ring. Needless to say, a number of ingredients were necessary to produce a combination that had all of these ideal characteristics. However, the low irritation properties of the ether sulfates and their good foaming characteristics in hard water contributed much to the overall performance of a bubble-bath preparation [177].

Surface-active agents may irritate the skin and remove nitrogenous materials from it. To reduce irritation and overcome the leaching of oils and proteins from the skin, water-soluble, partially hydrolyzed protein preparations were incorporated in lotions containing ether alcohol sulfates and alkyl sulfates [178]. Additives which were said to reduce the removal of fat from skin and the subsequent skin irritations were esters of aliphatic branched-chain alcohols and hydroxy substituted fatty acids. For example, 2,6-dimethyloctyl 12-hydroxystearate or 2,6-dimethyloctyl 12-hydroxyoleate were used in combination with sodium lauryl sulfate in formulations for bubble bath, toilet soap, cleansing cream, shampoo, facial cream, hand lotion, etc. [179]. Secondary alkyl sulfates prepared from olefins ranging in chain length from C_8 to C_{18} were used to prepare a synthetic washing agent. However, they were found to have an irritating effect on the skin of a test panel of 80 men [180].

4. Shaving Cream

Self-heating detergent aerosol sprays for shaving creams and shampoos containing a reducing agent such as 2-mercaptoimidazoline or 2-mercaptoben-
zoxazole, and an oxidizing agent such as peroxide, have been described [183].
The two components mixed as they passed simultaneously through the outlet
nozzle of the aerosol spray and became hot from the heat of reaction. The
reducing portion contained sodium lauryl sulfate and triethanolammonium
lauryl sulfate as the surfactant.

5. Medicinal Preparations

Surface-active agents can have a pronounced effect on the efficacy of medications because of their ability to lower interfacial tension which, in turn,
can affect potentiation and availability of medications. Surfactants were used
in the preparation of the creams, gels, and emulsions in which the medications were dispersed [184, 185]. Sodium lauryl sulfate itself was found to
have some antimicrobial activity. It also had a marked effect on the release
of antibiotics such as penicillin and tetracycline from a petroleum base [186].

The compatibility of 18 drugs with 20 different nonionic surfactants each
in combination with sodium lauryl sulfate in a hydrophilic ointment base was
evaluated. The combinations of polyoxyethylene octylphenol-formalin resin
and polyoxyethylene nonylphenol with sodium lauryl sulfate were most compatible with the drugs studied [187]. Products for use as local anesthetic
injectables or ointments were prepared from homologs and analogs of procaine and fatty alcohol sulfates. Sodium lauryl sulfate formed a typical
anionic-cationic salt complex with $p-NH_2C_6H_4CO_2CH_2CH_2N(CH_2CH_3)_2 \cdot HCl$
(procaine) which precipitated out of solution to give waxy crystals, mp 48-
52° C. The complex obtained with sodium stearyl sulfate melted at 71-75° C
and with sodium myristyl sulfate at 58-62° C. Sodium octyl sulfate gave a viscous oil [188]. Similarly, $p-CH_3CH_2CH_2NHC_6H_4CO_2CH_2CH_2N(CH_3)_2 \cdot HCl$
gave a complex with sodium lauryl sulfate melting at 53-54° C.

Heavy-metal salts of lauryl sulfuric acid were found to possess fungicidal and bacteriocidal activity. The alkyl sulfates showed greater activity
than other anionic surfactants such as dioctyl sulfosuccinate and isopropyl
naphthalenesulfonate with the same heavy-metal counterion [116]. The anti-
fungal activity of the heavy-metal ion surfactant salts was greatly increased
over that of corresponding chlorides, nitrates, or sulfates. Long-chain alkyl
sulfates were also used as wetting agents, foaming agents, and detergents in
denture cleaners [181] and compositions for oral hygiene in general [182].

Nonionic surfactants were established as effective complexers of iodine.
These carriers, when added to aqueous media, gradually released the iodine
so that it became germicidally effective. It was found that anionic surfactants

containing the polyoxyalkylene function, such as sodium lauryl ether sulfate, could also form complexes with iodine. The presence of the iodine in such compositions had little or no effect on foam properties and caused no staining [189].

B. Light-Duty Liquid Detergents

Liquid dishwashing compositions account for the largest volume of light-duty liquid detergents sold. The ether sulfates are the surfactants of choice in these preparations, followed by the alkyl sulfates and combinations with linear alkylaryl sulfonates, alkyl sulfonates, and to a minor degree ampho-teric surfactants. The ether sulfates have a good calcium-ion tolerance, and foam well in hard water. They are readily biodegradable and are less irri-tating to the skin than the alkyl sulfates. Alkyl sulfates such as n-hexadecyl sulfate in combination with β-branched methyl and ethyl fatty alcohol sulfates were claimed to give a better dishwashing compound [190-193] than either the straight- or branched-chain sulfates alone [191, 194].

A typical high-quality liquid dishwashing composition based on an ether alcohol sulfate contained, for example, 36 parts of the ammonium salt of a sulfated polyoxyethylated derivative of a middle-cut coconut fatty alcohol containing 3 moles of ethylene oxide, 8.6 parts of coconut fatty acid mono-ethanolamide, 19.5 parts of ethyl alcohol, and 35.9 parts by weight of water to make up the balance of the ingredients. The maximum detergency and mildness were obtained when the ether sulfate contained 2-4 moles of eth-ylene oxide. However, these properties deteriorated when the polyethoxy content was either increased or decreased [192, 195]. The ether alcohol sulfates can be used as either the sodium, potassium, ammonium, or al-kanolammonium salts [196, 205]. Oxyethylated tridecyl alcohol tolerated strong sulfating agents such as sulfur trioxide and chlorosulfonic acid better, with respect to product quality, than ethoxamers of other alcohols. The optimum ethylene oxide content for the ether sulfate in a liquid dishwashing detergent was found to be 4-5 moles per mole of alcohol, regardless of the water hardness or detergent concentration [197].

A light-duty liquid detergent suitable for washing dishes or fine fabrics was obtained from a blend of three types of anionic surfactants. Sodium dodecylbenzenesulfonate, ammonium dodecylphenoxyhexakis oxyethylene sulfate, and sodium xylenesulfonate were dissolved in water to give a liquid detergent containing about 45% solids [198].

An alkyl ether sulfate combined with a long-chain alkyl betaine was prepared as an excellent dishwashing detergent with an exceptionally low cloud point of -5° C [199]. A light-duty liquid detergent combination was pre-pared based on a secondary alkyl sulfate, urea, and ether alcohol sulfate and other ingredients [201, 202]. A liquid detergent composition which had strong bleaching properties and possessed an acceptable shelf life was pre-pared. Its primary surfactant component was an ether alcohol sulfate;

hydrogen peroxide was incorporated into the composition as the bleach [204]. Both alkyl sulfates and ether alcohol sulfates, although high foamers by nature, have been formulated into low-foaming detergent compositions [206].

H. E. Tschakert reviewed the literature for dishwashing detergents. Detergent theory, test methods, raw materials, requirements, and compositions were discussed [200]. The performance, processing, and biodegradability of alkyl ether sulfates were reviewed with respect to their suitability for use in shampoos, toiletries, and detergent compositions [203].

C. Heavy-Duty Detergents

Laundry-washing compounds are frequently combinations of alkyl sulfates or alkyl ether sulfates combined with an alkylaryl sulfonate for reasons of economy as well as good heavy-duty performance. The ether alcohol sulfates contribute greater foam, solubilizing properties, and improved fat-emulsifying properties [207-209]. However, alkyl sulfates have also been formulated into heavy-duty detergents as the sole surface-active agent [210-213]. One such composition contained an antimicrobial agent in suspension which was deposited on the washed fabric [214].

In a study of the potentiation of detergency in compositions containing secondary alkyl sulfates, it was found that the best detergency performance was obtained with a 1:1 ratio of alkyl sulfonates to secondary alkyl sulfates and a ratio of 3:1 for a combination of alkylaryl sulfonates with secondary alkyl sulfates [215].

A clear liquid heavy-duty detergent was formulated with 1 part of C_{14}-C_{18} secondary alkyl sulfates and 3 parts of C_8-C_{18} alkylbenzenesulfonates. The composition remained clear over a wide temperature range [216]. The foaming of detergents containing secondary alkyl sulfates was enhanced and greatly stabilized by the addition of 5-30% by weight of n-alkylglyceryl monoethers, such as the octyl, decyl, and dodecyl ethers of glycerol [217].

D. Textile Applications

Surfactants are commonly added to cotton-mercerizing solutions in order to improve their wetting properties. However, the usual agents have poor solubility in 30° Bé or more concentrated sodium hydroxide. Short-chain alkyl sulfates with a C_5-C_8 chain lengths are frequently used. Of the branched alcohols with eight carbon atoms, the sulfate of 2-ethylhexanol was found to be a better wetting agent than that from isooctanol [218]. A solvent such as butanol was frequently added as well as other alcohols or glycols, ethers, etc. [219, 220]. Tallow alcohol sulfates used at the rate of 5 g per liter were used in baths as softeners and antistatic agents for the treatment of synthetic yarns and filaments [221].

E. Rug Shampoos

Formulations for the cleaning of rugs and carpets on location are typically
high-foaming detergents whose foam is low in water content and which dries
to a hard friable powder that can be vacuumed with the dirt. The basic surfactant is frequently an alkyl or alkyl ether sulfate. Calcium, magnesium,
or ammonium salts have been used because they produce copious amounts of
foam of low moisture content. Other compounding ingredients must be
chosen carefully since nothing should be used that would leave a sticky,
gummy, dirt-gathering residue on the carpet fibers [222, 224]. A free-
flowing powder for dry-cleaning rugs contains sodium lauryl sulfate, along
with a high-flash-point petroleum distillate and infusorial earth [223].

F. Hard-Surface Cleaners

Surfactants are generally present in minor amounts in hard-surface detergent preparations. However, their contribution to the overall performance
of the composition is a major one. Inorganic salts and solvents usually are
the major ingredients. A careful analysis of the specific problem at hand is
the key to the formulation of good preparations.

A tertiary alkylamine polyoxyethylene sulfate with a C_{12}-C_{15} chain
length and 15 moles of ethylene oxide was found to be a good cleaner for
metal, glass, and plastics [225]. A built dishwashing detergent powder
based on sodium myristyl sulfate was prepared [226]. Alkyl sulfates were
used in combination with amphoteric sulfobetaines to prepare a general-
purpose detergent [227]. The amphoteric compound had a hydrotroping
effect, solubilizing the anionic surfactant in the presence of inorganic salts.
A detergent designed principally for the emulsification of greasy soil con-
sisted of a mixture of sodium dodecylsulfate and the amphoteric sodium β-
decylaminopropionate [228]. Alcohol sulfates and ether alcohol sulfates were
used in detergent preparations designed to clean glass such as windows and
automotive windshields. These surfactants were combined with polyglycol
monoethers to prevent fogging [229-231]. Small amounts of turpentine or
kerosene were added to aid in the removal of grease [232].

Stress cracking occurred with some plastics when they were washed
with detergents containing nonionic surfactants. Alkyl sulfates were included
in the preparations to inhibit these crazing tendencies [233]. A decontami-
nating solution for plastics contaminated with radioactive material was based
on an alkyl sulfate to which citric and oxalic acids were added as chelating
agent and precipitant, respectively [234].

G. Alkyl Sulfates as Chemical Intermediates

The principal application of the sulfates lies in their utility as surface-active
agents. Consequently, chemical operations generally stop at the point of

their manufacture. There exists, however, a small body of recent literature describing the use of sulfates (generally the alcohol sulfates rather than the ether alcohol sulfates) as intermediates. The treatment of sodium lauryl sulfate in an autoclave with ammonia or an amine gave mixtures of primary, secondary, and tertiary amines [235, 236]. Yields ranged up to 76% total amine. In the presence of alkali [235], a mixed amine was formed, containing mostly primary and secondary alkylamines, corresponding in chain length to the starting lauryl sulfate. When aqueous ammonia and a minimum of 25 wt % of an aromatic sulfonic acid salt, such as sodium xylenesulfonate, were used, 76% of laurylamine was obtained [236]. Omission of the sulfonate caused the yield to drop to 43%, and large amounts of lauryl alcohol and amine were formed. Similarly, $C_{12}H_{25}O(CH_2CH_2O)_2CH_2CH_2NH_2$ was prepared from the corresponding sulfate. Pure secondary ethoxyalcohol sulfates gave good yields of etheramines [244].

Alkyl sulfates have also been used elsewhere as alkylating agents. Under basic conditions and in an autoclave [237], sodium lauryl sulfate alkylated the following products to yield the corresponding derivatives:

Derivative prepared	From
Lauryl phenyl ether	Phenol
Lauryl mercaptan	Sodium hydrogen sulfide
Sodium lauryl sulfite	Sodium sulfite
Cyanododecane	Potassium cyanide

Under similar conditions, sodium triethoxylauryl sulfite is prepared from sodium triethoxylauryl sulfate and sodium sulfite.

V. LIST OF PRODUCTS BY TRADE NAMES [239]

A. Sodium Alkyl[a] Sulfates

Trade name	Form	Concentration, %	Manufacturer or supplier	Remarks
Akyposal NLS	Liquid	31	Chem-Y	
Akyposal SDS	Liquid	28	Chem-Y	
Alkasurf WAQ	Liquid	29	Alkaril	
Andinix LSS	Paste	20	Henkel/A	
Andinix LSS 30	Liq/paste	29	Henkel/A	

[a]"Alkyl" in lists A to I implies lauryl or other alkyl. Other chain lengths, where known, are noted under remarks.

A. <u>Sodium Alkyl Sulfates</u> (Continued)

Trade name	Form	Concen- tration, %	Manufacturer or supplier	Remarks
Andinix LSS 50	Paste	49	Henkel/A	
Andinix P 1090	Powder	83	Henkel/A	
Avirol 101 Special	Liquid	30	Henkel	
Avirol 106	Liquid	40	Henkel	2-Ethylhexyl
Avirol 110	Liquid	30	Henkel	n-Decyl
Avirol 113	Liquid	30	Henkel	Tridecyl
Avirol 115 Special	Liquid	30	Henkel	
Avitex AD	Liquid	--	Du Pont	
Avitex C	Paste	--	Du Pont	
Carsonol SHS	Liquid	30	Quad	2-Ethylhexyl
Carsonol SLS	Liquid	30	Carson	
Carsonol SLS-Paste B	Paste	29	Quad	
Carsonol SLS-Special	Liquid	30	Carson	
Cedepon LS 30	Liquid	30	C.D. Canada	
Cedepon L-S 30-P	Liquid	30	C.D. Canada	
Conco Sulfate C	Liquid	30	Conco	Cetyl
Conco Sulfate O	Liquid	30	Conco	Oleyl
Conco Sulfate T	Paste	30	Conco	Tallow
Conco Sulfate WA	Paste	30	Conco	
Conco Sulfate WA Special	Paste	30	Conco	
Conco Sulfate WA Dry	Powder	90	Conco	
Conco Sulfate WR	Powder	99	Conco	
Conco Sulfate WX	Powder	90	Conco	
Condanol NLS 35	Paste	35	D & R	
Condanol NLS 90	Powder	30	D & R	
Cosmopon 35	Paste	48	Tessilchimica	
Cycloryl MS	Liquid	27	Cyclo	Myristyl

A. Sodium Alkyl Sulfates (Continued)

Trade name	Form	Concen-tration, %	Manufacturer or supplier	Remarks
Cycloryl 21	Liq/paste	29	Witco/UK	
Cycloril 31	Paste	43	Witco/UK	
Cycloril 33	Paste	42	Witco/UK	
Cycloril 580	Powder	90	Witco/UK	
Cycloril 585N	Needles	85	Witco/UK	
Cycloril 599	Flakes	99	Witco/UK	
Cycloril OS	Liquid	18	Witco/UK	n-Octyl
Duponol 80	Liquid	--	Du Pont	
Duponol C	Powder	--	Du Pont	
Duponol D Paste	Paste	--	Du Pont	
Duponol LS Paste	Paste	--	Du Pont	
Duponol ME Dry	Powder	--	Du Pont	
Duponol QC	Liquid	--	Du Pont	
Duponol SN	Liquid	--	Du Pont	
Duponol WA Dry	Powder	--	Du Pont	
Duponol WA Paste	Paste	--	Du Pont	
Duponol WAQ	Liquid	--	Du Pont	
Duponol WAQA	Liquid	--	Du Pont	
Duponol WAQE	Liquid	--	Du Pont	
Duponol WN	Liquid	--	Du Pont	
Elfan 200	Powder	90	Akzo	
Elfan 260	Paste	30	Akzo	
Elfan 280	Paste	42	Akzo	
Elfan 280 Powder Conc.	Powder	90	Akzo	
Elfan KT 550	Paste	42	Akzo	Coco/tallow
Elfan KT 730	Paste	42	Akzo	Coco/tallow
Emal O	Powder	98	Kao-Atlas	
Emcol D5-10	Liquid	--	Witco	2-Ethylhexyl

A. Sodium Alkyl Sulfates (Continued)

Trade name	Form	Concentration, %	Manufacturer or supplier	Remarks
Emersal 6400	Liquid	30	Emery	
Emersal 6402	Liquid	30	Emery	
Emersal 6410	Liquid	30	Emery	
Empicol LS 30	Paste	30	A & W/A	
Empicol LS 30/E	Paste	30	A & W/A	
Empicol WAQ	Paste	35	A & W/A	
Empicol CHC–30	Paste	38	A & W/M	Cetyl/oleyl
Empicol LM	Powder	86	A & W/M	
Empicol LMV	Needles	82	A & W/M	
Empicol LX	Powder	89	A & W/M	
Empicol LX 28	Liquid	28	A & W/M	
Empicol LXV	Needles	84	A & W/M	
Empicol LY28S	Liq/paste	28	A & W/M	
Empicol LZ	Powder	88	A & W/M	
Empicol LZ 34	Paste	34	A & W/M	
Empicol LZ Pure	--	--	A & W/M	
Empicol LZG 30	Paste	30	A & W/M	
Empicol LZV	Needles	84	A & W/M	
Empicol LZVG	Needles	84	A & W/M	
Eufullon W 75	Paste	73	Z & S	Oleyl
Gardinol CA	Paste	30	A & W/A	Cetyl/oleyl
Lakeway 101–10	Liquid	30	Lakeway	
Lakeway 101–11	Liquid	30	Lakeway	
Lakeway 101–16 N	Needle	90	Lakeway	
Lakeway 101–16 P	Powder	90	Lakeway	
Lakeway 101–19	Liquid	45	Lakeway	
Laurylfoam	Liquid	30	Chemithon	
Lutensit AS 2233	Liquid	28	BASF	

A. Sodium Alkyl Sulfates (Continued)

Trade name	Form	Concentration, %	Manufacturer or supplier	Remarks
Maprofix 563	Powder	99	Onyx	
Maprofix LCP	Liquid	30	Onyx	
Maprofix LK	Powder	90	Onyx	
Maprofix NEU Powder	Powder	46	Onyx	
Maprofix Paste MM	Paste	41	Onyx	
Maprofix TAS	Paste	30	Onyx	Tallow
Maprofix WA Paste	Paste	30	Onyx	
Maprofix WAC	Liquid	30	Onyx	
Maprofix WAC-LA	Liquid	30	Onyx	
Maprofix WAQ	Liquid	30	Onyx	
Michelene LS-90	Powder	90	Michel	
Monogen	Paste	60	Dai-ichi	
Montopol LA Paste	Paste	45	Montagne	
Nikkol S.C.S.	Powder	100	Nikko	Cetyl
Nikkol S.L.S.	--	--	Nikko	
Nissan Persoft SK	Liquid	30	Nippon Oil	
Orvus WA Paste	Paste	--	P & G	
P & G Emulsifier No. 104	Liquid	30	P & G	
Polystep B-5	Liquid	29	Stepan	
Rewopol NLS 30	Paste	30	Rewo	
Rexowet 77	Liquid	30	Emkay	Heptadecyl
Richonol A	Liquid	30	Richardson	
Richonol A Powder	Powder	90	Richardson	
Richonol C	Paste	30	Richardson	
Sactipon 2 S 3	Paste	43	Lever	
Sactol 2 S 3	Paste	49	Lever	
Serdet DFK 40	Liquid	28	Servo	

A. <u>Sodium Alkyl Sulfates</u> (Continued)

Trade name	Form	Concen-tration, %	Manufacturer or supplier	Remarks
Serdet DSK 40	Liquid	38	Servo	2-Ethylhexyl
Sinnopon 12 Pure	Powder	100	S-S	
Sinnopon LS 35	Paste	35	S-S	
Sinnopon LS 55	Paste	55	S-S	
Sinnopon LCS 97	Bead	95	S-S	
Sinnopon LCSV 90	Needles	90	S-S	
Sinnopon LS 100	Powder	99	S-S	
Sipex BOS	Liquid	40	Alcolac	2-Ethylhexyl
Sipex EC 111	Paste	25	Alcolac	
Sipex MPS	Liquid	35	Alcolac	
Sipex OLS	Liquid	33	Alcolac	Octyl
Sipex OS	Liquid	26	Alcolac	Oleyl
Sipex SB	Liquid	29	Alcolac	
Sipex SD	Powder	93	Alcolac	
Sipex TDS	Liquid	25	Alcolac	Tridecyl
Sipex UB	Liquid	30	Alcolac	
Sipon LS	Liquid	28	Alcolac	
Sipon LSB	Liquid	29	Alcolac	
Sipon WD Crystals	Powder	98	Alcolac	
Solasol FFA	Powder	95	Aceto	
Solasol Needles	Needles	88	Aceto	
Solasol USP	Powder	92	Aceto	
Sole-Terge TS-2-S	Liquid	35	Sole	2-Ethylhexyl
Standapol E Conc.	Powder	90	Henkel	Cetyl/stearyl
Standapol WA-AC	Paste	28	Henkel	
Standapol WAQ Special	Liquid	28	Henkel	
Standapol WAS-100	Liquid	28	Henkel	

A. Sodium Alkyl Sulfates (Continued)

Trade name	Form	Concen- tration, %	Manufacturer or supplier	Remarks
Steinapol NLS 28	Liquid	28	Rewo/G	
Steinapol NLS 90	Powder	90	Rewo/G	
Stepanol ME Dry	Solid	90	Stepan	
Stepanol WA-100	Powder	99	Stepan	
Stepanol WA-C	Liquid	28	Stepan	
Stepanol WA Paste	Paste	28	Stepan	
Stepanol WA Special	Liquid	28	Stepan	
Stepanol WAQ	Liquid	28	Stepan	
Sterling WA	Paste	28	Can. Pack.	
Sterling WAQ	Liquid	28	Can. Pack.	
Sterling WAQ-CH	Liquid	28	Can. Pack.	
Sulfetal C 42	Paste	42	Z & S	
Sulfetal C 90	Powder	90	Z & S	
Sulfetal K 47	Paste	47	Z & S	
Sulfetal K 90	Powder	90	Z & S	
Sulfetal L 95	Powder	95	Z & S	
Sulfetal TC 50	Paste	50	Z & S	Tallow/coco
Sulfotex WA	Liquid	28	Textilana	
Sulfotex WA-6576	Liquid	28	Textilana	
Sulphonated "Lorol" Paste	Paste	40	R & M	
Sulfopon WA 1	Paste	30	Henkel/G	
Sulfopon WA 1 Special	Liq/paste	30	Henkel/G	
Sulfopon WA 2	Liquid	30	Henkel/G	
Sulfopon WA 3	Paste	30	Henkel/G	
Sulfopon WA 4	Paste	37	Henkel/G	Lauryl/cetyl
Surco SLS	Liquid	28	Surfact-Co.	
Surco SDS	Liquid	30	Surfact-Co.	Decyl

A. <u>Sodium Alkyl Sulfates</u> (Continued)

Trade name	Form	Concen-tration, %	Manufacturer or supplier	Remarks
Swascol 1 P	Powder	85	Swastik	
Synthetic Detergent 66	Liquid	--	P & G	
Teepol 610	Liquid	34	Shell and Shell/U.K.	Branched
Tergitol 08	Liquid	40	UCC	2-Ethylhexyl
Tergitol 4	Liquid	28	UCC	Branched C_{14}
Tergitol 7	Liquid	27	UCC	Branched C_{17}
Texapon CS Paste	Paste	57	Henkel/A	
Texapon DL Conc.	Paste	46	Henkel/G	
Texapon K 12	Powder	90	Henkel/A and Henkel/G	
Texapon K 12 Paste	Paste	54	Henkel/G	
Texapon K 14	Powder	90	Henkel/G	Myristyl
Texapon K 1294	Powder	94	Henkel/G	
Texapon K 1296	Powder	96	Henkel/G	
Texapon L 100	Powder	99	Henkel/G	
Texapon LS High Conc. Need.	Needles	90	Henkel/G	
Texapon V	Needles	90	Henkel	
Texapon V High Conc. Need.	Needles	88	Henkel/G	
Texapon Z	Powder	58	Henkel/G	
Texapon Z High Conc.	Powder	90	Henkel/G	
Texapon Z High Conc. Needles	Needles	88	Henkel/G	
Texapon Z Needles	Needles	90	Henkel/A	
Tylorol LS	Paste	42	Triantaphyllou	
Ultra Sulfate SL-1	Liquid	30	Witco	

B. Ammonium Alkyl Sulfates

Trade name	Form	Concen-tration, %	Manufacturer or supplier
Akyposal ALS 33	Liquid	33	Chem-Y
Alkasurf ALS	Liquid	27	Alkaril
Avirol 200	Liquid	28	Henkel
Carsonol ALS	Liquid	30	Carson
Carsonol ALS-Special	Liquid	30	Quad
Cedepon L-A 30	Liquid	30	C.D. Canada
Conco Sulfate A	Liquid	30	Conco
Cycloryl MA	Liquid	28	Witco/UK
Emersal 6410	Liquid	30	Emery
Empicol AL30	Liquid	30	A & W/A and A & W/M
Lakeway 101-20	Liquid	30	Lakeway
Maprofix NH	Liquid	30	Onyx
Maprofix NHL	Liquid	30	Onyx
Melanol LP 20	Liquid	35	Montagne
Montopol LA 20	Liquid	30	Montagne
Richonol AM	Liquid	30	Richardson
Richonol RS	Liquid	35	Richardson
Sactol 2 A	Liquid	25	Lever
Sinnopon LA 30	Liquid	27	S-S
Sipon L-22	Liquid	28	Alcolac
Standapol A	Liquid	28	Henkel
Stepanol AM	Liquid	30	Stepan
Sterling AM	Liquid	28	Can. Pack.
Sulfotex WAA	Liquid	28	Textilana
Sulphonated "Lorol" Liquid NH	Liquid	30	R & M

B. Ammonium Alkyl Sulfates (Continued)

Trade name	Form	Concen-tration, %	Manufacturer or supplier
Surco NH	Liquid	30	Surfact-Co.
Texapon Special	Liquid	44	Henkel/G
Texapon A 400	Liquid	33	Henkel/G

C. Alkanolamine Alkyl Sulfates

Trade name	Type[a]	Form	Concen-tration, %	Manufacturer or supplier
Akyposal MLS 33	MEA	Liquid	33	Chem-Y
Akyposal MLS 55	MEA	Liquid	55	Chem-Y
Akyposal TLS 42	TEA	Liquid	42	Chem-Y
Alkasurf DLS	DEA	Liquid	35	Alkaril
Alkasurf MLS	MEA	Liquid	32	Alkaril
Alkasurf TLS	TEA	Liquid	40	Alkaril
Avirol 300	TEA	Liquid	39	Henkel
Carsonol DLS	DEA	Liquid	35	Carson
Carsonol ILS	IPA	Liquid	90	Carson
Carsonol TLS	TEA	Liquid	40	Carson
Cedepon L-D-35	DEA	Liquid	35	C.D. Canada
Cedepon L-M-25	MEA	Liquid	25	C.D. Canada
Cedepon L-T-40	TEA	Liquid	40	C.D. Canada
Conco Sulfate EP	DEA	Liquid	37	Conco
Conco Sulfate TL	TEA	Liquid	42	Conco
Condanol DLS	DEA	Liquid	60	D & R

[a] DEA = Diethanolamine
IPA = Isopropanolamine
MEA = Monoethanolamine
TEA = Triethanolamine

C. Alkanolamine Alkyl Sulfates (Continued)

Trade name	Type[a]	Form	Concentration, %	Manufacturer or supplier
Condanol MLS	MEA	Liquid	33	D & R
Cosmopon TRC	TEA	Liquid	60	Tessilchimica
Cycloryl SA	MEA	Liquid	28	Cyclo and Witco/UK
Cycloryl TA	TEA	Liquid	40	Witco/UK
Cycloryl TAC	TEA	Liquid	40	Witco/UK
Cycloryl TAWF	TEA	Solid	76	Witco/UK
Duponol EP	--	Liquid	--	Du Pont
Duponol ST	--	Liquid	--	Du Pont
Elfan 240 M	MEA	Paste	34	Akzo
Elfan 240 T	TEA	Paste	40	Akzo
Emal TD	TEA	Liquid	40	Kao–Atlas
Emersal 6434	TEA	Liquid	30	Emery
Emersal 6440	--	Liquid	90	Emery
Empicol DA	DEA	Liquid	34	A & W/A
Empicol DLS	DEA	Liquid	34	A & W/M
Empicol LQ 27	MEA	Liquid	27	A & W/A and A & W/M
Empicol LQ 33	MEA	Liquid	33	A & W/M
Empicol MIPA	IPA	Liquid	30	A & W/A
Empicol TA 40	TEA	Liquid	40	A & W/A
Empicol TCR	TEA	Liquid	26	A & W/A
Empicol TDL 75	TEA/ DEA	Liquid	75	A & W/M
Empicol TL 40	TEA	Liquid	40	A & W/M
Empicol TLP	TEA	Liquid	38	A & W/M
Empicol TLP	TEA	Liquid	42	A & W/A
Empicol TLR	TEA	Liquid	36	A & W/M
Lakeway 101–30	TEA	Liquid	40	Lakeway

C. Alkanolamine Alkyl Sulfates (Continued)

Trade name	Type[a]	Form	Concen- tration, %	Manufacturer or supplier
Maprofix DLS–35	DEA	Liquid	35	Onyx
Maprofix TLS–65	TEA	Liquid	65	Onyx
Maprofix TLS–500	TEA	Liquid	40	Onyx
Maprofix TLS–513	TEA	Liquid	40	Onyx
Melanol LP 20 D	DEA	Liquid	35	Montagne
Melanol LP 20 M	MEA	Liquid	35	Montagne
Melanol LP 20 T	TEA	Liquid	40	Montagne
Melanol LP 1	IPA	Gel	60	Montagne
Product WAT	TEA	Liquid	--	P & G
Rewopol TLS–40	TEA	Liquid	40	Rewo
Richonol DLS	DEA	Liquid	40	Richardson
Richonol DTC	--	Liquid	90	Richardson
Richonol T	TEA	Liquid	40	Richardson
Sactipon 2 M	MEA	Liquid	29	Lever
Sactipon 2 T	TEA	Liquid	35	Lever
Sactol 2 M	MEA	Liquid	29	Lever
Sactol 2 T	TEA	Liquid	35	Lever
Sinnopon LT 42	TEA	Liquid	40	S–S
Sipon LT6	TEA	Liquid	40	Alcolac
Standapol DEA	DEA	Liquid	36	Henkel
Standapol T	TEA	Liquid	40	Henkel
Standapol TS–100	TEA	Liquid	40	Henkel
Steinapol MLS–35	MEA	Liquid	35	Rewo/G
Steinapol TLS 40	TEA	Liquid	40	Rewo/G
Stepanol DEA	DEA	Liquid	35	Stepan

C. Alkanolamine Alkyl Sulfates (Continued)

Trade name	Type[a]	Form	Concen- tration, %	Manufacturer or supplier
Stepanol WAT	TEA	Liquid	40	Stepan
Sterling WADE	DEA	Liquid	35	Can. Pack.
Sterling WAME	MEA	Liquid	31	Can. Pack.
Sterling WAT	TEA	Liquid	40	Can. Pack.
Sulfetal KT 400	TEA	Liquid	40	Z & S
Sulfotex WAD	DEA	Liquid	34	Textilana
Sulfotex WAT	TEA	Liquid	40	Textilana
Sulphonated "Lorol" Liquid DA	DEA	Liquid	34	R & M
Sulphonated "Lorol" Liquid MA	MEA	Liquid	32	R & M
Sulphonated "Lorol" Liquid TA	TEA	Liquid	42	R & M
Surco DEA-LS	DEA	Liquid	35	Surfact-Co.
Surco TEA-LS	TEA	Liquid	40	Surfact-Co.
Swascol 4L	TEA	Liquid	40	Swastik
Texapon DLS	DEA	Liquid	30	Henkel/G
Texapon MLS	MEA	Liquid	30	Henkel/A
Texapon MLS	MEA	Liquid	33	Henkel/G
Texapon T	TEA	Liquid	50	Henkel/G
Texapon T-35	TEA	Liquid	35	Henkel/G
Texapon T-42	TEA	Liquid	42	Henkel/G
Texapon T Paste	TEA	Paste	80	Henkel/G
Tylorol LM	MEA	Liquid	30	Triantaphyllou
Tylorol LT 50	TEA	Liquid	42	Triantaphyllou

D. Miscellaneous Alkyl Sulfates

Trade name	Cation[a]	Form	Concen- tration, %	Manufacturer or supplier
Alkasurft MGLS	Magnesium	Liquid	27	Alkaril
Andinix L	Sodium/TEA	Liquid	30	Henkel/A
Cedepal S-K 404	Potassium	Liquid	--	C.D. Canada
Cedepon L-K 30	Potassium	Liquid	30	C.D. Canada
Conco Sulfate M	Magnesium	Liquid	30	Conco
Conco Sulfate P	Potassium	Liquid	30	Conco
Cycloryl GA	Magnesium	Liquid	28	Witco/UK
Cycloryl KA-45	Potassium	Paste	45	Witco/UK
Cycloryl LA	Lithium	Liquid	30	Witco/UK
Cycloryl LC	Lithium	Liquid	30	Cyclo and Witco/UK
Cycloryl TMA	TEA and ammonia	Liquid	30	Witco/UK
Emalox C-60	Potassium	Paste	30	Yoshimura
Empicol HL 25	Lithium	Liquid	25	A & W/A
Empicol ML 26	Magnesium	Liquid	26	A & W/M
Empicol TC 34	TEA and ammonia	Liquid	35	A & W/M
Empicol TC 34	TEA and ammonia	Liquid	33	A & W/A
Maprofix MG	Magnesium	Liquid	28	Onyx
Sinnopon LL 30	Lithium	Liquid	30	S-S
Sipon LM	Magnesium	Liquid	27	Alcolac
Steinapol TLS 90 F	"Triamine"	Liquid	90	Rewo/G
Stepanol Mg	Magnesium	Liquid	30	Stepan
Sterling WAM	Magnesium	Liquid	30	Can. Pack.
Sulphonated "Lorol" Liquid TN	TEA and ammonia	Liquid	36	R & M

[a]TEA = Triethanolamine

D. Miscellaneous Alkyl Sulfates (Continued)

Trade name	Cation[a]	Form	Concentration, %	Manufacturer or supplier
Surco MG-LS	Magnesium	Liquid	30	Surfact-Co.
Texapon AT	Ammonia and TEA	Liquid	28	Henkel/G
Texapon K High Conc. Needles	Potassium	Needles	90	Henkel/G

E. Sodium Alkyl Ether Sulfates

Trade name	Form	Concentration, %	Manufacturer or supplier	Remarks
Akyposal DS 28 L	Liquid	28	Chem-Y	
Akyposal DS 56	Liquid	56	Chem-Y	
Akyposal 23 ST 70	Paste	70	Chem-Y	
Akyposal EO 20	Liquid	28	Chem-Y	
Akyposal EO 20 MW	Liquid	28	Chem-Y	
Akyposal EO 20 PA	Liquid	56	Chem-Y	
Akyposal EO 20 SF	Liquid	28	Chem-Y	
Akyposal LM 70	Paste	70	Chem-Y	
Akyposal MS	Liquid	25	Chem-Y	
Akyposal MS Conc.	Liquid	60	Chem-Y	
Alfonic 1412-S	Liquid	58	Conoco	
Alkasurf ES-60	Liquid	58	Alkaril	
Andinix M 40 B	Liquid	25	Henkel/A	
Aremsol SLES	Liquid	27	R & M	
Avirol 100-E	Liquid	25	Henkel	
Avirol 116-E	Liquid	28	Henkel	
Avirol BOD 153	Liquid	60	Henkel	
Calimulse ES-30	Liquid	30	Pilot	
Carsonol SES-S	Liquid	60	Carson	

E. Sodium Alkyl Ether Sulfates (Continued)

Trade name	Form	Concen-tration, %	Manufacturer or supplier	Remarks
Carsonol SLES	Liquid	30	Carson	
Cedepal S-S 406	Liquid	60	C.D. Canada	
Conco Sulfate 219	Liquid	60	Conco	
Conco Sulfate WE	Liquid	30	Conco	
Condanol NL 2	Liquid	28	D & R	
Condanol NL 2 (56)	Liquid	56	D & R	
Cosmopon LE 50	Paste	50	Tessilchimica	
Cycloryl NA	Liquid	27	Witco/UK	
Cycloryl NA60	Liquid	60	Witco/UK	
Dobanol 25-3S/27	Liquid	27	Shell N.V. and Shell U.K.	
Dobanol 25-3S/60	Liquid	59	Shell N.V. and Shell U.K.	
Elfan NS 242	Liquid	28	Akzo	2 Moles E.O.[a]
Elfan NS 243 S	Liquid	28	Akzo	3 Moles E.O.
Elfan NS 243 S Conc.	Liq/paste	70	Akzo	3 Moles E.O.
Elfan NS 252 S	Liquid	28	Akzo	2 Moles E.O.
Elfan NS 252 S Conc.	Liq/paste	70	Akzo	2 Moles E.O.
Elfan NS 423 SH	Liquid	27	Akzo	3 Moles E.O.
Empicol ESB	Liquid	28	A & W/A	
Empicol ESB-3	Liquid	27	A & W/M	
Empicol ESB-30	Liquid	27	A & W/M	
Empicol ESB-50	Paste	50	A & W/M	
Empicol ESC-30	Liquid	28	A & W/M	

[a]The term "moles E.O." refers to the average number of moles of ethylene oxide added to the alcohol before it is sulfated.

E. Sodium Alkyl Ether Sulfates (Continued)

Trade name	Form	Concentration, %	Manufacturer or supplier	Remarks
Empimin KSN 27	Liquid	27	A & W/M	
Empimin KSN 60	Liquid	60	A & W/M	
Empimin SQ25	Liquid	25	A & W/A	
Empimin SQ60	Liquid	60	A & W/A	
Hyonic JN-400-SA	Liquid	60	Nopco	
Lakeway 201-10	Liquid	60	Lakeway	
Lakeway 201-11	Liquid	30	Lakeway	
Levenol WX	Liquid	30	Kao-Atlas	
Levenol WZ	Liquid	30	Kao-Atlas	
Lutensit AS 2230	Liquid	28	BASF	
Lutensit AS 3330	Liquid	28	BASF	
Lutensit AS 3333	Liquid	28	BASF	
Lutensit AS 3334	Liquid	28	BASF	
Lutensit AS 4330	Liquid	28	BASF	
Lutensit AS 4333	Liquid	28	BASF	
Maprofix 60S	Liquid	60	Onyx	
Maprofix ES	Liquid	25	Onyx	3.5 Moles E.O.
Maprofix ESY	Liquid	25	Onyx	1 Mole E.O.
Maprofix OS	Liquid	30	Onyx	Tridecyl
Merpoxal LM 3028	Liquid	28	Kempen	
Merpoxal LM 3030	Liquid	30	Kempen	
Merpoxal LM 3070	Paste	70	Kempen	
Merpoxal TR 4060	Paste	60	Kempen	iso-Tridecyl
Neodol 25-3S	Liquid	58	Shell	$C_{12}-C_{15}$
Neodol 25-3S-40	Liquid	40	Shell	$C_{12}-C_{15}$
Nikkol NES-203	Liquid	20	Nikko	
Nissan Persoft EK	Liquid	30	Nippon Oil	
Nyapon DG Liquid	Liquid	30	Nyanza	

E. <u>Sodium Alkyl Ether Sulfates</u> (Continued)

Trade name	Form	Concen-tration, %	Manufacturer or supplier	Remarks
Polystep B-12	Liquid	60	Stepan	
Polystep B-16	Liquid	25	Stepan	
Polystep B-19	Liquid	26	Stepan	
Polystep B-23	Liquid	60	Stepan	
Rewopol NL 2	Liquid	28	Rewo	
Richonol S-1285C	Liquid	60	Richardson	
Richonol S-5260	Liquid	30	Richardson	
Sactipon 2 OS	Liquid	25	Lever	
Sactipon 2 OS 28	Liquid	28	Lever	
Sactol 2 OS	Liquid	25	Lever	
Sactol 2 OS 28	Liquid	28	Lever	
Sandet EN	Liquid	--	Sanyo	
Sandet END	Liquid	30	Sanyo	
Sandet ENM	Liquid	30	Sanyo	
Serdet DCK 30	Liquid	28	Servo	
Serdet DPK 30	Liquid	28	Servo	
Serdet DPK 3/60	Paste	58	Servo	
Sinnopon LES 228	Liquid	25	S-S	
Sinnopon WES 35	Paste	35	S-S	Cetyl
Sinnopon WS 30	Paste	33	S-S	Cetyl/stearyl
Sipex EST-30	Liquid	29	Alcolac	Tridecyl
Sipex EST-60	Liquid	57	Alcolac	Tridecyl
Sipon ES	Liquid	27	Alcolac	
Sipon ESY	Liquid	25	Alcolac	
Standapol ES-1	Liquid	25	Henkel	1 Mole E.O.
Standapol ES-2	Liquid	26	Henkel	
Standapol ES-3	Liquid	28	Henkel	

E. <u>Sodium Alkyl Ether Sulfates</u> (Continued)

Trade name	Form	Concen-tration, %	Manufacturer or supplier	Remarks
Standapol ES-40 Conc.	Liquid	60	Henkel	Myristyl
Steinapol NL 2-28	Liquid	28	Rewo/G	
Steinapol NL 3-28	Liquid	28	Rewo/G	
Steol CS 460	Liquid	60	Stepan	
Steol KS 460	Liquid	60	Stepan	
Sterling Super A	Liquid	60	Can. Pack.	
Sterling Super 3E	Liquid	28	Can. Pack.	
Sulfotex EL	Liquid	30	Textilana	
Sulfotex LMS-E	Liquid	62	Textilana	
Sulfotex RIF-S	Liquid	47	Textilana	
Surco LES 60 C	Liquid	60	Surfact-Co.	
Texapon BS	Liquid	22	Henkel/G	
Texapon EVR	--	35	Henkel/G	
Texapon K 14S	Liquid	30	Henkel/G	Lauryl/ myristyl
Texapon N 40	Liquid	25	Henkel/A	
Texapon N 70	Paste	70	Henkel/G	
Texapon ND Paste	Paste	57	Henkel/G	
Texapon NSF	Liquid	28	Henkel/G	
Texapon Q	Liquid	30	Henkel/G	
Texapon Q Conc.	Paste	65	Henkel/G	
Tex-Wet 1158	Liquid	60	Intex	
Tylorol 2S	Liquid	28	Triantaphyllou	
Tylorol 1000	Liquid	28	Triantaphyllou	
Tylorol BS	Liquid	22	Triantaphyllou	
Ultra Sulfate 60	Liquid	60	Witco	

E. Sodium Alkyl Ether Sulfates (Continued)

Trade name	Form	Concentration, %	Manufacturer or supplier	Remarks
Ultra Sulfate 1050	Liquid	45	Witco	
Zetesol Conc.	Liq/paste	73	Z & S	
Zetesol NV	Paste/liq	27	Z & S	

F. Ammonium Alkyl Ether Sulfates

Trade name	Form	Concentration, %	Manufacturer or supplier	Remarks
Alfonic 1412-A	Liquid	58	Conoco	
Alipal AB-436	Liquid	58	GAF	
Alipal CD-128	Liquid	58	GAF	
Alkasurf EA-60	Liquid	58	Alkaril	
Avirol BOD 253	Liquid	60	Henkel	
Calfoam NEL-60	Liquid	60	Pilot	
Carsonol ALES	Liquid	30	Carson	
Carsonol SES-A	Liquid	60	Carson	
Cedepal S-A 406	Liquid	60	C.D. Canada	
Conco Sulfate 216	Liquid	60	Conco	
Conco Sulfate WM	Liquid	30	Conco	
Cosmopon AEC	Paste	80	Tessilchimica	
Cycloryl M3D60	Liquid	60	Witco/UK	
Cycloryl MD	Liquid	27	Witco/UK	
Dobanol 25-3A/60	Liquid	60	Shell N.V. and Shell U.K.	
Empicol EAB	Liquid	23	A & W/M	
Empimin AQ60	Liquid	60	A & M/A	
Lakeway 201-20	Liquid	60	Lakeway	
Lakeway 201-29	Liquid	30	Lakeway	

F. Ammonium Alkyl Ether Sulfates (Continued)

Trade name	Form	Concentration, %	Manufacturer or supplier	Remarks
Maprofix MB	Liquid	30	Onyx	3.5 Moles E.O.
Maprofix MBO	Liquid	30	Onyx	1 Mole E.O.
Merpoxal LMA 3045	Liquid	45	Kempen	
Merpoxal LMA 3060	Liquid	60	Kempen	
Neodol 25-3A	Liquid	58	Shell	$C_{12}-C_{15}$
Polystep B-11	Liquid	59	Stepan	
Polystep B-20	Liquid	26	Stepan	
Polystep B-22	Liquid	28	Stepan	
Rewopol AL3	Liquid	60	Rewo	
Richonol S-1300C	Liquid	60	Richardson	
Sactipon 2 O A	Liquid	25	Lever	
Sactol 2 O A	Liquid	25	Lever	
Serdet DCN 30	Liquid	28	Servo	
Serdet DPN 3/30	Liquid	28	Servo	
Sinnopon LEA 225	Liquid	25	S-S	
Sipon EA	Liquid	27	Alcolac	
Standapol EA-40 Conc.	Liquid	60	Henkel	Myristyl
Steinapol AL 3	Liquid	60	Rewo/G	
Steol CA 460	Liquid	60	Stepan	
Steol KA 460	Liquid	60	Stepan	
Sterling ES 600	Liquid	60	Can. Pack.	
Sulfotex LMT	Liquid	58	Textilana	
Sulfotex OT	Liquid	58	Textilana	
Sulfotex PAI	Liquid	48	Textilana	
Sulfotex PAI-S	Liquid	47	Textilana	
Sulfotex RIF	Liquid	47	Textilana	
Sulfotex SAL	Liquid	58	Textilana	

F. Ammonium Alkyl Ether Sulfates (Continued)

Trade name	Form	Concentration, %	Manufacturer or supplier	Remarks
Sulfotex SAT	Liquid	58	Textilana	
Surco LES 60 A	Liquid	60	Surfact-Co.	
Surflo-S30	Liquid	72	Baroid	
Surflo-S302	Liquid	94	Baroid	
Surflo-S36	Liquid	75	Baroid	
Surflo-S362	Liquid	100	Baroid	
Texapon NA	Liquid	24	Henkel/A and Henkel/G	
Ultra Sulfate 58	Liquid	58	Witco	
Ultra Sulfate AE-3	Liquid	58	Witco	
Zetesol A 64	Liquid	60	Z & S	
Zetesol AP	Liquid	60	Z & S	

G. Miscellaneous Alkyl Ether Sulfates

Trade name	Cation[a]	Form	Concentration, %	Manufacturer or supplier
Cycloryl GD	Magnesium	Liquid	27	Witco/UK
Cycloryl TD	TEA	Liquid	35	Witco/UK
Nikkol NES-303	TEA	Solid	30	Nikko
Sactipon 2 O M G	Magnesium	Liquid	28	Lever
Sactipon 2 O T	TEA	Liquid	35	Lever
Sactol 2 O M G	Magnesium	Liquid	28	Lever

[a] MEA = Monoethanolamine
TEA = Triethanolamine

G. Miscellaneous Alkyl Ether Sulfates (Continued)

Trade name	Cation[a]	Form	Concen- tration, %	Manufacturer or supplier
Sactol 2 O T	TEA	Liquid	35	Lever
Sandet ET	TEA	Liquid	--	Sanyo
Sinnopon LEG 325	Magnesium	Liquid	25	S-S
Texapon M	MEA	Liquid	28	Henkel/G
Texapon NT	TEA	Liquid	35	Henkel/G
Tylorol SM	MEA	Liquid	28	Triantaphyllou

H. Sodium Alkylphenoxy Ether Sulfates

Trade name	Form	Concen- tration, %	Manufacturer or supplier
Alipal CO-433	Liquid	28	GAF
Alipal EO-526	Liquid	58	GAF
Cycloril NZ (series)	Liquid	30	Witco/UK
Emal 20C	Liquid	25	Kao-Atlas
Gardisperse AC	Liquid	30	A & W/A
Lutensit A-ES	Liquid	40	BASF
Newcol 560 SN	Liquid	30	Nippon N.
Newcol 861 S	Paste	30	Nippon N.
Newcol 1305 SN	Liquid	30	Nippon N.
Newcol 1310 SN	Liquid	30	Nippon N.
Triton W-30 Conc.	Liquid	27	R & H
Triton X-200	Liquid	28	R & H
Triton X-202	Liquid	30	R & H
Triton X-301	Paste	20	R & H

I. Ammonium Alkylphenoxy Ether Sulfates

Trade name	Form	Concen-tration, %	Manufacturer or supplier
Alipal CO-436	Liquid	58	GAF
Alipal EP-110	Liquid	30	GAF
Alipal EP-120	Liquid	30	GAF
Concopal A	Liquid	60	Conco
Concopal AS	Liquid	60	Conco
Merpoxal NOA 4080	Paste	80	Kempen
Merpoxal NOA 4558	Liquid	60	Kempen
Neutronyx S-60	Liquid	60	Onyx
Newcol 560 SF	Liquid	50	Nippon N.
Polystep B-1	Liquid	59	Stepan
Sulfotex NTS	Liquid	60	Textilana

J. Index of Manufacturers or Suppliers

A & W/A	Albright & Wilson (Australia) Ltd., Melbourne, Australia
A & W/M	Albright & Wilson Ltd., Marchon Div., Whitehaven, Cumberland, England
Aceto	Aceto Chemical Co., Inc., Flushing, New York
Akzo	Akzo Chemie GmbH, Duren (Rhld), West Germany
Alcolac	Alcolac, Inc., Baltimore, Maryland
Alkaril	Alkaril Chemicals, Ltd., Mississauga, Ontario, Canada
Baroid	Baroid Div., N.L. Industries, Inc., Houston, Texas
BASF	BASF Aktiengesellschaft, Ludwigshafen, West Germany
Can. Pack.	Canada Packers Ltd., Chemical Div., Toronto, Ontario, Canada
Carson	Carson Chemical, Inc., Long Beach, California
C.D. Canada	Chemical Developments of Canada Ltd., Pointe Claire, Quebec, Canada
Chemithon	Chemithon Corp., Seattle, Washington

Chem-Y	Chem-Y, Fabriek van Chemische Producten NV, Bodegraven, Holland
Conco	Continental Chemical Co., Clifton, New Jersey
Conoco	Conoco Chemicals, Continental Oil Co., Saddle Brook, New Jersey
Cyclo	Cyclo Chemicals Corp., Miami, Florida
Dai-ichi	Dai-ichi Kogyo Seiyaku Co., Ltd., Tokyo, Japan
D & R	Dutton & Reinisch Ltd., London, England
Du Pont	E.I. duPont de Nemours & Co., Wilmington, Delaware
Emery	Emery Industries, Inc., Mauldin, South Carolina
Emkay	Emkay Chemical Co., Elizabeth, New Jersey
GAF	GAF Corp., Chemical Div., New York, New York
Henkel	Henkel, Inc., Chemical Specialties Div., Teaneck, New Jersey
Henkel/A	Henkel Argentina S.A.C.I.F. y de M., Buenos Aires, Argentina
Henkel/G	Henkel International GmbH, Dusseldorf, West Germany
Intex	Intex Products, Inc., Greenville, South Carolina
Kao-Atlas	Kao-Atlas Co., Ltd., Tokyo, Japan
Kempen	Elektrochemische Fabrik Kempen GmbH, Kempen-Niederrhein, Holland
Lakeway	Lakeway Chemicals, Inc., Muskegon, Michigan
Lever	Lever Industriel, Noisy-le-Sec, France
Michel	M. Michel & Co., Inc., New York, New York
Montagne	Produits Chimiques de la Montagne Noire, Paris, France
Nikko	Nikko Chemicals Co., Ltd., Tokyo, Japan
Nippon N.	Nippon Nyukazai Co., Ltd., Tokyo, Japan
Nippon Oil	Nippon Oils & Fats Co., Ltd., Tokyo, Japan
Nopco	Nopco Chemical Div., Diamond Shamrock Chemical Co., Morristown, New Jersey
Nyanza	Nyanza, Inc., Lawrence, Massachusetts
Onyx	Onyx Chemical Co., Div. of Millmaster Onyx Corp., Jersey City, New Jersey
P & G	Procter & Gamble Co., Cincinnati, Ohio
Pilot	Pilot Chemical Co., Los Angeles, California
Quad	Quad Chemical Corp., Long Beach, California
R & H	Rohm and Haas Co., Philadelphia, Pennsylvania
R & M	Ronsheim & Moore Ltd., Castleford, Yorkshire, England
Rewo	Rewo Chemicals, Inc., East Farmingdale, New York
Rewo/G	Rewo Chemische Fabrik GmbH, Steinau, West Germany
Richardson	The Richardson Co., Organic Chemicals Div., Des Plaines, Illinois

S–S	Sidobre-Sinnova, Neuilly-sur-Seine, France
Sanyo	Sanyo Chemical Industries, Ltd., Kyoto, Japan
Servo	Servo Chemische Fabriek B.V., Delden, Holland
Shell	Shell Chemical Co., Industrial Chemicals Div., Houston, Texas
Shell N.V.	Shell Nederland Chemie N.V., Gravenhage, Holland
Shell U.K.	Shell Chemicals U.K., Ltd., London, England
Sole	Sole Chemical Corp., Div. of Hodag Chemical Corp., Skokie, Illinois
Stepan	Stepan Chemical Co., Northfield, Illinois
Surfact-Co.	Surfact-Co., Inc., Blue Island, Illinois
Swastik	Swastik Oil Mills, Bombay, India
Tessilchimica	La Tessilchimica SpA, Bergamo, Italy
Textilana	Textilana Corp., Hawthorne, California
Triantaphyllou	Th. G. Triantaphyllou & Co., Athens, Greece
UCC	Union Carbide Corp., Chemicals and Plastics, New York, New York
Witco	Witco Chemical Corp., Organics Div., New York, New York
Witco/UK	Witco Chemical, Cyclo Div., London, England
Yoshimura	Yoshimura Oil Chemical Co., Ltd., Toyonakaski, Japan
Z & S	Zschimmer & Schwarz, Lahnstein-2/Rhein, West Germany

REFERENCES

1. W. Schrauth, Chem. Ztg., 55, 3, 17 (1931); Ber., 65B, 93 (1932); Angew. Chem., 46, 459 (1933); W. Schrauth, O. Schenck, and K. Stickdorn, Ber., 64B, 1314 (1931); see also O. Schmidt, Ber., 64B, 2051 (1931).
2. D. B. Hatcher, J. Am. Oil Chemists' Soc., 34, 175 (1957).
3. B. G. Wilkes and J. N. Wickert, Ind. Eng. Chem., 29, 1234 (1939).
4. C. Q. Sheely, Jr. and R. G. Rose, Ind. Eng. Chem. Prod. Res. Develop., 4, 24 (1965).
5. R. C. Hurlburt, R. F. Knott, and H. A. Cheney, Soap Chem. Specialties, 43 (5), 122 (1967).
6. R. C. Hurlburt, R. F. Knott, and H. A. Cheney, Soap Chem. Specialties, 43 (6), 88 (1967).
7. A. Pryce, So. African Pat. 67-07, 709 (1969), to Unilever Ltd.; Chem. Abstr., 72, 50717 (1970).
8. M. Klang and R. Stoica, Rev. Chim. (Bucharest), 9, 23 (1958); Chem. Abstr., 53, 200 (1959).
9. A. E. Sowerby, Fr. Pat. 1,556,042 (1969), to Marchon Products Ltd.
10. J. W. Lohr, U.S. Pat. 3,232,976 (1966), to Andrew Jergens Co.
11. A. Davidsohn and J. Manor, Isr. Pat. 12,536 (1959), to "Dahlia" Kibbutz Hashomer Hazir.

12. Proctor and Gamble Co., Br. Pat. 1,111,208 (1968).
13. J. Vahala, J. Sponar, and J. Weigner, Chem. Prumysl., 13 (5), 236 (1963); Chem. Abstr., 59, 8579 (1963).
14. L. A. Potolovskii et al., Tr., Vses. Nauch.-Issled. Inst. Pererab. Nefti, 1970 (13), 355-358; Chem. Abstr., 73, 100285 (1970).
15. Kao Soap Co., Ltd., Fr. Pat. 1,576,526 (1969).
16. S. Holzman and U. Zoller, Soap Chem. Specialties, 45, 43 (1969).
17. W. Schenk, J. Starrger, and R. Irnich, Ger. Pat. 1,286, 026 (1969), to Badische Anilin-und Soda-Fabrik A.-G.
18. M. Ballestra, Fr. Pat. 1,451,626 (1966).
19. Stepan Chemical Co., Br. Pat. 974,298 (1964).
20. E. A. Knaggs and M. L. Nussbaum, U.S. Pat. 3,544,613 (1970), to Stepan Chemical Co.
21. G. Place, Br. Pat. 799,038 (1958), to Thomas Hedley & Co., Ltd.
22. R. J. Brooks and B. Brooks, U.S. Pat. 3,069,242 (1962), to Chemithon Corp.
23. S. Popovic and J. Michic, J. Tehnika (Belgrade), 20, 963 (1965); Chem. Abstr., 66, 4103 (1967).
24. G. Szekely, Rev. Chim. (Bucharest), 21 (8), 458 (1970); Chem. Abstr., 74, 32906 (1971).
25. P. Roele, Belg. Pat. 638,577 (1964), to Unilever N.V.
26. A. M. Shiman and P. L. Shkurenko, Maslo-Zhir. Prom., 36 (6), 38 (1970); Chem. Abstr., 73, 78813 (1970).
27. A. M. Shiman and P. L. Shkurenko, Neftepererab, Neftekhim (Moscow), 1970 (7), 37-40; Chem. Abstr., 73, 89454 (1970).
28. R. W. Meritt and L. B. Graham, Br. Pat. 1,005,065 (1965), to Procter & Gamble, Ltd.
29. Atlas Powder Co., Br. Pat. 766,706 (1957).
30. E. E. Gilbert and B. Veldhuis, J. Am. Oil Chemists' Soc., 36, 208 (1959).
31. Colgate Palmolive Co., Belg. Pat. 617,968 (1962).
32. E. E. Gilbert and B. Veldhuis, J. Am. Oil Chemists' Soc., 37, 298 (1960).
33. G. L. Broussalian, U.S. Pat. 3,287,389 (1966), to Monsanto Co.
34. L. G. Nunn Jr., U.S. Pat. 3,150,161 (1965), to General Aniline and Film Corp.
35. Esso Research and Engineering Co., Br. Pat. 942,130 (1963).
36. A. J. Rutkowski and A. F. Turbak, U.S. Pat. 3,133,949 (1964), to Esso Research and Engineering Co.
37. A. F. Turbak, U.S. Pat. 3,168,547 (1965), to Esso Research and Engineering Co.
38. J. L. Smith and R. C. Harrington, Jr., U.S. Pat. 2,957,014 (1960), to Eastman Kodak Co.
39. J. Vyskocil and A. Lollok, Czech. Pat. 108,690 (1963); Chem. Abstr., 60, 10932 (1964).
40. V. I. Babaev et al., Russ. Pat. 215,383 (1968); Chem. Abstr., 69, 51577 (1968).

41. A. Ujhidy et al., Chem. Tech. (Berlin), 15 (9), 554 (1963).

42. F. Weiss, East Ger. Pat. 12,783 (1957).

43. B. Herold, Ger. Pat. 1,150,673 (1963).

44. E. Maurer, Rev. Chim. (Bucharest), 17 (11), 665 (1966); Chem. Abstr., 66, 67027 (1967).

45. A. I. Kudryashov et al., Russ. Pat. 196,788 (1967); Chem. Abstr., 68, 51188 (1968).

46. A. P. Dienekhovskaya and A. M. Karnaukh, Sbornik Statei o Rabot, Ukrain. Nauch.-Issledovatel. Inst. Maslozhir. Prom., 1958 (3), 75-78; Chem. Abstr., 55, 10300 (1961).

47. C. E. Redemann, U.S. Pat. 3,332,979 (1967), to Purex Corp. Ltd.

48. L. I. Slominski and M. K. Yakubov, Masloboino-Zhirovaya Prom., 26 (11), 22-25 (1960); Chem. Abstr., 55, 4410 (1961).

49. G. P. Touey and J. E. Kiefer, U.S. Pat. 3,075,965 (1963), to Eastman Kodak Co.

50. R. C. Harrington, Jr., U.S. Pat. 2,928,860 (1960), to Eastman Kodak Co.

51. L. D. Volkova and L. V. Yatsenko, Masloboino-Zhirovaya Prom., 35, 25 (1969); Chem. Abstr., 71, 5139 (1969).

52. I. S. Sukhoterin, Russ. Pat. 210,151 (1968); Chem. Abstr., 69, 51578 (1968).

53. I. I. Tokarskaya et al., Neftekhimiya, 3 (5), 758 (1963); Chem. Abstr., 60, 1572 (1964).

54. E. E. Gilbert, Sulfonation and Sulfation, in Kirk-Othmer Encyclopedia of Chemical Technology (A. Standen, ed.), 2nd ed., Vol. 19, John Wiley & Sons, Inc., New York, 1969, pp. 301 ff.

55. D. D. Whyte, J. Am. Oil Chemists' Soc., 32, 313 (1955).

56. G. Bozzetto, U.S. Pat. 3,055,929 (1962); G. Bozzetto, Ger. Pat. 1,118,195 (1961); G. Bozzetto, Ital. Pat. 592,223 (1959).

57. F. Meyer, East Ger. Pat. 42,927 (1966).

58. L. D. Volkova and L. U. Yatsenko, Maslozhir Prom., 33, 18 (1967); Chem. Abstr., 67, 63659 (1967).

59. N. Whitman, U.S. Pat. 3,337,601 (1967), to E.I. Du Pont de Nemours & Co., Inc.

60. V. Blinoff and G. Braude, U.S. Pat. 2,975,141 (1961), to American Alcolac Corp.

61. I. K. Getmanskii, Russ. Pat. 154,360 (1963); Chem. Abstr., 60, 6747 (1964).

62. K. Shimokai and M. Fukushima, Yukagaku, 12, 516 (1963); Chem. Abstr., 60, 7032 (1964).

63. H. J. Krause, U.S. Pat. 3,484,474 (1969), to Henkel & Cie., G.M.B.H.

64. H. J. Krause, Ger. Pat. 1,248,034 (1967), to Henkel & Cie., G.M.B.H.

65. Fr. Pat. 1,337,681 (1963), to Shell Internationale Research Maatschappij N.V.

66. Neth. Application 6,600,934 (1966), to Union Carbide Corp.; Chem. Abstr., 66, 20243 (1967).

67. S. Ropuszynski et al., Przem. Chem., 47, 83 (1968); Chem. Abstr., 68, 115893 (1968).
68. M. C. Rozzi, U.S. Pat. 3,313,839 (1967), to General Aniline & Film Corp.
69. S. M. Loktev and E. L. Vulakh, Maslozhir. Prom., 33, 15 (1966); Chem. Abstr., 67, 65735 (1967).
70. D. Santmyers and R. Aarons, Sulfamic Acid and Sulfamates, in Kirk-Othmer Encyclopedia of Chemical Technology (A. Standen, ed.), 2nd ed., Vol. 19, John Wiley & Sons, Inc., New York, 1969, pp. 242 ff.
71. J. M. Walts and L. M. Schenck, U.S. Pat. 3,395,170 (1968), to General Aniline & Film Corp.
72. E. L. Vulakh et al., Neftekhimiya, 4, 780 (1964); Chem. Abstr., 62, 2701 (1965).
73. S. M. Loktev et al., Nefteperab, Neftekhim (Moscow), 1970 (8), 32-34; Chem. Abstr., 73, 132250 (1970).
74. H. Bruschek, Ger. Pat. 1,154,460 (1963), to Dr. Th. Boehme A.-G., Chem. Fabrik.
75. J. M. Walts and L. M. Schenck, U.S. Pat. 3,392,185 (1968), to General Aniline & Film Corp.
76. I. S. Litvak et al., Zavodskaya Laboratoriya, 33, 607 (1967); Chem. Abstr., 67, 73098 (1967).
77. R. Grimmer and G. Moser, East Ger. Pat. 51,853 (1966); Chem. Abstr., 66, 65097 (1967).
78. E. L. Vulakh et al., Russ. Pat. 181,224 (1966), to Chemical Combine, Novomoskovsk; Chem. Abstr., 65, 10814 (1966).
79. H. Yamaguchi and K. Yamane, Hiroshima Daigake Kogakubu Kenkyu Hokou, 16 (2), 287 (1968); Chem. Abstr., 70, 37091 (1969).
80. K. Nakano and H. Yamaguchi, Kogyo Kagaku Zasshi, 67, 2055 (1964); Chem. Abstr., 65, 583 (1966).
81. H. Yamaguchi and T. Oomori, Hiroshima Daigaku Kogakubu Kenkyu Hokoku, 16, 283 (1968); Chem. Abstr., 70, 28326 (1969).
82. H. C. Borghetty and C. A. Bergman, J. Am. Oil Chemists' Soc., 27, 88 (1950).
83. K. Miyamoto et al., Jap. Pat. 464 (1964), to Nitto Chemical Industry Co.; Chem. Abstr., 60, 11899 (1964).
84. H. Yamaguchi et al., Jap. Pat. 23,333 (1965), to Nitto Chemical Industry Co., Ltd., and Nitto Physico-Chemical Research Institute; Chem. Abstr., 64, 4941 (1966).
85. R. O. Mumma, Biochim. Biophys. Acta, 165, 571 (1968); Chem. Abstr., 70, 11909 (1969).
86. K. Takiura et al., Yakugaku Zasshi, 87, 1262 (1967); Chem. Abstr., 68, 48966 (1968).
87. J. Bohunek, Ger. Pat. 1,201,334 (1965), to Farbwerke Hoechst, A.-G.
88. D. L. Kass and T. W. Martinex, U.S. Pat. 3,158,639 (1965), to The Pure Oil Co.
89. B. Babos et al., Veszpremi Vegyip, Egyet. Kozlemen, 9, 119 (1965); Chem. Abstr., 66, 2302 (1967).

90. B. S. Nazarenko et al., Gazokondensaty Nefti, Mater. Srendeaziat. Nauch. Soveshch, Neftekhim. Khim. Pererab. Uglevodorodov, 2nd, 1967, 361-364; Chem. Abstr., 71, 82903 (1969).

91. I. Velea et al., Roman. Pat. 48,911 (1967), to Romania, Ministry of the Chemical Industry; Chem. Abstr., 69, 20639 (1968).

92. I. S. Sukhoterin and I. G. Agranovich, Mashlozhir Prom., 31, 39 (1965); Chem. Abstr., 66, 4113 (1967).

93. R. J. Brooks and B. Brooks, U.S. Pat. 3,058,920 (1962), to The Chemithon Corp.

94. S. S. Nazrova and V. K. Tsyskovskii, U.S.S.R. Pat. 133,855 (1960); Chem. Abstr., 55, 10929 (1961).

95. Henkel & Cie., G.M.B.H., Neth. Appl. 6,603,475 (1966); Chem. Abstr., 66, 37432 (1967).

96. Romania Ministry of Petroleum and Chemical Industry, Neth. App. 6,407,289 (1966); Br. Pat. 884,656 (1961); Chem. Abstr., 66, 4123 (1967).

97. P. B. Kooyman, Neth. Pat. 81,602 (1956); Chem. Abstr., 51, 9188 (1957).

98. F. H. van Heel, Neth. Pat. 84,711 (1957); Chem. Abstr., 52, 14650 (1958).

99. E. Clippinger, Ind. Eng. Chem. Prod. Res. Develop., 3, 3 (1964).

100. A. T. Kyll et al., Izvest. Akad. Nauk Eston, S.S.R., Ser. Tekh. i Fizmat. Nauk, 7 (2), 105-116 (1958); Chem. Abstr., 53, 5709 (1959).

101. P. L. Kooijman and R. W. Kreps, Neth. Pat. 86,626 (1957); Chem. Abstr., 53, 14002 (1959).

102. N.V. de Bataafsche Petroleum Maatschappij, Br. Pat. 782,466 (1957).

103. N. V. Milovidova et al., Tr. Vses. Nauchn.-Issled. Inst. po Pererabotke Nefti, 9, 81 (1963); Chem. Abstr., 60, 1938 (1964).

104. Shell Internationale Research Maatschappij N.V., Br. Pat. 465,435 (1964).

105. J. S. Berber and R. V. Rahfuse, U.S. Bur. Mines, Rep. Invest. 1968, No. 7115.

106. M. Buis, Br. Pat. 655,459 (1951), to Naamlooze Venootschap de Bataafsche Petroleum Maatschappij.

107. P. J. Garner and H. N. Short, Br. Pat. 656,064 (1951), to Shell Refining and Marketing Co., Ltd.

108. Societe Anon. des Produits Chimiques Shell—Saint Gobain, Fr. Pat. 1,236,165 (1960).

109. J. Hill and J. Vyskoeil, Abhandl. Deut. Akad. Wiss. Berlin, Kl. Chem. Geol. Biol., 1963 (2), 13-25; Chem. Abstr., 60, 10530 (1964).

110. E. Szmidtgal and H. Pasternak, Fette, Seifen, Anstrichmittel, 66, 225 (1964).

111. T. F. Rutledge, Ger. Pat. 1,403,963 (1969), to Atlas Chemical Industries.

112. E. Koenig, P. Hahn, K. Stickdorn, and H. G. Braun, Ger. Pat. 1,274,118 (1968), to Veb Deutsches Hydrierwerk Rodleben.

113. J. K. Weil, A. J. Stirton, and E. B. Leardi, J. Am. Oil. Chemists' Soc., 44, 522 (1967).
114. J. E. Gotte, Fette Seifen Anstrichmittel, 56, 583 (1954).
115. A. Lottermoser and F. Stoll, Kolloid Z., 63, 19 (1933).
116. J. N. Bone and D. W. O'Day, J. Am. Pharm. Assoc., 47, 795 (1958).
117. E. W. Maurer and A. J. Stirton, U.S. Pat. 3,305,578 (1967), to United States of America.
118. E. W. Maurer, A. J. Stirton, and J. K. Weil, J. Am. Oil Chemists' Soc., 37, 34 (1960).
119. E. W. Maurer and A. J. Stirton, U.S. Pat. 3,133,946 (1964), to United States of America.
120. E. W. Maurer and A. J. Stirton, U.S. Pat. 3,291,750 (1966), to United States of America.
121. R. G. Bistline, Jr., A. J. Stirton, and E. W. Maurer, J. Am. Oil Chemists' Soc., 34, 516 (1957).
122. G. Braude, R. R. Egan, M. Warren, and L. Galitzin, Chem. Specialties Mfrs. Assoc. Proc. Ann. Meeting, 43, 174-178 (1956).
123. J. K. Weil, A. J. Stirton, and E. A. Barr, J. Am. Oil Chemists' Soc., 43, 157 (1966).
124. F. J. Gohlke and H. Berghausen, Soap Chem. Specialties, 43, 60 (1968).
125. S. P. Harrold, J. Colloid Sci., 15, 280 (1960).
126. J. K. Weil, A. J. Stirton, and R. G. Bistline, Jr., J. Am. Oil Chemists' Soc., 31, 444 (1955).
127. J. K. Weil, A. J. Stirton, and E. W. Maurer, J. Am. Oil Chemists' Soc., 32, 148 (1955).
128. A. J. Stirton, E. W. Maurer, and J. K. Weil, J. Am. Oil Chemists' Soc., 33, 290 (1956).
129. J. K. Weil, A. J. Stirton, and R. G. Bistline, Jr., J. Am. Oil Chemists' Soc., 36, 241 (1959).
130. J. C. Cowan, J. K. Weil, and A. J. Stirton, U.S. Pat. 2,938,872 (1960), to United States of America.
131. S. H. Lenher, Am. Dyestuff Reptr., 22, 663 (1933).
132. B. M. Finger, G. A. Gillies, G. M. Hartwig, W. W. Ryder, Jr., and W. M. Sawyer, J. Am. Oil Chemists' Soc., 44, 525-530 (1967).
133. V. V. R. Subrahamanyam and K. T. Achaya, J. Chem. Eng. Data, 6, 38 (1961).
134. K. Shimokai and M. Fukushima, Yukagaku, 12, 516 (1963); Chem. Abstr., 60, 7032 (1964).
135. F. Püschel, Tenside, 3, 71 (1966).
136. M. Kashiwagi and H. Ezaki, Bull. Chem. Soc. Japan, 32, 624 (1959); Chem. Abstr., 54, 8115 (1960).
137. J. Pearson and A. Lawrence, Trans. Faraday Soc., 63, 488 (1967).
138. T. P. Matson, Soap Chem. Specialties, 34, 52 (1963).
139. M. Matsuda, N. Kawamura, W. Yamo, and W. Kumura, Yukagaku, 18, 132 (1969).

140. J. K. Weil, A. J. Stirton, and M. V. Nunez-Ponzoa, J. Am. Oil Chemists; Soc., 43, 603 (1966).
141. A. M. Schwartz, J. W. Perry, and J. Berch, Surface Active Agents, Vol. 2, Interscience Publishers, Inc., New York, 1958, pp. 622 ff.
142. Anon., Chem. Eng. News, 31, 5289 (1953).
143. H. W. Hibbott, Handbook of Cosmetic Science, The MacMillan Co., New York, 1963.
144. I. K. Getmanskii et al., Tr. Nauchn.-Issled. Inst. Sintetich. Zhiro-zamenitelei i Moyushchikh Sredstv., 1962 (3), 95; Chem. Abstr., 60, 9096 (1964).
145. J. R. Livingston, Jr., R. Drogin, and R. J. Kelly, Ind. Eng. Chem. Prod. Res. Develop., 4, 28 (1965).
146. C. K. von Fenyes, R. C. Johnson, and D. G. Norton, Drug Cosmetic Ind., 107 (2), 36 (1970).
147. J. M. Allen and G. V. Scott, U.S. Pat. 2,879,231 (1959), to Colgate Palmolive Co.
148. H. W. McCune, U.S. Pat. 3,313,735 (1967), to Procter & Gamble.
149. Unilever N.V., Neth. Appl. 6,608,258 (1967).
150. J. Schneider and N. S. Blodgett, U.S. Pat. 3,267,039 (1966), to N. S. Blodgett.
151. Yu. M. Volkov et al., Russ. Pat. 251,765 (1968); Chem. Abstr., 72, 35684 (1970).
152. F. Zschimmer, E. Zschimmer, R. Schwarz, and W. Schwarz, Br. Pat. 1,027,898 (1966).
153. Gillette Co., Br. Pat. 843,379 (1960).
154. March Products, Ltd., Fr. Pat. 1,403,213 (1965).
155. R. K. Lehne and L. J. Murphy, Fr. Pat. 1,409,328 (1965), to Colgate-Palmolive Co.
156. G. Proserpio, Riv. Ital. Essenze-Profumi, Piante Offic., Aromi-Saponi, Cosmet.-Aerosol., 50 (5), 250-251 (1968).
157. G. Barber, Ger. Pat. 2,025,481 (1971), to Witco Chemical Co.
158. K. Owada, Jap. Pat. 4024 (1962), to Marumiya Co., Ltd.
159. Tran Anh Tuan, Ger. Pat. 1,807,338 (1969), to Helene Curtis Indus-tries, Inc.; Chem. Abstr., 71, 114480 (1969).
160. F. J. Gohlke and H. Bergerhausen, Parfum., Cosmet. Savons, 12, 143 (1969).
161. J. N. Masci and Normand A. Poirier, Br. Pat. 850,514 (1960).
162. K. Bergwein, Drug Cosmetic Ind., 81, 163-165, 236-237 (1957).
163. J. F. L. Chester, Perfumery Essent. Oil Record, 58, 539-546 (1967).
164. F. Roesch, U.S. Pat. 2,865,811 (1958), to Irval Cosmetics Inc.
165. Veb Farbenfabrik Walten, East Ger. Pat. 11,984 (1956).
166. A. Shansky, Am. Perfumer Cosmet., 82 (7), 43-45 (1967).
167. Unilever Ltd., Fr. Pat. 1,553,432 (1969).
168. K. R. Dutton, and Dutton and Reinisch Ltd., Br. Pat. 1,040,011 (1966).
169. F. E. Boettner and J. L. Rainey, U.S. Pat. 2,982,737 (1961), to Rohm and Haas Co.
170. T. Sano, Jap. Pat. 8576 (1956), to Kao Soap Co.

171. J. D. Barnhurst, G. M. Leigh, and J. A. Monick, U.S. Pat.
     3,442,812 (1964), to Colgate-Palmolive Co.
172. G. Hewitt, U.S. Pat. 3,350,320 (1967), to Colgate-Palmolive Co.
173. K. A. Schumann, K. H. Schumann, and J. Bruecki, East Ger. Pat.
     32,812 (1965).
174. D. P. Barrett, J. P. Parke, and J. O. Murray, Ger. Pat. 2,007,883
     (1970), to Unilever N.V.
175. A. Alsbury, B. H. Hampson, and H. Moore, Ger. Pat. 2,007,557
     (1970), to Unilever N.V.
176. A. Alsbury and D. P. Barrett, Ger. Pat. 1,809,034 (1969), to Uni-
     lever N.V.
177. P. Alexander, Mfg. Chemist Aerosol News, 36, 41-43 (1965).
178. E. Eigen and S. Weiss, U.S. Pat. 3,548,056 (1970), to Colgate-
     Palmolive Co.
179. O. K. Jacobi, Ger. Pat. 1,952,057 (1970), to Kolmar Laboratories,
     Inc.
180. K. A. Loperkhova and V. Zabolivaniza, Khim. Prom., 1965, 308.
181. B. Puetzer, A. V. Finn, and L. Mackles, U.S. Pat. 3,337,466 (1967),
     to Tintex Corp.
182. Peter Riethe, J. Soc. Cosmet. Chem., 18 (5), 291-302 (1967).
183. G. Lemetre, L. Porta, and P. Tomassini, Ger. Pat. 2,013,994 (1970).
184. P. H. Ekvorthy, Pharm. J., 199, 107 (1967).
185. E. Nuernberg, Deut. Apotheker-Ztg., 1968 (26), 907-913.
186. S. N. Sharma and S. D. Gupta, Indian J. Pharm., 29, 309 (1967).
187. J. Miyazaki and K. Arikawa, Yakuzaigaku, 1968, 28 (1), 69-73; Chem.
     Abstr., 69, 89687 (1968).
188. I. Utsumi and K. Harada, Jap. Pat. 99 (1956), to Tanabe Drug Co.,
     Chem. Abstr., 51, 5831 (1956).
189. H. E. Tschakert, Seifen Oele Fette Wachse, 79, 673 (1953).
190. G. Weder, Textil-Rundschau, 13, 637-642 (1958).
191. A. J. Stirton, E. W. Maurer, and J. K. Weil, J. Am. Oil Chemists'
     Soc., 33, 290-291 (1956).
192. H. D. Weiss, O. Gellner, and G. W. Panzer, U.S. Pat. 3,256,202
     (1966), to Alcolac Chem. Corp.
193. W. J. Dewitt and R. C. Taylor, U.S. Pat. 3,480,556 (1969).
194. T. G. Jones and D. W. Stephens, U.S. Pat. 3,281,367 (1967), to Lever
     Bros. Co.
195. Thomas Hedley & Co., Ltd., Br. Pat. 791,704 (1958).
196. L. Fernandez, U.S. Pat. 2,861,956 (1958).
197. W. C. Schar, M. H. Paulson, Jr., F. N. Baumgartner, and E. J.
     Wickson, J. Am. Oil Chemists' Soc., 37, 427-430 (1960).
198. J. H. Wilson, U.S. Pat. 3,150,098 (1964).
199. Kaekawa, Sadao, Naganuma, and Yoshisumi, Jap. Pat. 18,987 (1970).
200. Hans E. Tschakert, Seifen Oele Fette, Wachse, 93 (13), 421-427 (1967).
201. A. V. Matrenina, M. Martuzane, M. Smagare, and D. Spinga, Russ.
     Pat. 216,889 (1968), to Riga Chemical Plant, "Aerozol"; Chem. Abstr.,
     69, 53081 (1968).
202. L. Shapiro, Am. Dyestuff Reptr., 39, 38 (1950).

203. F. J. Gohlke and H. Bergerhausen, Soap Chem. Spec., 43, 47–49 (1967).
204. Unilever Ltd., Br. Pat. 855,679 (1960).
205. N. S. Smith, U.S. Pat. 3,179,598 (1965), to Procter & Gamble.
206. H. Y. Lew, U.S. Pat. 3,231,508 (1966), to Chevron Research Co.
207. Charles F. Jelinek and Raymond L. Mayhew, U.S. Pat. 2,941,951 (1955).
208. N. V. de Bataafsche Petroleum Maatschappij, Neth. Pat. 89,559 (1958).
209. E. Gotte and W. Gundel, Ger. Pat. 1,071,873 (1959), to Henkel & Cie., G.M.B.H.
210. A. J. Stirton, E. W. Maurer, and J. K. Weil, J. Am. Oil Chemists' Soc., 33, 290–291 (1956).
211. H. D. Weiss, D. Gellner, and G. W. Panzer, U.S. Pat. 3,256,202 (1966), to Alcolac Chemical Corp.
212. L. E. Cohen, U.S. Pat. 3,039,971 (1962), to FMC Corp.
213. V. Dvorkovitz and J. A. Goldman, U.S. Pat. 2,982,736 (1961), to Diversey Corp.
214. John J. Parran, Jr., So. African Pat. 68 07,526 (1969), to Procter & Gamble Co.; Chem. Abstr., 72, 33547 (1970).
215. S. M. Fadeeva and T. V. Yapryntseva, Kul't Byt. Izdeliya, 1967, 77–82; Chem. Abstr., 70, 12839 (1969).
216. Shell Internationale Research Maatschappij, N.V., Neth. Pat. 97,122 (1961).
217. Frederick M. Fowkes, Webster M. Sawyer, Jr., and Martin J. Shick, U.S. Pat. 2,900,346 (1959).
218. E. Maurer and H. Schuster, Bul. Inform. Lab. Central Colorist, 1966, 327–341.
219. Kurt Lindner, Ger. Pat. 955,857 (1957).
220. C. Panaltide, M. Frenkel, A. Avachian, and S. Danaila, Bul. Inform. Lab. Central Colorist, 1966, 343–347.
221. Chemische Werke Stockhausen & Cie., and Vereinigte Glanzstoff-Fabriken Akt.-Gas., Ger. Pat. 1,056,576 (1959).
222. Paranosa of London, Ltd., Br. Pat. 1,056,850 (1967).
223. O. D. Hoxie, Belg. Pat. 668,163 (1965), to Bissell Inc.
224. W. Hackett, J. Deterg. Spec., 1970, 16–17.
225. Rohm & Haas Co., Ger. Pat. 1,112,602 (1959).
226. W. L. St. John, U.S. Pat. 2,985,592 (1961).
227. Henkel & Cie., G.M.B.H., Belg. Pat. 614,604 (1962).
228. G. O. Funderburk, R. C. Johnson, and R. H. Smith, U.S. Pat. 2,833,722 (1958).
229. N. M. Elias, U.S. Pat. 3,075,288 (1963).
230. M. E. Stonebraker and S. P. Wise, U.S. Pat. 3,463,735 (1969).
231. C. O. Durbin and Gert G. Levy, U.S. Pat. 3,304,264 (1967).
232. Theodore A. Ruemele, Br. Pat. 805,768 (1958).
233. C. M. Bodach, U.S. Pat. 3,352,787 (1967), to W. R. Grace & Co.
234. J. Vyskocil and A. Lollok, Czech. Pat. 108,690 (1963); Chem. Abstr., 60, 10932 (1964).

235. S. H. Shapiro, U.S. Pat. 3,379,763 (1968), to Armour & Co.
236. Marchon Products, Ltd., Br. Pat. 1,067,762 (1967).
237. Marchon Products, Ltd., Fr. Pat. 1,346,673 (1963).
238. A. L. Kudryashov and I. K. Getmanskii, Russ. Pat. 159,594 (1963); Chem. Abstr., 61, 2066 (1964).
239. McCutcheon's Detergents and Emulsifiers, North American Ed. and International Ed. (2 Vol.), McCutcheon's Div., Allured Publishing Corp., Ridgewood, N.J., 1974.
240. A. E. Blood and J. D. Heller, Fr. Pat. 1,462,888 (1966), to Eastman Kodak Co.
241. H. S. Bloch, G. E. Illingworth, and G. W. Lester, U.S. Pat. 3,681,424 (1972), to Universal Oil Products Co.
242. H. M. Muijs, J. Meisner, and C. Kortland, Fette Seifen Anstrichmittel, 73, 315 (1971).
243. M. Nakamura, W. Yano, and W. Kimura, Yukagaku, 20, 165 (1971).
244. A. Struve, W. Stein, and W. Umbach, Fette Seifen Anstrichmittel, 74, 331 (1972).
245. M. J. Rosen, J. Am. Oil Chemists' Soc., 51, 461 (1974).
246. W. C. Griffin and R. W. Behrens, Anal. Chem., 24, 1076-1077 (1952).
247. T. R. Briggs, J. Phys. Chem., 24, 120-126 (1920).

CHAPTER 6

SULFATED MONOGLYCERIDES AND
SULFATED ALKANOLAMIDES

James K. Weil and Alexander J. Stirton[*]

Eastern Regional Research Center
U.S. Department of Agriculture
Philadelphia, Pennsylvania

I. INTRODUCTION

In the development of synthetic detergents, many methods have been investigated for attaching hydrophilic groups to hydrophobic carbon chains. Fatty alcohol sulfates were shown to have excellent detergent properties but their growth was limited by the expense involved in converting fats to fatty alcohols. In a search for a more economical route to the alcohol sulfate type of synthetic detergent, fatty acids or their esters were reacted with glycols or

---

[*]Deceased.

219

hydroxyalkylamines to produce detergents having the general formula,

$$RCOXR'OSO_3Na$$

where X is O, NH, or N-alkyl, and R' is $-CH_2CH_2-$, $-CH_2CHOHCH_2-$, or other alkylene, alkoxyalkyl, and hydroxyalkyl groups.

This type of compound has an auxiliary hydrophilic structure which imparts desirable surface-active properties but whose utility may be limited by the susceptibility of the ester or amide to hydrolysis. Those compounds where R' is $-CH_2CH_2-$ are particularly unstable because of anchimeric assistance in hydrolysis.

The development of sulfated monoglycerides and sulfated alkanolamides naturally followed the technology of monoglycerides and alkanolamides. Sulfated monoglycerides therefore presented an early achievement in the development of synthetic detergents and a great deal of work was done by one of the major soap companies in the development and marketing of products containing this type of compound. Interest in the sulfated alkanolamide followed the later development of fatty amides. Although the Ninols invented by Kritchevsky [1] were diethanolamides used as foam stabilizers and viscosity improvers, much of the technology developed for them is applicable to the production of monoalkanolamides for sulfation.

II. SULFATED MONOGLYCERIDES

A. Synthesis

Thieme first reported investigation on the sulfation of monoglycerides [2]. He studied the equilibrium of triglycerides at different temperatures and strengths of sulfuric acid. He proposed that the shift toward greater hydrolysis was aided by the formation of sulfated glycerol esters, shown in Eq. (1) for the hydrolysis of α, α'-diglyceride,

$$H_2SO_4 + RCO_2CH_2CHOHCH_2O_2CR \longrightarrow$$
$$RCO_2CH_2CHOHCH_2OSO_3H + RCO_2H \qquad (1)$$

Because of its low cost, this is the best method for the production of commercial sulfated monoglycerides. The reaction is carried out under conditions which permit the formation of monoglycerides and sulfation to take place simultaneously. Early processes [3, 4] have called for the reaction of one mole of triglyceride, two moles of anhydrous glycerine, and an excess of sulfuric acid. Gray [5] obtained a product high in active ingredient and low in ether soluble compounds by the reaction of coconut oil with oleum and glycerol sulfuric acid. His method recommends the preparation of glycerol sulfuric acid by passing a 1:20 molar mixture of SO_3 in air through 99.5%

glycerol at 65° C until 2 moles of SO_3 are absorbed per mole of glycerol. The glycerol sulfuric acid was then mixed with a larger quantity of oleum, allowed to stand for 20 min, and then mixed with the triglyceride and stirred for one and a half hours at 60° C.

The inability of glycerine to form homogeneous solutions with fats or fatty acids gives rise to an adverse distortion of product distribution in the preparation of pure monoglycerides. As monoglyceride is formed, it is carried into the fatty layer where it is more highly acylated to di- or triglyceride. Hilditch and Rigg [6] have shown that the monoglyceride content does not exceed 50% when fatty acids react with excess glycerine without solvent, but the use of phenol as a cosolvent can give a product with 92% monoglyceride when 3 moles of glycerine per mole of fatty acid are used, and 98% monoglyceride when the mole ratio is increased to ten to one. A disadvantage of this method is the reactivity of the solvent phenol leading to undesirable by-products which are difficult to separate. The necessary contact of glycerol and fatty reactant has also been accomplished by continuous countercurrent flow systems [7] or other high-temperature mixing methods [8]. High-vacuum distillation has been used as a method for purifying monoglycerides [9]. Kurt [10] increased the monoglyceride content from 30-40%, usually found in such reaction mixtures, to 92-94% by a molecular distillation process.

The sulfation step in the preparation of sulfated monoglycerides may be accompanied by hydrolysis and rearrangement unless special precautions are taken. A complex formed from chlorosulfonic acid and sodium chloride has been used to reduce hydrolysis in laboratory preparations where high purity is desired [11]. Sulfur dioxide has been used [12] as a combination solvent and cooling agent to minimize hydrolysis by keeping the temperature low during sulfation and neutralization. Gebhart and Mitchell [13] have recommended the use of carbonates to neutralize the sulfonation mass in a mixed aqueous-alcoholic solvent. Excessive heat build up is prevented by the limited solubility of the carbonate and the cooling effect of carbon dioxide as it is evolved. Muncie's early patent [14] on the process for making sulfated monoglycerides describes equipment for the efficient dissipation of heat and careful monitoring of pH and temperature in the neutralization step.

Hydrolysis of some of the glycerol ester may not always be a disadvantage. It has been shown [15] that sodium stearate and monostearin are a particularly effective pair for forming stable emulsions, and it is unlikely that small amounts of these surface-active agents would adversely affect the detergent properties of sulfated monoglycerides.

The most popular starting material for the preparation of sulfated monoglycerides has been coconut oil. For a number of years coco monoglyceride sulfate was the active ingredient for the product Vel marketed by the Colgate-Palmolive Co. It had high foaming power, good solubility and adequate detergency properties which are typical of other derivatives of coconut oil. Coco monoglyceride sulfate may still be of interest in the

Philippines [16] where coconut oil has a local advantage over other raw materials. Other fatty acids which have been suggested for the manufacture
of monoglyceride sulfates are rosin acids [17] and synthetic fatty acids [18].
Since sulfated monoglycerides are more soluble than many other derivatives,
tallow and other long-chain, low-cost, starting materials may be worthy of
consideration.

B. Properties and Uses

The first broad basic patent [19] covering sulfated monoglycerides was issued to the Colgate-Palmolive-Peet Company in 1936 and subsequently this
company has developed and marketed sulfated monoglycerides for a number
of years. In spite of industrial interest and success, however, very little
can be found in the literature on the characterization of pure sulfated monoglycerides. The instability of these materials toward hydrolysis and ester
rearrangement makes preparation of pure individual surfactants difficult.

Biswas and Mukherji prepared pure sulfated monoglycerides from
lauric, myristic, palmitic, stearic, oleic, and linoleic acids, and determined their colloidal [11] and surface active [20] properties (see Table 1).
Purified monoglycerides were sulfated with chlorosulfonic acid-sodium
chloride or puridine-sulfur trioxide complexes. Critical micelle concentration for sulfated monolaurin was found to be 0.19% by the conductance method
and 0.098% by dye titration which may be compared with values ranging from
0.17 to 0.25% for sodium dodecyl sulfate.

Table 1 lists surface tension, foam height, and emulsifying power for
the sulfated monoglycerides. Sulfated monomyristin exhibits the greatest
surface-tension lowering effect, and the greatest lowering effect on the
water-heptane interfacial tension. Sulfated monomyristin also showed the
best foaming power and the greatest emulsion volume.

Jedlinski [21] evaluated the detergency of built and unbuilt preparations
of sulfated monoglycerides from lauric acid and synthetic fatty acids on
three different test cloths. The compound made from a mixture of C_{12}-C_{13}
synthetic fatty acids had nearly the same detergency as that made from
lauric acid but not as good as that prepared from a mixture of C_{13}-C_{14} synthetic acids. Method of preparation of the monoglyceride and builder formulation had little effect on the detergency of the final product.

Sulfated monoglycerides have been marketed under the trade name Vel
as a fully formulated household detergent. It has been claimed that the addition of small amounts of fatty alcohols [22] or amides [23] increases the
detergency and foaming power of the formulation. The patent literature
describes detergent bars made from both sulfated monoglycerides [24] and
sulfated diglycerides [25], as well as the use of sulfated monoglycerides in
soap-syndet Combo bars [26, 27]. Although sulfated glycerol esters have
the disadvantage of being susceptible to hydrolysis, liquid shampoo

TABLE 1

Surface-Active Properties of Sulfated Monoglycerides[a] [20]

Fatty acid	Surface tension, dyn/cm	Interfacial tension,[b] dyn/cm	Foam height,[c] mm	Emulsion volume,[d] ml
Lauric	34	39	7	1.1
Myristic	26	19	190	5.4
Palmitic	46	21	20	3.6
Stearic	46	26	21	4.6
Oleic	36	22	180	5.0
Linoleic	37	25	9	4.9

[a]Concentration, 0.1% at 25°C.
[b]Heptane-aqueous interface, measured by du Nouy tensiometer.
[c]Ross-Miles pour test.
[d]Volume of emulsion remaining after shaking 15 ml of 0.1% solution and 5 ml of heptane and allowing to stand 30 min.

formulations have been prepared with the aid of buffering agents such as urea [28], polyacrylamides [29], and iminodipropionates [30]. Sulfated monoglycerides have also been proposed as an ingredient of an aerosol shampoo [31]. Probably the most unusual use suggested for these materials is as an ingredient of a composition to reduce the desire to smoke [32].

III. OTHER SULFATED POLYOL ESTERS

Harris' basic patent for sulfated monoglycerides [19] includes many types of sulfated esters and ethers, and another broad British patent has been granted to I.G. Farbenindustrie [33]. Although sulfated monoglycerides have been most popular because they are low-cost, efficient detergents, a word should be said about the related ester sulfates which have been considered. The simplest member of this series, the sulfated monoester of ethylene glycol has been mentioned by Harris and elsewhere [34] in the patent literature but has never become commercially important. Hydrolytic stability may be expected to be poor if we consider the analogy between these compounds and the sulfated hydroxyethylamides which will be discussed later. Sulfated pentaerythritol esters have also been proposed for this type of detergent [17, 18].

Similar compounds have been made by opening heterocyclic rings. The Boehme Fettchemie made several such compounds in the laboratory during World War II [35]. Pentamethylene oxide was allowed to react with lauric acid and chlorosulfonic acid at 80°C for 5-6 hours [Eq. (2)].

$$
\begin{array}{c}
CH_2 \\
/ \;\; \backslash \\
CH_2 \;\; CH_2 \\
|\;\;\;\;\;\;\; | \\
CH_2 \;\; CH_2 \\
\backslash \;\; / \\
O
\end{array}
+ C_{11}H_{23}COOH + ClSO_3H \longrightarrow C_{11}H_{23}COO(CH_2)_5 OSO_3H \quad (2)
$$

Another variation [36] of this type of product is made from sulfuric acid and a fatty tetrahydrofurfuryl ester [Eq. (3)].

$$
\begin{array}{c}
CH_2{-}CH_2 \\
|\;\;\;\;\;\;\;\; | \\
RCOOCH_2CH \;\;\;\; CH_2 \\
\backslash \;\; / \\
O
\end{array}
+ H_2SO_4 \longrightarrow RCOOCH_2 CH(OH)(CH_2)_5 OSO_3H \quad (3)
$$

A product from the dimerization of acrolein was also esterified with fatty acids of paraffin origin and sulfated.

IV. SULFATED ALKANOLAMIDES

A. Synthesis of Alkanolamides

Although the preparation of fatty alkanolamides has been covered more completely in an earlier volume of this series [37, 38] the subject will be reviewed briefly here for continuity.

Hydroxyalkylamides may be prepared by the direct reaction of hydroxyalkylamines with fatty acids or esters, or by the reaction of epoxides with fatty amides. Since the latter process yields a mixture of products, the former is preferred when a single derivative is desired.

Alkanolamides were first made by the reaction of fatty acids with an excess of diethanolamine [1]. Although this type of product, in this case a dialkanolamide, has found a market as an ingredient in detergent formulations, it was later found that alkanolamides, particularly the monohydroxyalkylamides, could be prepared in higher yields and better purity by the alkali catalyzed reaction [Eq. (4)] of hydroxyalkylamine with fatty esters [39-45].

$$
RCOOR' + HOCH_2CH_2NH_2 \xrightarrow{\;\;NaOCH_3\;\;} RCONHCH_2CH_2OH + R'OH \quad (4)
$$

This method is applicable for any fatty ester, including the natural glycerol esters. Naudet [43] studied the glycerol ester composition changes in the partial ammonolysis of triglyceride. The composition of free glycerine, monoglyceride, diglyceride, triglyceride, and amide ester was close to that predicted by random distribution of the remaining nonamidated acyl groups on the available hydroxyl groups.

Esters of low-molecular-weight alcohols, such as methanol and ethanol, are the preferred starting material for the laboratory preparation of pure hydroxyalkylamides because the low boiling by-product alcohols are easily removed. Although most of the strong bases such as sodium methoxide, sodamide, sodium, or sodium hydroxide have been used, metallic sodium has been recommended to ensure the necessary dryness in small-scale laboratory runs. A number of pure alkanolamides have been prepared by this method [44, 45].

To illustrate this procedure, the preparation of 2-hydroxyethylpalmitamide is described. Freshly cut sodium, 0.25 g was added to a mixture of 86 g (0.32 mole) of methyl palmitate and 21 g (0.35 mole) of ethanolamine at 100°C. As the sodium dissolved, the stirred reaction mixture was heated to 115-125°C for 40 min, at which time the distillation of methanol was completed. Crystallization from ethanol at room temperature after clarification by hot filtration gave a 91% yield of pure product.

Amides may be oxyalkylated [Eq. (5)] with alkaline catalysts to give hydroxyalkylamides along with products having different degrees of oxyalkylation,

$$RCONH_2 + nCH_2\underset{O}{\overset{}{\diagdown\diagup}}CH_2 \xrightarrow{\overline{OH}} RCONH(CH_2CH_2O)_nH \tag{5}$$

where n is an average number whose components are distributed in accordance with statistical theory. Evidence for the equation shown above, rather than an equation involving both amide hydrogens, has been presented by Knaggs [46]. He showed that the infrared spectra of the alkali-catalyzed polyoxyalkylation product was the same as that for an authentic sample of diglycolamide, $RCONH(C_2H_4O)_2H$, rather than that of the diethanolamide, $RCON(C_2H_4OH)_2$. This structural difference is important in achieving hydrolytic stability of the sulfation product. Uncatalyzed oxyalkylations have been reported [38] to lead to substitution of both amido hydrogens; one might expect similar results with acid catalysis.

B. Sulfation

The monohydroxyalkylamides may be sulfated by common sulfating agents very much as one would sulfate a fatty alcohol. The laboratory procedure [45, 47], carried out in chloroform or carbon tetrachloride solution at room temperature using chlorosulfonic acid as the sulfating agent, gives nearly quantitative yields of sulfate with no noticeable hydrolysis of the amide linkage [Eqs. (6) and (7)].

$$RCONHC_2H_4OH + ClSO_3H \longrightarrow RCONHC_2H_4OSO_3H + HCl \qquad (6)$$

$$RCONHC_2H_4OSO_3H + NaOH \longrightarrow RCONHC_2H_4OSO_3Na + H_2O \qquad (7)$$

Sulfation may be carried out with equal ease on the simple alkanolamides, those having higher degrees of oxyalkylation, or those with alkyl side chains. The only exceptions are compounds with fully substituted nitrogen where instability makes it difficult to isolate a pure product from the sulfation process.

An unusual behavior is observed in the sulfation of unsaturated alkanolamides. Instead of the usual double-bond reaction encountered in the sulfation of unsaturated fatty materials like oleyl alcohol [48], hydroxyalkyloleamides can be sulfated by the conventional reagents without double-bond attack [49]. Spada and Gavioli [50] reported the sulfation of hydroxyethyloleamide with chlorosulfonic acid, although their sulfation mixture did contain a small amount of sodium chloride which is known to spare unsaturation when used in greater proportions. The double-bond protection found here is probably related to that found for different types of amine and ether complexes [48]. It has been shown [51] that unsaturated ether alcohols are sulfated with about 60% double-bond retention.

Commercially sulfations have been carried out with conventional sulfating agents such as sulfuric acid, oleum, chlorosulfonic acid, or urea-sulfuric acid [52, 53]. The reaction is carried out on the neat alkanolamide or dissolved in a chlorinated solvent at temperatures below 75° C but preferably at 30-35° C. Complete sulfation is not always desirable. Coconut alkanolamides have the best properties when not sulfated an extent greater than 75-85% because the unsulfated product contributes to detergency [54]. Products containing 50% unsulfated material have excellent lathering and cleaning power.

Swern [55] made sulfated hydroxypropylmyristamide by the reaction of N-allylmyristamide with sulfuric acid containing small amounts of sulfur trioxide at room temperature [Eq. (8)].

$$RCONHCH_2CH{=}CH_2 + H_2SO_4 \longrightarrow RCONHCH_2CH(OSO_3H)CH_3 \qquad (8)$$

Darker products were obtained when the concentration of sulfur trioxide was increased up to 22%.

C. Effect of Structure on Properties and Uses

Table 2 shows the effect of changes in chemical structure on surface-active properties. Unlike the ether alcohol sulfates [56] or the fatty acids [57],

TABLE 2

Surface-Active Properties of Sulfated Alkanolamides [45]

Sulfated alkanolamide	Melting point of parent alkanolamide, °C	Krafft point, 1%, °C	CMC,[a] millimoles/liter	Detergency,[b] ΔR
$C_{11}H_{23}CONHCH_2CH_2OSO_3Na$	89	14	10.1	16
$C_{15}H_{31}CONHCH_2CH_2OSO_3Na$	99	42	0.55	30
$C_{17}H_{35}CONHCH_2CH_2OSO_3Na$	103	53	0.16	32
$C_{17}H_{35}CONHCH_2CH(CH_3)OSO_3Na$	86	27	0.14	30
$C_{17}H_{35}CONHCH_2CH_2OSO_3Na$	97	57	0.13	31
$C_{17}H_{35}CON(CH_3)CH_2CH_2OSO_3Na$	55	21	0.11	22

[a]Critical micelle concentration.
[b]Active ingredient (0.05%) + builder (0.20%) at 60°C, 300 ppm.

there is little relationship between the melting point of alkanolamides and Krafft point for the sulfated alkanolamides. Side-chain substitution lowers the Krafft point nearly by 30° C while an increase in the length of the alkyl chain between the amide and sulfate gives a slight raising of the Krafft point and lowering of the critical micelle concentration. The close correspondence of the critical micelle concentration of sulfated hydroxyethylamides with alkyl sulfates of the same fatty chain length indicated that the hydrophilicity of the amide group is sufficient to counterbalance the additional chain length-ening from the hydroxyalkylamine. All of the sulfated alkanolamides showed good lime-soap dispersing properties, and showed good detergency and foam-height [45, 58]. Recent work [58a, 58b] demonstrates the good detergency obtained when these materials are combined with soap.

Sulfated alkanolamides have been suggested for use in shampoos [59, 60] because they are not irritating to the skin. Toilet bars have been made from mixtures of sulfated coconut monoethanolamides and soap [61].

Sulfated alkanolamides were found [62, 62a] to be readily biodegradable under aerobic conditions at 25° C as well as under anaerobic conditions at 35° C.

D. Stability

1.   Theory

Desnuelle and Micaelli [63] observed that sulfated ethanolamides showed con-siderably less stability to alkaline hydrolysis than sulfated 3-hydroxypropyl-amides or sodium dodecyl sulfate and proposed that an oxazoline ring inter-mediate facilitated the hydrolysis of sulfated ethanolamides.

More recently [45], it has been found that the N-methyl-(hydroxyethyl)-amide sulfate is more rapidly hydrolyzed than the unsubstituted hydroxyethyl-amide sulfate and that the main product is a hydroxyalkylamide with only smaller amounts of fatty acid [Eq. (9)]. The methyl-substituted compound could not form the oxazoline ring and the electron-donating effect of the methyl group would aid in the decomposition of the ring structure formed by the participation of the carbonyl group.

$$\text{(9)}$$

TABLE 3

Hydrolysis of Sulfated Alkanolamides [45]

	0.05 N amide in 0.05 N NaOH at 100° C		0.05 N amide in 0.05 N HCl at 80° C	
Sulfated alkanolamide	Half-life, min	Fatty acids, %	Half-life, min	Fatty acids, %
$C_{15}H_{31}CONHCH_2CH_2CH_2OSO_3Na$	500	35	180	50
$C_{17}H_{35}CONHCH_2CH(CH_3)OSO_3Na$	126	25	85	85
$C_{11}H_{23}CONHCH_2CH_2OSO_3Na$	83	16	142	40
$C_{15}H_{31}CONHCH_2CH_2OSO_3Na$	83	--	83	65
$C_{17}H_{35}CON(CH_3)CH_2CH_2OSO_3Na$	9	4	45	22

The ring compound formed from the nonmethyl substituted amide may dehydrate to the oxazoline structure, slowing the reaction by formation of a more stable intermediate. A similar intermediate may be proposed for the acid-catalyzed reaction.

2. Effect of Structure

Table 3 lists the time required for 50% hydrolysis and composition of the hydrolyzed product for the hydrolysis of 0.05 N amide in 0.05 N NaOH at 100° C and in 0.05 N HCl at 80° C. The 3-hydroxypropylamide shows the greatest stability in both acid and alkaline systems. Methyl substitution near the sulfate gives a small increase in stability to alkaline hydrolysis while substitution at the nitrogen atom greatly reduces stability in acid or alkaline systems and most of the hydrolysis occurs at the sulfate ester linkage.

Stability is greatly increased by separation of the amide from the sulfate by more than two atoms. The use of diglycolamine has been suggested [59] to obtain a stable sulfated product suitable for use in shampoos. Igepon B is a compound prepared from 3-hydroxy-1-aminobutane [35] and a fatty 1,3-propanolamide has also been patented [64].

V. OTHER SULFATED AMIDES

The potential stability of the amide linkage and the effective contributions to detergency made by the amide and sulfate groups has inspired other workers to prepare related compounds for use as detergents. An example is the product obtained from the reaction of a fatty acid or aliphatic sulfonyl chloride with tetrahydrofurfurylamine and sulfation of the resulting amide [35].

Amide and sulfate groups are sometimes used in a different order. Tetrahydrofurfuryl alcohol was reacted with dodecyl isocyanate and the resulting urethane was sulfated with an excess of sulfuric acid at 60-80° C [35]. Mehltretter [65] reacted fatty amines with d-gluconic acid and one or more of the free hydroxy groups were sulfated.

A new polymeric amide sulfate has been made [66] by the reaction of butadiene, urea, and sulfuric acid.

REFERENCES

1. W. Kritchevsky, U.S. Pat. 2,089,212 (1937).
2. B. W. Van E. Thieme, K. Akad. Wetenschappen, Amsterdam, 10, 855 (1909); Chem. Abstr., 4, 754 (1910).
3. Colgate-Palmolive-Peet Co., Fr. Pat. 810,847 (1937).
4. K. Brodersen and M. Quaedvlieg, Ger. Pat. 702,598 (1941), to I. G. Farbenindustrie Akt.-Ges.
5. F. W. Gray, U.S. Pat. 2,868,812 (1959), to Colgate-Palmolive Co.
6. T. P. Hilditch and J. G. Rigg, J. Chem. Soc., 1935, 1774.
7. M. H. Ittner, U.S. Pat. 2,474,740 (1949), to Colgate-Palmolive-Peet Co.
8. A. C. Bell and W. G. Alsop, U.S. Pat. 2,496,328 (1950), to Colgate-Palmolive-Peet Co.
9. C. J. Arrowsmith and J. Ross, U.S. Pat. 2,383,581 (1945), to Colgate-Palmolive-Peet Co.
10. N. H. Kuhrt, E. A. Welch, and F. J. Kovarik, J. Am. Oil Chemists' Soc., 27, 310 (1950).
11. A. K. Biswas and B. K. Mukherji, J. Phys. Chem., 64, 1 (1960).
12. G. H. Weinreich and J. L. Salvat, Afinidad, 23, 485 (1966); Chem. Abstr., 67, 11124 (1966).
13. A. I. Gebhart and J. E. Mitchell, U.S. Pat. 2,660,588 (1953), to Colgate-Palmolive-Peet Co.
14. F. W. Muncie, U.S. Pat. 2,242,979 (1941), to Colgate-Palmolive-Peet Co.
15. H. H. G. Jellinek and H. A. Anson, J. Soc. Chem. Ind., 68, 108 (1949).
16. V. P. Arida, F. C. Borlaza, and W. J. Schmitt, Philippine J. Sci., 94, 311 (1965); Chem. Abstr., 66, 106166 (1967).
17. W. F. Carson, U.S. Pat. 2,362,882 (1944), to Hercules Powder Co.
18. L. I. Slominskii and I. M. Kozhevnikova, Sbornik Statei Rabot. Ukrain. Nauch.-Issledovatel. Inst. Maslozhir. Prom., 1958 (3), 70-83 (Pub. 1960); Chem. Abstr., 56, 10309 (1962).
19. B. R. Harris, U.S. Pat. 2,023,387 (1935), to Colgate-Palmolive-Peet Co.

20. A. K. Biswas and B. K. Mukherji, J. Am. Oil Chemists' Soc., 37, 171 (1960).
21. Z. Jedlinski, Proc. 3rd Int. Congr. Surface Activity, Cologne, 1, 51 (1960) (Pub. 1961); Chem. Abstr., 57, 7405 (1962).
22. J. Ross, U.S. Pat. 2,731,422 (1956), to Colgate-Palmolive Co.
23. P. T. Vitale and M. E. Liftin, U.S. Pat. 2,733,213 (1956), to Colgate-Palmolive Co.
24. I. R. Schmolka, U.S. Pat. 2,945,816 (1960), to Colgate-Palmolive Co.
25. E. E. Dreger and A. C. Bell, U.S. Pat. 2,385,614 (1945), to Colgate-Palmolive-Peet Co.
26. J. A. V. Turck, Jr., U.S. Pat. 3,030,310 (1962).
27. R. M. Anstett, U.S. Pat. 3,076,766 (1963), to Colgate-Palmolive Co.
28. A. F. Anderson, U.S. Pat. 2,773,835 (1956), to Colgate-Palmolive Co.
29. K. R. Hansen, U.S. Pat. 3,001,949 (1957), to Colgate-Palmolive Co.
30. L. Wei, U.S. Pat. 3,341,460 (1967), to Colgate-Palmolive Co.
31. J. M. Allen and G. V. Scott, U.S. Pat. 2,879,231 (1959), to Colgate-Palmolive Co.
32. D. E. McCarthy, Belg. Pat. 662,976 (1965).
33. I. G. Farbenindustrie A.-G., Br. Pat. 499,144 (1939).
34. L. A. Mikeska, U.S. Pat. 2,293,265 (1942), to Standard Oil Development Co.
35. J. D. Brandner, W. H. Lockwood, R. H. Nagel, and K. L. Russell, PB Rept. 81819, FIAT Final Rept. No. 1141 (1947).
36. K. L. Russell and A. C. Bell, U.S. Pat. 2,235,534 (1941), to Colgate-Palmolive-Peet Co.
37. L. W. Burnette in Nonionic Surfactants (M. J. Schick, ed.), Chaps. 11 and 12, Marcel Dekker, Inc., New York, 1967, pp. 372-418.
38. E. Jungermann and D. Taber in Nonionic Surfactants (M. J. Schick, ed.), Chap. 8, Marcel Dekker, Inc., New York, 1967, pp. 208-239.
39. E. M. Meade, U.S. Pat. 2,464,094 (1949), to Lankro Chemicals Ltd.
40. Chimictechnic union chimique du nord et du Rhone (Soc. anon.), Fr. Pat. 979,000 (1951).
41. J. A. Monick, J. Am. Oil Chemists' Soc., 39, 213 (1962).
42. L. J. Garrison, J. H. Paslean, and M. S. Edmondson, Detergent Age, 5 (1), 27 (1968).
43. M. Naudet, E. Sombuc, P. Desnuelle, and G. Reutenauer, Bull. Soc. Chim. France, 1952, 476.
44. G. F. D'Alelio and E. E. Reid, J. Am. Chem. Soc., 59, 111 (1937).
45. J. K. Weil, N. Parris, and A. J. Stirton, J. Am. Oil Chemists' Soc., 47, 91 (1970).
46. E. A. Knaggs, Soap Chem. Specialties, 40 (12), 79 (1964).
47. C. Paquot, J. Rech. Centre Natl. Rech. Sci., 13, 169 (1950); Chem. Abstr., 45, 10617 (1951).
48. J. K. Weil, A. J. Stirton, and R. G. Bistline, J. Am. Oil Chemists' Soc., 31, 444 (1954).

49. J. K. Weil, N. Parris, and A. J. Stirton, J. Am. Oil Chemists' Soc., 48, 35 (1971).

50. A. Spada and E. Gavioli, Farm. Sci. Tech. (Pavia), 7, 441 (1952); Chem. Abstr., 47, 894 (1953).

51. J. K. Weil, A. J. Stirton, and E. B. Leardi, J. Am. Oil Chemists' Soc., 44, 522 (1967).

52. J. W. Orelup, Br. Pat. 450, 672 (1936).

53. H. Manneck, Seifen-Öle-Fette-Wachse, 82, 649 (1956).

54. K. R. Dutton and W. B. Reinisch, Mfg. Chemist, 28, 124 (1957).

55. E. T. Roe, J. M. Stutzman, and D. Swern, J. Am. Chem. Soc., 73, 3642 (1951).

56. J. K. Weil, A. J. Stirton, and A. N. Wrigley, 5th Int. Congr. Detergency, Barcelona, A/II, 45 (1968).

57. M. Demarcq and D. Dervichian, Bull. Soc. Chim. France, 12, 939 (1945).

58. K. R. Dutton and W. B. Reinisch, Mfg. Chemist, 28, 176 (1957).

58a. R. G. Bistline, W. R. Noble, J. K. Weil, and W. M. Linfield, J. Am. Oil Chemists' Soc., 49, 63 (1972).

58b. R. G. Bistline, W. R. Noble, and W. M. Linfield, J. Am. Oil Chemists' Soc., 50, 294 (1973).

59. V. Zorayam and G. Vanlerberghe, Br. Pat. 1,107,441 (1966); U.S. Pat. 3,562,170 (1971).

60. H. I. Bernstein and C. P. Fuchs, U.S. Pat. 2,588,197 (1952), to Emulsol Corp.

61. K. Hennig, Ger. Pat. 1,108,368 (1961), to Henkel & Cie.

62. E. W. Maurer, T. C. Cordon and A. J. Stirton, J. Am. Oil Chemists' Soc., 48, 163 (1971).

62a. T. D. Cordon, E. W. Maurer, and A. J. Stirton, J. Am. Oil Chemists' Soc., 49, 174 (1972).

63. P. Desnuelle and O. Micaelli, Bull. Soc. Chim. France, 1950, 671.

64. E. Plötz and K. Matschat, Ger. Pat. 1,072,251 (1958).

65. C. L. Mehltretter, M. S. Furry, R. L. Mellies, and J. C. Hankin, J. Am. Oil Chemists' Soc., 29, 202 (1952).

66. T. F. Rutledge, F. A. Hughes, T. J. Galvin, and J. D. Zech, J. Am. Oil Chemists' Soc., 44, 367 (1967).

CHAPTER 7

SULFATED FATS AND OILS

Bernard A. Dombrow[*]

Nopco Chemical Division
Diamond Shamrock Chemical Company
Morristown, New Jersey

[*]Present address: Shetland Drive, Lakewood, New Jersey.

I. INTRODUCTION

Basically, a sulfated oil is the reaction product of concentrated sulfuric
acid and a fatty oil (glyceride). The active component of sulfated fats and
oils is a salt of a half ester of sulfuric acid where the alcoholic moiety is a
hydroxy fatty acid which may or may not be unsaturated. Furthermore, the
carboxyl group may be free or esterified with glycerine, as in the case of
natural fats and oils, or with monohydric alcohols. Commercially, these
products are named according to the substrate from which they are derived.
Thus, sulfated oleic acid has as its main active component the salt of the
sulfate ester of 10-hydroxystearic acid. Another method [1] of differentiating
them from other sulfuric acid monoesters is to refer to them as having
"internally esterified sulfo groups" in contrast to the external type where
the sulfuric acid ester group is at the end of a hydrophobic chain. Some
of the latter can be found in Chapter 5 (Alcohol and Ether Sulfates) and
Chapter 6 (Sulfated Monoglycerides and Alkanolamides). Other compounds
such as sulfated allyl and methallyl esters [2] and ethers [3] of fatty
acids and alcohols, respectively, are not discussed elsewhere in this
volume. Hence, this chapter covers the products obtained by the sulfation of unsaturated and hydroxy fatty acids and natural fats and oils
containing their glycerides, as well as monohydric alcohol esters of these
acids.

Sulfated oils were the first nonsoap organic surfactants, dating back to
the turn of the last century. It has been claimed that Papillon, in 1790,
used a mixture of olive oil and sulfuric acid as a mordant [4]. However, in
1834, F. F. Runge prepared what he termed a "sulfuric acid oil" from the
same ingredients and demonstrated it to be superior to untreated olive oil
(then being used) as a mordant (see Ref. 1, pp. 321-322 for a verbatim ex-
cerpt). Experimentation continued along these lines based upon olive oil
until about 1875 when the sulfated castor oil was introduced as the first
commercial sulfated-type textile assistant [4]. Since this product was used
as a mordant for Turkey-red dyeing, it acquired the commercial name of
Turkey-red oil. Details of these studies, as well as those carried out before the 1930s when other surfactants appeared upon the market supplanting
sulfated oils, can be found in Refs. 1 and 4-6. During this 100-year period,
attempts were made to adapt these oils to the entire scope of surfactants.
Patents can be cited of uses ranging from wetting agents through detergents
to emulsifiers. Discussions, as well as bibliographies, may be found in
these references covering chemical studies of mechanisms of sulfation and
compositions. In recent times, only comparatively few studies may be
found in the literature. The bulk of the later work in this field appears to
cover attempts to use native fats and oils in sulfation processes in underde-
veloped countries.

In contrast to most anionic surfactants such as sulfated alcohols, sulfosuccinates, etc., which are usually available in fairly pure form, the sulfated fats and oils are mostly, if not entirely, represented by chemically highly impure materials. As a matter of fact, the sulfation process is usually carried out deliberatively in such a manner as to produce this type of mixture. The commercial process is more of an art, where the product is tailor-made for utilitarian end uses, rather than to produce a chemical. Because of this fact, it is found that, in general, at most about 30% of active sulfo ingredients are present in the final water-free organic portions. The entire complex can be considered to be a solution of a salt of the half ester of sulfuric acid dissolved in unreacted oil (or fat) fluidized by water to form a clear liquid. Other important components are soap and fatty acids. As water is usually present, inorganic salts are also found as impurities. It has been possible to purify these products to some extent. Thus, sulfated oils of higher strengths may be obtained by removing unreacted oil (in the broadest sense) by means of extraction and solvent partition using halogenated solvents [7] and alcohols and ketones [8]. Methods are also available for the removal of inorganic salts [9]. It must be recognized that these purified products still cannot be considered to be as chemically pure as would be sulfated alcohols or sulfosuccinates. Hence, a typical sulfated oil is a unique complex of sulfoesters, soap, water, fatty acids, and neutral oil tailored for an end use. They are often misnamed sulfonated oils.

During the period between 1928 to 1935 [10], many new surfactants were developed (isopropylnaphthalenesulfonate dates back to 1917), such as sulfated fatty alcohols, sulfated monoglycerides, ethylene oxide adducts, sulfonated alkylbenzenes and Igepon types. Before World War II, they were already supplanting the sulfated oils in many applications. Though the total production of surface active agents has steadily increased, that of sulfated oil remains constant.

Table 1 presents a good picture of the raw materials (substrates) used in the manufacture of sulfated oils. The drying oils are not represented here and of the semidrying variety only two, namely, soya bean oil, which is used in small quantities, and cod oil. The latter, though apparently in fair use, is being slowly phased out, since it cannot be used with chrome-tanned leather, but only with the vegetable-tanned type. As to coconut oil (with an iodine value quite low), it is doubtful that it is sulfated as such. A most interesting point is the large quantity of sperm oil utilized. It is the only liquid wax available and its derivative has shown much promise as an emulsification additive. Since the sperm whale has become a threatened species, whose killing is now banned by most Western nations, this usage is bound to decrease. Except for drying and semidrying oils, as well as oils with a very low iodine value, practically every natural fat and oil has been a successful candidate for the sulfation process.

TABLE 1

Surface–Active Agents: U.S. Production and Sales of Sulfated Esters,
Acids, and Oils, 1973

Compound	Production, 1,000 lb	Sales		
		Quantity, 1,000 lb	Value, 1,000 $	Unit value, per lb
Sulfuric acid esters (and salts thereof), total		222,817	57,559	0.26
Acids, amides, and esters, sulfated, total	--	16,385	4,620	0.28
Esters of sulfated oleic acid, total	4,759	4,761	1,631	0.34
Butyl oleate, sulfated, sodium salt	1,458	1,577	459	0.29
Propyl oleate, sulfated, sodium salt	517	515	161	0.31
All other	2,784	2,669	1,011	0.38
Tall oil, sulfated, sodium salt	3,492	3,372	680	0.20
Other acids, amides, and esters, sulfated	--	8,252	2,309	0.28
Natural fats and oils, sulfated, total	31,853	30,149	8,287	0.28
Castor oil, sulfated, sodium salt	5,920	5,664	2,707	0.48
Coconut oil, sulfated, sodium salt	871	774	280	0.36
Cod oil, sulfated, sodium salt	1,666	1,656	274	0.17
Herring oil, sulfated, sodium salt	690	688	139	0.20
Mixed fish oils, sulfated, sodium salt	4,023	3,728	747	0.20
Neatsfoot oil, sulfated, sodium salt	2,066	1,407	333	0.24
Ricebran oil, sulfated, sodium salt	9	9	2	0.22

TABLE 1 (Continued)

Compound	Production, 1,000 lb	Sales		
		Quantity, 1,000 lb	Value, 1,000 $	Unit value, per lb
Natural fats and oils, sulfated (Continued)				
Soybean oil, sulfated, sodium salt	614	566	101	0.18
Sperm oil, sulfated, sodium salt	778	688	184	0.27
Tallow, sulfated, sodium salt	5,860	5,869	1,014	0.17
All other	9,356	9,100	2,506	0.28

Source: U.S. International Trade Commission, Synthetic Organic Chemicals, 1973, Washington, D.C.: U.S. Govt. Print. Off., 1975.

II. PREPARATORY METHODS

There are two basic methods for the synthesis of sulfuric acid esters of natural fats and oils. The first one is based upon the addition reaction of sulfuric acid across a double bond of an unsaturated fatty acid chain (olefinic addition), as shown in Eq. (1):

$$CH_3-(CH_2)_7-\overset{\overset{\displaystyle H}{|}}{C}=\overset{\overset{\displaystyle H}{|}}{C}-(CH_2)_7-COOH \ + \ H_2SO_4 \longrightarrow$$

$$CH_3-(CH_2)_7-\overset{\overset{\displaystyle H}{|}}{\underset{\underset{\displaystyle OSO_3H}{|}}{C}}-(CH_2)_8-COOH \qquad (1)$$

The second method, namely the esterification of the hydroxy fatty acid chains by sulfuric acid, is illustrated by Eq. (2).

$$CH_3(CH_2)_5 \overset{\overset{\displaystyle H}{|}}{\underset{\underset{\displaystyle OH}{|}}{C}} - \overset{\overset{\displaystyle H}{|}}{\underset{\underset{\displaystyle H}{|}}{C}} - \overset{\overset{\displaystyle H}{|}}{C} = \overset{\overset{\displaystyle H}{|}}{C} - (CH_2)_7 - COOH + H_2SO_4 \rightleftharpoons$$

$$CH_3(CH_2)_5 \overset{\overset{\displaystyle H}{|}}{\underset{\underset{\displaystyle SO_3H}{|}}{\underset{O}{|}}{C}} - CH_2 - \overset{\overset{\displaystyle H}{|}}{C} = \overset{\overset{\displaystyle H}{|}}{C} - (CH_2)_7 - COOH + H_2O \qquad (2)$$

It should be noted that water is split off in this reaction, diluting the reagent sulfuric acid. These two methods will be discussed in the above order, using a typical general procedure each time as the basis for our discussions. Sunderland [15] mentions various practical sulfation methods which can be used as historical background (he used the misnomer of sulfonated oils).

A. Olefinic Addition

1. Procedure

In a typical run, 25 lb of 98% sulfuric acid is slowly added to 100 lb of olive oil at such a rate that the temperature range of the reaction mass can be kept below 25° C. Normally, an addition time of about one hour is to be expected. A total of two hours, including acid addition, is needed for the entire sulfation reaction. To remove the unreacted sulfuric acid, the reaction mass is washed by adding it to a 7-10% aqueous solution of sodium chloride at 40° C. On standing, the initially viscous opaque emulsion separates into two layers in approximately one hour. The lower aqueous layer is drawn off and discarded. Aqueous alkali, usually sodium hydroxide, is stirred into the upper layer until an end point of slightly alkaline to methyl orange is obtained. The oil is ready now for the panning stage. A water layer is allowed to bleed out of the mass while the latter is held at 40° C overnight or some other suitable period. The new upper layer is cleared by a further addition of alkali and adjusted for its ultimate use. As relatively viscous reaction masses are involved, good agitation is essential, especially in the first two stages.

2. Chemistry of Olefinic Addition

In the case of oleic acid, the indications are that the sulfo group is found at the carbon-10 position [11]. Theoretically [12], this would be the preferred addition point according to "Markovnikov's Rule." Much of the confusion on this point is most likely due to the raw material that is sulfated. Chemically speaking, commercial oleic acid is a not-too-pure grade of cis-9,10-octa-decenoic acid. It is believed that many of the side reactions cited in the old literature may be avoided by keeping the temperature of sulfation in the specified range [13]. In fact, the claim is made that at elevated temperatures (near that of boiling water) sulfonic acids are formed with oleic acid [14]. A comparatively short reaction time can be employed. The addition reaction is very fast, taking place almost instantaneously [15]. With good mixing, the reaction mass is ready for washing as soon as a good intimate mixture has been obtained. The viscosity of the reaction mass and agitation are the determining factors [16], as long as the temperature is kept low.

The concentration of sulfuric acid used in the above procedure is 98%, yielding an SO_3 content of the sulfated oil of 8-9% (which corresponds to a conversion to sulfate ester of roughly 50% based upon the sulfuric acid employed). If acid of lower strength is used, it has been found, that in order to obtain the same rate of conversion to sulfo esters, somewhat more acid is required than might be expected from a straight linear extrapolation (calculated on 100% strength base). Thus for 94% sulfuric acid, about 20% more acid is necessary. Indications are that acid strengths below 85% do not yield significant quantities of sulfate esters [17]. The studies in Ref. 17 are based upon a reaction temperature too high (90° C) and hence can only be used for derivative information. However, it is demonstrated that by an increase of acid strength, a definite and increasing formation of sulfonic acids occurs, most, if not all, probably due to the high reaction temperature. The use of more sulfuric acid than required by the molar ratios of the process has its limitations with regard to increasing the yield of sulfate esters. The indications from one study [18] are that even with 3 moles of sulfuric to one of oleic acid, over 15% of unreacted unsaturation was observed. Furthermore, increases in mineral acid content cause viscosities to build up, which may cause mixing problems.

One of the side reactions suppressed by addition of sulfuric acid at low temperature is hydrolysis of the carboxylic esters. No significant increase in acid value (organic) is found before the wash step. Some hydrolytic action can be found to occur during washing. However, this may be minimized by lowering the temperature to 25° C. The higher temperatures are used to speed up the separation into two layers of the wash emulsion. Sodium chloride is employed as the salting-out agent, in order to eliminate as much as possible the presence of sodium sulfate as an inorganic impurity. The latter tends to crystallize out with water of crystallization.

The panning stage is usually used to tailor-make the sulfated oil for its ultimate use. By controlling the alkalinity or mineral acidity of this operation, variations in acid value, soap content, etc., can be obtained. The end use determines the quantity of alkali for saponification. In order to produce clear products, some free acid must be developed. It has been found that if too low an acid value is attained, some of the carboxylic ester linkages must be saponified for this purpose. Such a combination of water, sulfate esters, soap, neutral oils, inorganic salts, etc., required adjustment for proper internal miscibility. Of course, these operations are limited by the fact that sulfate esters are also susceptible to mineral acid hydrolysis.

B. Esterification

1. Procedure

In a typical run, 25 lb of 98% sulfuric acid is added slowly to 100 lb of castor oil, holding the top temperature of the reaction mass at 20° C. After all the mineral acid has been added, the exotherm of the reaction can be permitted to raise the reaction-mass temperature to a maximum of 35° C. A longer total reaction time (including acid addition time) is employed here than in the previous method. The wash solution used is again aqueous sodium chloride. Both the wash and panning stages are similar to those of the olefinic addition sulfation. Again, the conditions of panning and adjusting the upper layer from that stage are dependent upon the final use for the sulfated oil in question.

2. Chemistry of Esterification

Equation (2) represents the general reaction of this procedure. The need for higher temperatures and longer reaction times over those of the olefinic addition method is due to the nature of the esterification reaction in the sulfation of castor oil. In addition, these conditions overcome the higher viscosities encountered in this process. Table 2 illustrates a 25% sulfation of castor oil with 92% sulfuric acid at 29.5° C.

TABLE 2

Sulfation of Castor Oil [19]

Minutes	SO_3 combined, %
30	1.02
75	1.62
120	2.14
165	2.59
210	2.98
255	3.29
300	3.54
345	3.71

In contrast to the rapid addition of sulfuric acid to a double bond, the esterification reaction is quite slow, especially with less concentrated sulfuric acid. Incidentally, Ref. 19 discloses that the use of mercury or its sulfate tends to enhance the sulfate esters content to 5.5% SO_3 after 345 minutes. Despite the large difference in reactivity between the double bond and hydroxyl group towards sulfuric acid, only the hydroxyl group is attacked significantly during sulfation. The iodine value of hydrolyzed sulfated castor oil fatty acids does not vary much, if any, from that of the original castor oil fatty acids [20]. The preservation of the double bond during the sulfation was noted very early in studies with castor oil [21]. A detailed analysis [22] of commercial turkey-red oil led one analyst to conclude that only about 5% of the total reaction had occurred at the double bond; hence it could be ignored. In a recent study [21], it was found that a similar condition exists when alkyl ricinoleates are sulfated with chlorosulfonic acid. In this case, even a 200% excess of reagent could not raise the yield over 80-90%. Many old literature studies and patents cover the blocking of hydroxy group before sulfation [1, 4-6].

Since water is liberated during the esterification and the reaction rate is slow even at somewhat elevated temperatures, it is not surprising to find that before the wash stage some hydrolysis takes place. Acid values of 25-30 may be encountered which is indicative of roughly 15% hydrolysis of the carboxylic ester. Otherwise, the comments in Sec. II,A as to the washing and panning steps hold for this type of sulfation also. A lower range (7-8%) of combined SO_3 is obtained, which is about one unit less than that for olefinic addition. The presence of water of reaction thus appears to reduce the efficiency of the sulfation. The data in Table 3 demonstrate that use of somewhat weaker sulfuric acid reduces the yield of combined SO_3 about in half. Hence the presence of water is more detrimental to the sulfation of castor oil than to that of a straight unsaturated natural oil such as olive oil.

III. GENERAL PROPERTIES

Even when we consider only those sulfated products which are obtained directly from the sulfation process, ignoring possible blends with raw oils, we are faced with a group of materials which are not homogeneous chemical compounds. Rather, we have complex mixtures of known chemical entities blended with compounds of uncertain chemical structures. Many variables complicate the picture for this class of anionic surfactants more than for any other class. To lay a base line for any discussion of these products, we look upon pure sulfated oleic acid, i.e., the monosodium salt of 10-sulfostearic acid, as a model of the principal active ingredient. This compound would have an SO_3 content of 20% and an acid value of 139. Commercially sulfated products are available in the SO_3 content range of 3-17% (based upon total solids). Since water is normally present, some of the acid must be converted to soap in order to obtain a pH high enough for hydrolytic stabilization of the sulfate group. Soap derived from the sulfo ester, as well as from unreacted fatty acid are present. Thus, the total amount of soap in the sulfated oils can be varied from a few percent to complete saponification. Similarly both the carboxylic acidity and the degree of esterification can be varied. Neutral oil is an important ingredient in these products and not merely an impurity. It aids in the blending process with raw oils, etc. Often the raw oil is added after panning just before adjustment. The interplay of the various constituents determines the utility of the final sulfated product [23].

Besides a minimum of the three variables mentioned above, two additional variables are also used to control composition. The first is the water content, which is obvious. But what is not apparent is that some products require a minimum amount of water content for good final water dispersion. The other is the type of alkali used for neutralization which can affect the blending characteristics. Sodium and potassium hydroxides are the most common. Ammonia has been used for special purposes. A combination of two alkalis can be used, especially where one of them is ammonium hydroxide. The fixed alkali is used to neutralize the sulfo group while the other (ammonia) may be used to finish, that is to react with the carboxylic groups. These sulfated products may be considered to be a class of complex materials designed by empirical means to perform certain commercial tasks.

A. Solubility in Aqueous Media

Since we are dealing with a very broad class of materials, we cannot expect to be too specific in our comments as to their solubilities in aqueous media. There are two solutizing groups present, hence it is the interaction of the sulfate groups and the soaps with the rest of the organic components which determines the type of emulsion obtained. As these products are in reality clear, self-emulsifiable oils, acid value and alkalinity must first be adjusted to produce homogeneous liquids. When these are subsequently stirred in water, a range of emulsions are obtained whose appearance varies from clear through translucency to varying degrees of milkiness to deep opacity. The emulsion stability can vary from infinite life (i.e., no oil separation for days) through creaming to free oil separation on standing for a period of time. All these have their utility. Thus, in some uses such as agricultural spray oils, water is used as a spreader for the emulsified oil but the oil must separate out on the foliage to prevent washoff. A general rule is that the more soap and sulfate esters can be blended clear in the sulfated product, the better the emulsion type and stability.

One of the early uses for sulfated castor oil was as a dye assistant in acid dye baths. This was made possible because it was found to dissolve clear in dilute acetic acid. In fact, a standard test requiring a clear stable solution in about 5.6% acetic acid was set up. An approximate SO_3 content of 7.0% or better was sufficient for this purpose. Higher SO_3 contents were found necessary to prevent salting out from dilute mineral acid solutions.

Thus, an SO_3 content of 13% enables a sulfated oleic acid to dissolve clear in 1% hydrochloric acid at room temperature [7]. The standard method for SO_3 determination is based upon hydrolysis in 1N-sulfuric acid. The products of higher SO_3 initially dissolve clear in the acid solution. Hydrochloric acid, both on a weight percentage or normality basis is a stronger salting out agent than sulfuric acid.

B. Hydrolytic Stability

Since the sulfate ester is a half ester of a dibasic mineral acid, it has a negative charge. Because of this factor, the sulfated oils have a certain amount of resistance to alkaline hydrolysis. However, care must be taken during saponification of the carboxylic esters (such as glycerides) to have good agitation. Localized overheating and localized alkali excess should be avoided. The picture as to mineral acid hydrolysis is quite different. The sulfate ester group is quite susceptible to mineral acid conditions, even for high SO_3 products which initially may be clearly soluble. It is only a matter of time and temperature before complete hydrolysis occurs. The sulfate ester group is so unstable that for good shelf life some soap must be present (alkaline to methyl orange). Usually a total alkali content of at least 0.5% (expressed as potassium hydroxide) is sufficient. It should be recognized that the hydrolysis is autocatalytic since sulfuric acid is split off. A hydroxy fatty acid is generated by this process.

IV. ANALYSIS

There are two good sources for the official methods of analysis for sulfated oils, the American Oil Chemists' Society (AOCS) and the American Society for Testing Materials D-500-55 (ASTM). The methods will not be discussed here in detail, but are discussed only in so far as they pertain to the meaning of that method in explaining the composition of sulfated oils. More detailed procedures may be found in Ref. 24. An interesting investigation is described in Ref. 25, in which the author separated several commercial sulfated oils into a number of fractions for a detailed study. Besides the confusion of nomenclature where "sulfate" and "sulfonate" are used interchangeably in the old literature, these sources indiscriminately report results on the oils on the "as is" basis or on dry solids as well as on the fat basis. The official methods use the "as is" basis; in our discussions, we shall use the dry-solids base as more appropriate for comparative studies.

A. Water

This determination is required to obtain the dry-solids content. Free water as well as water of crystallization of the sodium sulfate are determined.

Water has the function of producing fluid clear liquids which, as will be noted later, was found necessary for good emulsification of the various blends. (AOCS - F 1a-44; ASTM, Sec. 3-8.)

B. Sulfur Trioxide Content

As noted above, acid hydrolysis works excellently with sulfate esters liberating the esterified sulfuric acid which is easily titrated. If the SO_3 content is above 10%, the oil is more resistant and hence it is advisable to heat somewhat longer. (AOCS - F 2a-44; ASTM, Sec. 14-18.)

C. Inorganic Salts

This is a measure of chlorides and inorganic sulfates present. (AOCS - F6-44; ASTM, Sec. 41-45.)

D. Total Alkali

The alkali value is reported arbitrarily in terms of percentages of potassium hydroxide equivalents. It represents the amount of alkali used to react with the carboxylic groups to form soap but does not represent any neutralization of sulfate esters. (AOCS F 7-44; ASTM Sec. 46-48.)

E. Acid Value

This procedure is supposed to determine unreacted carboxylic acid groups. However, in this method, ammonia soaps and ammonia neutralized sulfate ester titrate as free acidity and correction must be made for this factor. (AOCS F 9a-44; ASTM Sec. 52-55.)

F. Unsaponifiables

This determination is usually run when mineral oil may be present. Of course, if the extract has a significant hydroxyl value, the presence of sperm oil may be demonstrated and corrections can be made in the analysis. (AOCS F5-44; ASTM Sec. 36-40.)

G. Neutral Fat

This comparatively simple method is not official but is often used to identify part of the total composition. The method as found in literature is applicable to sulfated oils other than castor oil [26]. It is based upon the extraction with petroleum ether of an aqueous alcoholic solution of neutralized sulfated oil. Thus, it represents all fatty material in the sulfated oil except soap, fatty acids, and sulfate esters. Hence, it may include many side-reaction products mentioned in the old literature as well as some unsaponifiable components. The latter are usually mineral oils. These may be further isolated by determining the unsaponifiable content of the extract. In the case of castor oil [27] which is insoluble in petroleum ether, the extraction is run by dissolving the sulfated oil in water, lowering the pH to the methyl orange end point and extracting with diethyl ether. A correction is made on the extract by allowing for the fatty acid content.

V. APPLICATIONS

With the advent of the newer surfactants in the 1930s, many uses for which sulfated oils were being specially tailored, were taken over by these new products. Most of the early research and patent work on sulfated oils was shunted aside. At present, emulsification appears to be fundamentally the most important surface activity remaining for these products.

A. Emulsifiers

Self-emulsifiable oils represent the largest group of applications for sulfated oils. Commercial products are available for solubilizing many water-immiscible liquids. These clear blends can be found to disperse in water, without need for mechanical means such as homogenizers, to form emulsions which may vary in clarity from clear through translucence and opalescence to opacity with varying degrees of emulsion stability.

1. Nonfatty Liquids

Specific emulsifiers may be purchased which are miscible with either mineral oils, pine oil, or chlorinated solvents, etc., up to the ratio of 1 to 4. These blends can be stirred into water-forming stable emulsions whose opacity depends upon the above ratio. The entire range of mineral oils up to and including paraffin wax [28] have been solubilized. Water is an important ingredient in the final blend. It must be present in sufficient amounts for proper hydration [28] but too much will interfere with the miscibility of an emulsifier with the water-insoluble liquid in question. The sulfated oil is usually adjusted to the individual mineral oil; the source, fraction, etc., have a bearing upon blendability and subsequent emulsification. These liquids are used as cutting oils for other metal working compositions and as oil sprays for insecticides, etc., in the form of aqueous emulsions. In the case of textile oils, they can be employed also neat, being easily removable by water rinses. A self-emulsifiable paraffin wax has been used as a paper size [28].

Liquids containing pine oil in concentrations up to 80% have found use in cleaners, wetting agents, and disinfectants. Similarly chlorinated solvents, for example orthodichlorobenzene, have been employed as degreasers, drain cleaners, etc.

2. Fats and Oils

Sulfated oils can be blended clear with raw fats and oils to yield self-emulsifiable liquids. However, a lower carrying power is found for this group; roughly a ratio of 1.5 is the most fatty oil that can be emulsified in this manner. Perhaps, the most important application for these materials is in spinning oils for textiles. The yarns may be oiled either neat or from an aqueous emulsion. After processing, the resulting fabric may be rinsed free of oil retaining a good hand (softness). Of course, mineral oils may also be blended into these textile oils if desirable.

Leather after the tanning operations and before drying must have so-called fat liquor oils incorporated into it. Otherwise, on drying, the internal fibers cement together producing a stiff, weak, and harsh material. When the oils are present, a supple leather results with good tensile strength, good stitch strength, etc. Sulfated oils are excellent for aiding in the penetration of raw oils into the interior of the leather. The degree of penetration may be controlled by the sulfation process as well as by blending raw oils with sulfated products. Initially, sulfated oils displaced competing products because of their reproducibility and their ease of handling. In the case of chrome-tanned leather, an extra advantage is found in the fact that sulfated oils, being anionic, are fixed by the chrome tan inside the leather. This tends to prvent oil exudation to the surface of the leather.

B. Wetting Characteristics

The classical use for Turkey-red oil was as a dye assistant for turkey red (alizarin) helping to produce a level color. Of course, the use of sulfated oils expanded from this dye to other classes of dyes, especially acid dyes where their tolerance to dilute acetic acid permitted their use. The action of these oils in the dye batch is not a simple one. At least, three different functions have been ascribed to them [29]. They reduce the surface tension, enabling water, which is the vehicle for the dyestuff, to penetrate the yarn better. They act as retarders (or inhibitors) competing with the dyestuffs for dye sites followed by being gradually displaced by the dye. Finally, they dissociate the dye aggregates so that single molecules tend to reach the dye site. All these activities aid in the production of a level color on the textile. Rewetting represents another important application for the textile industry. It is important in textile operations that when a dry cloth is passed into another aqueous operation, the fabric is wetted out quickly. Thus, even though a dye batch may contain wetting agents, retardation of wetting as the cloth is fed into the bath, could result in uneven color. If a coating of a latex or some other aqueous solution is applied to a cloth, good rewetting is required for uniformity of application. An added advantage in all of these operations is usually a soft feel (hand) for the treated textile.

C. Miscellaneous

Besides the above applications, there are several which do not fall clearly into the above classes. Over the years, many uses have been developed, in some of these the newer surfactants may work better. Economic factors, ease of manufacture of sulfated oils as well as technical inertia have had the combined effect to keep the sulfated oils on the market. In the cosmetic industry, latherless shampoos, neutral hand soaps (of pH close to 7), water-soluble perfumes, etc., are a few of these applications. Sulfated oils exhibit promise as plasticizers for glue. In addition to their being employed as general wetting agents, they are good dispersion aids for powders. One big advantage that occurs in many applications appears to be a general softness conferred upon the substrates employed.

D. Commercial Products

Tables 3 and 4, listing sulfated products with their manufacturers, were compiled from McCutcheon's Detergents and Emulsifiers, 1969 Annual [30]. They provide a list of trade names (trademarks) and manufacturers active in this field, as well as available products.

TABLE 3

Commercial Sulfated Fatty Esters

Trade name	Manufacturer
Ahcowet RS	ICI America, Inc.
Amawet E	American Aniline Products, Inc.
Atlasol P	Atlas Refining, Inc.
Avirol Supra	Gardinol Div., Albright & Wilson (Australia), Ltd.
Calsolene Oil HSA	ICI America, Inc.
Chemkal 65, 70	Synthron, Inc.
Cinwet BOS	Cindet Chemical, Inc.
Detersol T-1621	Finetex, Inc.
Duofol AS, L	Hart Products Corp.

TABLE 3 (Continued)

Trade name	Manufacturer
Dynesol LD	Amalgamated Chemical Corp.
Emkafol D, OT	Emkay Chemical Co.
Gaftex 288	GAF Corporation, Textile Chemical Div.
Glycopen N	Arkansas Co., Inc.
Kara Wet 70	Refined—Onyx Div., Millmaster Onyx Corp.
Lebcol Series	Branchflower Co.
Marvanol SP	Marlowe—Van Loan Corp.
Morowet 3	Moretex Chemical Products
Nopco 2272R	Nopco Chemical Div., Diamond Shamrock Chemical Co.
Procastol	Soluol Chemical Co., Inc.
Sanfodex	Dexter Chemical Corp.
Sanol 50	Bryant Chemical Corp.
SD-71	Laurel Products Corp.
Stantex 322	Standard Chemical Products, Inc.
Surfax 1410, WO	E. F. Houghton & Co.
Tetranol	Arkansas Co., Inc.
Textrapen	Tex-Chem. Co., Inc.
Tri-A-Nol FW	Scholler Brothers, Inc.
Wixol 8091	Wica Chemical, Inc.

TABLE 4

Commercial Sulfated[a] Oils and Fatty Acids

Trade name	Manufacturer	Raw material[b]
Ahcol Series	ICI America, Inc.	Various
Aquasol Series	American Cyanamid Co., Industrial Chemicals Div.	Castor oil
Arizona Surface Active Agent 351, 352	Arizona Chemical Co.	Tall-oil fatty acids

TABLE 4 (Continued)

Trade name	Manufacture	Raw material[b]
Atlasol G	Atlas Refinery, Inc.	Glyceryltrioleate
Avirol 130	Standard Chemical Products, Inc.	Castor oil
Avirol T	Gardinol Div., Albright & Wilson (Australia), Ltd.	Castor oil
Calsolene Oil HS	ICI (England)	--
Chemoil 412	Standard Chemical Co.	Castor oil
Concental 75	ICI America, Inc.	Tallow
Depcolevel JDS	DePaul Chemical Co.	--
Emka Finishing Oil	Emkay Chemical Co.	Vegetable oil
Eureka 102-H	Atlas Refinery, Inc.	Rincinoleic acid
Eureka 392	Atlas Refinery, Inc.	Vegetable fatty acids
Laurel Textile Oil	Laurel Products Corp.	Castor oil
Mello-Neats K	Kehew-Bradley Co.	Animal fats
Monopol Oil 48	GAF Corporation, Dyestuff & Chemical Div.	Castor oil
Monopole Oil	Nopco Chemical Div., Diamond Shamrock Chemical Co.	Castor oil
Monosulph	Nopco Chemical Div., Diamond Shamrock Chemical Co.	Castor oil
Morofin SFA	Morotex Chemical Products	Fatty acid
Nopco 1408	Nopco Chemical Div., Diamond Shamrock Chemical Co.	Castor oil
Nopco 1471	Nopco Chemical Div., Diamond Shamrock Chemical Co.	Vegetable oil
Para Soap	Scholler Brothers, Inc.	Animal and vegetable oil blends
Pli-A-Tex	Scholler Brothers, Inc.	Long-chain fatty acid glyceride
Prestabit Oil V	GAF Corporation, Dyestuff & Chemical Div.	Fatty acids

TABLE 4 (Continued)

Trade name	Manufacturer	Raw material[b]
Sulfated Castor Oil V63	Hart Products Corp.	Castor oil
Sulphol	Kehew–Bradley Co.	Cod oil
Supratol VF	Hart Products Corp.	Castor oil
Titan Castor No. 75	Titan Chemical Products, Inc.	Castor oil
Trisco Oil No. 894	Scholler Brothers, Inc.	Castor oil
Trisulphoil XX	Scholler Brothers, Inc.	Castor oil

[a]Some of these products are sometimes called sulfonated oils rather than sulfated oils. In the author's opinion, Tables 3 and 4 contain only sulfated oils.

[b]Fatty oil or fatty acid, as indicated in McCutcheon's Annual 1969.

REFERENCES

1. K. Linder, in Chemie und Technologie der Fette, Vol. VI (H. Schonfeld and G. Hefter, eds.), Julius Springer Verlag, Wien, 1937, p. 317.
2. D. Price and R. Kapp, U.S. Pat. 2,341,060 (1944), to National Oil Products Co.
3. D. Price and B. Dombrow, U.S. Pat. 2,241,421 (1941), to National Oil Products Co.
4. J. P. Sisley, Am. Dyestuff Reptr., 43, 741 (1954).
5. W. Herbig, Die Öle und Fette in der Textil Industrie, 2nd ed., Wissenschaftliche Verlagsgesellschaft m.b.H., Stuttgart, 1929, pp. 249–322.
6. A. M. Schwartz and J. W. Perry, Surface Active Agents, Interscience Publishers, Inc., New York, 1949, Vol. I, pp. 44–52; Vol. II, pp. 54–56.
7. B. Dombrow and R. Beach, U.S. Pat. 2,203,524 (1940), to National Oil Products Co.
8. B. Dombrow, U.S. Pat. 2,280,118 (1942), to National Oil Products Co.
9. R. Kapp, L. J. Mosch, and E. T. Woods, U.S. Pat. 2,285,337 (1942), to National Oil Products Co.
10. B. Blaser, Vortrage in Originalfassung, III, Intern. Kongr., Grenzflächenaktive Stoffe, Köhn 1960, Band I, Verlag der Universitatsdruckerei, Mainz, p. 35.
11. M. De Groote, B. Keiser, A. F. Wirtel, and L. J. Monson, Ind. Eng. Chem. Anal. Ed., 3, 244 (1931).

12. N. Isenberg and M. Grdinic, J. Chem. Educ., 46, 602 (1969).
13. T. N. Mehta and B. Y. Rao, J. Indian Chem. Soc. Ind. News Ed., 18, 195 (1955).
14. Reference 11, p. 247.
15a. A. E. Sunderland, Soap, 11, 63-64 (Oct. 1935).
15b. A. E. Sunderland, Soap, 11, 63-64 (Nov. 1935).
16. R. H. Trask, J. Am. Oil Chemists' Soc., 33, 570 (1956).
17. Reference 13, p. 196.
18. E. T. Roe, B. B. Schaeffer, J. A. Dixon, and W. C. Ault, J. Am. Oil Chemists' Soc., 24, 45 (1947).
19. S. Rangarajan and N. P. Palaniappan, Ind. Eng. Chem., 50, 1787 (1958).
20. P. Soderdahl, J. Am. Oil Chemists' Soc., 24, 70 (1947); Reference 5, p. 273.
21. H. Bertsch, H. Reinheckel, and G. Czichouki, Fette, Seifen, Anstrich-mittel, 67, 677-682 (1965).
22. R. M. Koppenhoefer, J. Am. Leather Chemists' Assoc., 34, 627 (1930).
23. A. E. Sunderland, Soap, 11, 67 (Dec. 1935).
24. M. J. Rosen and H. A. Goldsmith, Systematic Analysis of Surface-Active Agents, in Chemical Analysis (P. J. Elving and I. M. Kolthoff, eds.), Vol. XII, Interscience Publishers Inc., New York, 1960.
25. R. M. Koppenhoefer, J. Am. Leather Chemists' Assoc., 34, 622-639 (1939).
26. A. E. Sunderland, Soap, 11, 68 (Dec. 1935).
27. A. DeCastro, Private communication.
28. R. M. Cobb, D. S. Chamberlin, and B. A. Dombrow, Paper Trade J., 97 (10), 35-38 (1933).
29. E. R. Trotman, Dyeing and Chemical Technology of Textile Fibres, 3rd ed., Griffen, London, 1964, p. 355.
30. McCutcheon's Detergents and Emulsifiers, 1969 Annual, John W. McCutcheon, Inc., Morristown, New Jersey.

CHAPTER 8

ALKYLARYLSULFONATES

George C. Feighner[*]

Petrochemical Department
Continental Oil Company
Saddle Brook, New Jersey

[*]Present affiliation: Scientific Services, Franklin Lakes, New Jersey.

253

I. INTRODUCTION

A. Scope

This chapter on alkylarylsulfonates includes recent developments in the
field with emphasis on the surfactant aspects. Effects of structure on properties, applications, and manufacture are the central theme. Since the recent history of anionic surfactants includes the switch from nonbiodegradable
to biodegradable surfactants, much of the subject matter is related to linear
alkylbenzene sulfonates.

To better define the topic, alkylarylsulfonates, it is necessary to re-
member that we are discussing primarily anionic surfactants. Since most
applications are in aqueous solutions we shall consider not only chemicals
like dodecylbenzenesulfonates, but hydrotropes and coupling agents like the
xylenesulfonates. Some consideration will be given to the oil-soluble alkylarylsulfonates of high molecular weight when emulsifiers are discussed;
however, the truly oil-soluble sulfonate surfactants are the subject of a
separate chapter of this book.

B. Definition

Alkylarylsulfonate molecules consist of three functional parts. The aryl
group serves to join a nonpolar lipophyllic alkyl group to the ionic hydro-
phyllic sulfonate radical. The most common aryl group is derived from benzene, but naphthalene, diphenyl ether, and other aromatic structures may
be employed. The size of the alkyl group and configuration provide the
greatest source of variation of surfactant properties. The aromatic ring
may be substituted with more than one alkyl group which may be short or
long; branched, linear, or cyclic; and may even contain hereto atoms or
functional groups. The sulfonate group is inserted into the molecule in commercial practice by sulfonation with sulfuric acid, sulfur trioxide, or mixtures thereof. Since alkylarylsulfonate surfactant are salts, variation from
this part of the molecule depends on use of different organic and inorganic
cations. Sodium and ammonium salts are used for water-soluble products;
but, where surface activity involves solid or second-liquid interfaces, relatively covalent salts of amines and polyvalent cations are used.

C. History

Probably the earliest manufacture and usage of alkylarylsulfonate came
about as a result of by-products of sulfuric acid treatment of petroleum
fractions. The so-called "natural sulfonates" result from sulfonation of
relatively complex alkyl aromatics found in petroleum, shale oil, coal-tar

fractions, and other naturally occurring hydrocarbons [1-6]. As usage of these crude products increased, synthetic materials were developed. However, the natural sulfonates are still important today and are much improved in quality. They are also manufactured selectively for specialty applications. In some cases natural sulfonates have proceeded from by-product status to the primary product of manufacture.

Alkylation of naphthalene and conversion to the sulfonic salts have been practiced for many decades. Today, the most important naphthalenesulfonates have relatively short alkyl groups and are often complex molecules with more than one alkyl substituent and crosslinks derived from formaldehyde. At the early stage of development, anionic surfactants were primarily used for industrial applications. The advent of the kerylbenzenesulfonates started the usage of these chemicals in household products.

Kerylbenzenesulfonates were developed in the late 1930s. Although most usage was in textile and other industrial applications, household formulations were soon made. Kerylbenzenesulfonates were synthesized by alkylating benzene with chlorinated kerosene and then sulfonating the resulting alkylbenzene [7]. These derivatives of kerylbenzene probably represent the first commercial product called by the generic name alkylarylsulfonate.

After World War II, a greatly improved alkylarylsulfonate appeared. This was ABS or dodecylbenzenesulfonate. It was derived from propylene tetramer which was a by-product of gasoline manufactured by the process of propylene polymerization. Because of the ability of petrochemical companies to convert propylene tetramer to high quality dodecylbenzene in large volumes at low prices, a revolution occurred in the soap industry. Synthetic detergents were rapidly substituted for soap. Dodecylbenzenesulfonates quickly became the largest organic surfactant used in the United States.

Although ABS and its variations were excellent surfactants, they had one serious drawback. Their rate of biodegradation in surface waters and sewage treatment plants was slow and incomplete. To solve this problem the detergent industry in the early 1960s converted to linear alkylbenzenesulfonate or LAS, which has since proved to have adequate biodegradability. The detergent-foam problem we heard so much about in the 1950s has been substantially eliminated.

In the preceding paragraphs I have outlined the development of the largest and best known alkylarylsulfonate, but numerous other compounds are available and are used in a multitude of fashions. The short-chain alkylarylsulfonates, described in the last section of this chapter, function as hydrotropes, couplers, and anticaking agents. Higher-molecular-weight products are available for emulsification. Corrosion inhibition, crystal modification, foaming, improvement of electroplating and numerous other useful functions are effected by alkylarylsulfonates.

II. MANUFACTURE OF ALKYLARYLSULFONATES

A. Production of Alkylating Agents

Since processes for manufacturing alkylbenzenes and sulfonation to produce
surfactants are covered in other chapters, only sufficient discussion of
these topics for understanding the properties and applications of alkylarylsulfonates will be covered here. The chemistry involved in producing alkylating agents and condensing these with the aryl compound are important to
the properties of the final surfactant.

A large variety of alkylating agents is available for the synthesis of
alkylbenzenes. Also, the various catalysts and processes used for alkylation of aromatic rings can cause variations in the alkylbenzene. Olefin-
alkylating agents constitute the largest source of alkyl groups. Although
recent emphasis has been upon production of biodegradable linear alkylbenzenesulfonates [8], many branched and cyclic olefins have also been used
for alkylation. Propylene tetramer is the best known branched olefin alkylating agent [9-13]. The propylene tetramer of commerce is a very complex
mixture which probably contains none of the expected head-to-tail polymers.
Rather, it is a mixture of olefins of almost every carbon chain lengths,
many branched isomers and a variety of carbon-carbon double-bond positions. Structures [1], [2], and [3] are typical.

$$
\begin{array}{ccc}
\underset{|}{R}\ \ \underset{|}{R} & \underset{|}{R} & \underset{|}{CH_3}\ \ \underset{|}{CH_3} \\
R-C\!=\!C-H & R-C\!=\!CH_2 & R-C-CH\!=\!C-R \\
 & & \underset{CH_3}{|}
\end{array}
$$

$$
\quad [1] \qquad\qquad [2] \qquad\qquad\qquad [3]
$$

Isobutene polymers have also received extensive study because of their
low cost [15]. However, the presence of a quaternary carbon atom adjacent
to the double bond leads to extreme fragmentation and production of low-
molecular-weight alkylbenzene by-products [Eq. (1)].

(1)

Other branched olefins that have received serious study are the dimers and trimers [16] of linear α-olefins such as hexene-1 [17-20]. Dimerization with aluminum alkyl and similar catalysts give only a minimum of branching [Eq. (2)]. Alkylbenzenesulfonates derived from these olefins are much more biodegradable than those obtained from propylene tetramer.

$$CH_3(CH_2)_3 CH=CH_2 \longrightarrow CH_3(CH_2)_3 - \underset{\underset{CH_2}{\|}}{C} - (CH_2)_5 CH_3$$

$$CH_3(CH_2)_3 - \underset{\underset{\bigcirc}{|}}{\overset{\overset{CH_3}{|}}{C}} - (CH_2)_5 - CH_3$$

(2)

Thermal cracking of petroleum [18-21], shale oil [22, 23], and lignite produce olefinic streams. These have substantial branching as well as other impurities and produce alkylbenzenesulfonates with inferior biodegradability. The Fischer-Tropsch process yields an olefin-paraffin stream which is fairly linear. This stream is used in Germany to produce linear alkylbenzene (LAB) [24].

The thermal cracking of purified petroleum wax can result in high purity, linear α-olefins. These olefins can be converted to highly biodegradable alkylarylsulfonates. However, removal of branched and cyclic components of wax requires expensive equipment for urea adduction, and other sources of linear alkylating agents compete with the wax cracking process.

High purity α-olefins are obtained by ethylene polymerization using aluminum alkyls in the Ziegler process. Although the purity of these olefins is excellent, too broad a homolog range results; more competitive processes for linear alkylating agents are available.

In addition to thermal cracking of the normal paraffins in petroleum wax, methods of converting normal paraffins to olefin alkylating agents have been adopted. The high-purity normal paraffins obtained by molecular-sieve separation of kerosene fractions are the materials of preference. Chlorination followed by dehydrochlorination is a commercial process [25]. Recently, dehydrogenation catalysts have been developed for directly converting normal paraffins to linear olefins [26-28]. Oxidation to secondary alcohols followed by dehydration also produces linear olefins [29, 30], but purification costs and low yields have prevented commercialization of this process.

Under certain conditions, the Fischer-Tropsch process produces fairly linear olefins and these have provided a route to linear alkylate sulfonate (LAS) in Europe and South Africa.

A historic process of introducing reactivity into paraffinic molecules is chlorination. This process was used to produce keryl benzene, the original alkylarylsulfonate surfactant, and is widely employed today for the synthesis of LAS. However, there is a difference between the chemistry of the chlorination of kerosene and that of normal paraffins. Kerosene contains aromatics [31], cycloalkanes, isoparaffins, n-paraffins, and nonhydrocarbons. The chlorination-alkylation sequence with n-paraffins leads to secondary phenylalkanes and small amounts of the 1-phenylalkanes. On the other hand with kerosene, refractory aromatic chlorides, cycloalkylbenzenes, isoalkylbenzenes, diphenylalkanes, tetralins, indans, dehydrogenation products of cycloalkanes, and mixed structures are also formed. In one example only 23.5% of linear products were obtained from chlorinated kerosene [32].

The most widely used process for making linear alkylbenzene employs chlorination of n-paraffins followed by condensation with benzene catalyzed by $AlCl_3$. The n-paraffins are chlorinated to a low degree of conversion per pass to minimize undesirable di- and polychlorides. Usually the chlorides are condensed with benzene directly [33, 34], but at least one company dehydrohalogenates and uses the olefin for alkylation [35].

A modification of propylene tetramer to produce a more biodegradable product involved hydrocarbonylation [36]. The idea was to introduce a linear linkage which would be subject to degradation by bacterial processes.

B. Production of Alkylate

Although the choice of the alkylating agent determines to a large extent the molecular composition of the resulting linear alkylbenzenesulfonate, the alkylation reaction results in a considerable number of possible variations. The chemistry of production of linear alkylbenzenesulfonates depends on both the alkylating agent and the catalyst used for the condensation. Table 1 shows the different results obtained with $AlCl_3$ and HF [39] on dodecene-1.

TABLE 1

Alkylation of Benzene with Dodecene-1 with Different Catalysts

Catalyst	Isomer distribution, %					
	1-phenyl	2-phenyl	3-phenyl	4-phenyl	5-phenyl	6-phenyl
$AlCl_3$[a]	0	32	22	16	15	15
HF[a]	0	20	17	16	23	24
H_2SO_4, 98%	0	41	20	13	13	13

[a]Anhydrous.

The rate of alkylation and heat of reaction is greater with olefins than with alkyl chlorides [524]. With linear olefins use of anhydrous fluoride gives a wide distribution of secondary phenylalkanes favoring the internal isomers. Aluminum chloride, on the other hand, causes formation of a larger amount of the 2-phenylalkane isomer than would be expected from a random attack. Considerable industrial research was devoted to understanding and controlling isomer distribution because there are measurable differences in surfactant properties between the internal (3-, 4-, 5-, 6-, and 7-phenylalkane) and external (1- and 2-phenylalkane) isomers of linear alkylbenzenesulfonates [37-40, 527].

With an α-olefin, under mild alkylation conditions and with aluminum chloride catalyst, high concentrations of 2-phenylalkanes are obtained. Equilibration with aluminum chloride results in a higher than statistical amount of 2-phenyl isomer. Internal or α-olefins give higher than statistical amounts of internal isomers with HF. However, if HF is contacted with linear alkylbenzene at 100-150°C, the isomer distribution is typical of that obtained with $AlCl_3$.

Two methods have been used to control isomer distribution of LAB. The choice of catalyst is most important. Aluminum chloride and chloro-paraffins give increased amounts of 2-phenylalkane isomers. Use of HF and olefins favors the internal phenylalkane isomers. The other method is to separate the lower-boiling internal isomers from the higher-boiling external ones by fractional distillation. The unwanted fraction is isomerized to the equilibrium mixture which is fractionated again [38].

Since most commercial alkylating agents also contain small amounts of difunctional compounds and branched structures, other reactions occur. Some of these are shown in Eqs. (3) to (5).

$$\text{(3)}$$

DIPHENYLALKANE   DIALKYLTETRALIN

$$\text{(4)}$$

$$\text{(5)}$$

The chemistry of branched alkylbenzenesulfonates is much less understood than that of the linear products. Because propylene tetramer is such a complex mixture, its chemistry is understood only in general terms. Some of the salient features are outlined in Eqs. (6) to (9).

ALKYLATION

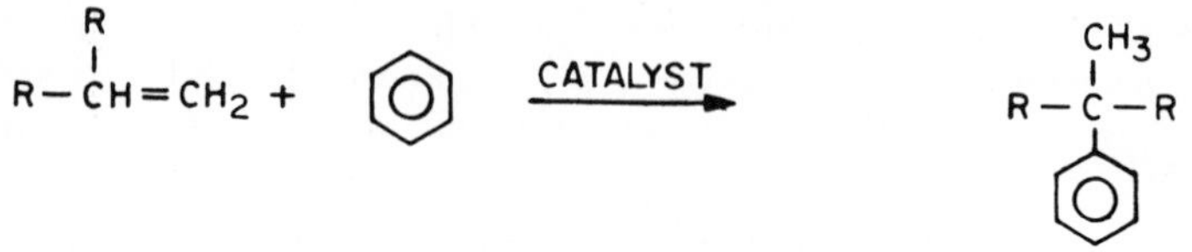

$$R-\overset{\overset{\textstyle R}{|}}{C}H=CH_2 + \bigcirc \xrightarrow{\text{CATALYST}} R-\overset{\overset{\textstyle CH_3}{|}}{\underset{\bigcirc}{C}}-R \qquad (6)$$

FRAGMENTATION

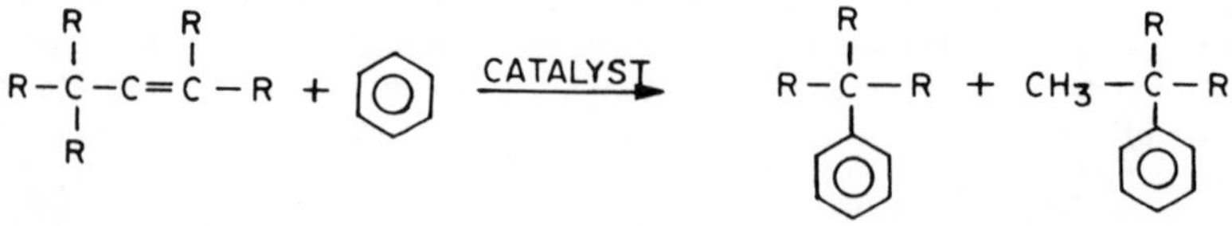

$$R-\overset{\overset{\textstyle R}{|}}{\underset{\underset{\textstyle R}{|}}{C}}-C=\overset{\overset{\textstyle R}{|}}{C}-R + \bigcirc \xrightarrow{\text{CATALYST}} R-\overset{\overset{\textstyle R}{|}}{\underset{\bigcirc}{C}}-R + CH_3-\overset{\overset{\textstyle R}{|}}{\underset{\bigcirc}{C}}-R \qquad (7)$$

DIMERIZATION – ALKYLATION

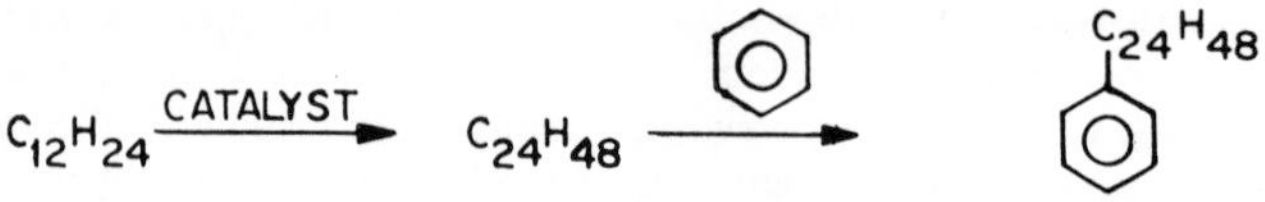

$$C_{12}H_{24} \xrightarrow{\text{CATALYST}} C_{24}H_{48} \quad \bigcirc \longrightarrow \overset{C_{24}H_{48}}{\bigcirc} \qquad (8)$$

FRAGMENTATION– ADDITION– ALKYLATION

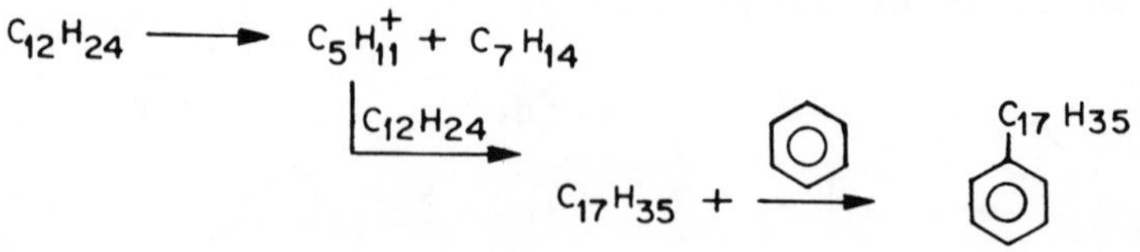

$$C_{12}H_{24} \longrightarrow C_5H_{11}^+ + C_7H_{14}$$
$$\quad\quad\quad\quad \downarrow C_{12}H_{24}$$
$$\quad\quad\quad\quad C_{17}H_{35} + \bigcirc \longrightarrow \overset{C_{17}H_{35}}{\bigcirc} \qquad (9)$$

Equation (6) schematically depicts the synthesis of branched "dodecylbenzene." The alkyl groups are highly branched, with multiple methyl groups. The total carbon number of the dodecyl group in a typical branched detergent alkylate contains 10, 11, 12, 13, 14, and 15 carbon atoms, but averages C_{12} or a little less.

Use of olefins resulting from dimerizationof linear α-olefins is much more straightforward. Outlined in Eqs. (10 and (11) are the reactions involved in converting hexene-1 to alkylbenzene. The Phillips process produces linear internal olefins by catalytic elimination of ethylene over NiO and the chemistry of alkylation with these olefins is without complication [520].

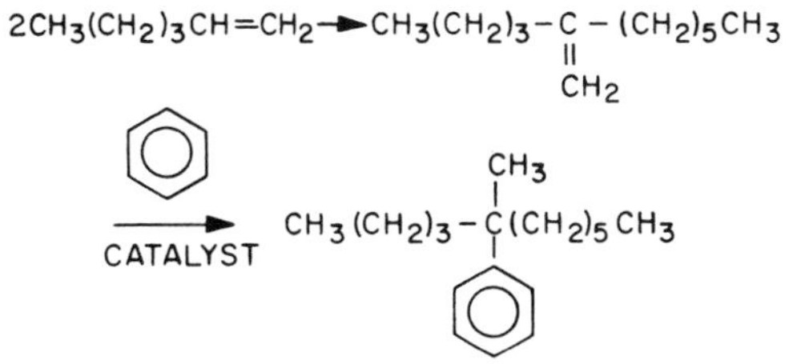

$$2CH_3(CH_2)_3CH=CH_2 \longrightarrow CH_3(CH_2)_3-\underset{\underset{CH_2}{\|}}{C}-(CH_2)_5CH_3$$

(10)

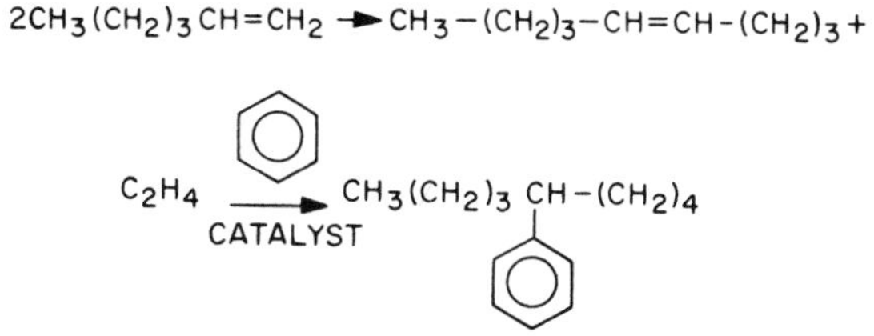

$$\xrightarrow[\text{CATALYST}]{} \quad CH_3(CH_2)_3-\underset{\underset{\bigcirc}{|}}{\overset{\overset{CH_3}{|}}{C}}(CH_2)_5CH_3$$

$$2CH_3(CH_2)_3CH=CH_2 \longrightarrow CH_3-(CH_2)_3-CH=CH-(CH_2)_3+$$

(11)

$$C_2H_4 \xrightarrow[\text{CATALYST}]{\bigcirc} CH_3(CH_2)_3 \underset{\bigcirc}{CH}-(CH_2)_4$$

When impure feedstocks such as kerosene for chlorination, olefins from cracking of petroleum fractions, or lignite carbonization products are used, the chemistry is so complicated that one must use his imagination to count the possibilities. Some structures which are important to the properties of the resultant alkylbenzene sulfonates are outlined in the Eqs. (12) to (14).

AROMATIC RING CHLORINATION

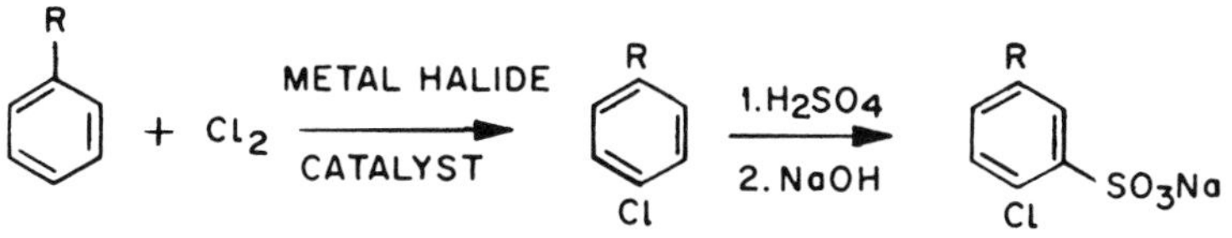

$$\underset{}{\overset{R}{\bigcirc}} + Cl_2 \xrightarrow[\text{CATALYST}]{\text{METAL HALIDE}} \underset{Cl}{\overset{R}{\bigcirc}} \xrightarrow[\text{2. NaOH}]{\text{1. H}_2\text{SO}_4} \underset{Cl}{\overset{R}{\bigcirc}}SO_3Na$$

(12)

POLYCHLORINATION

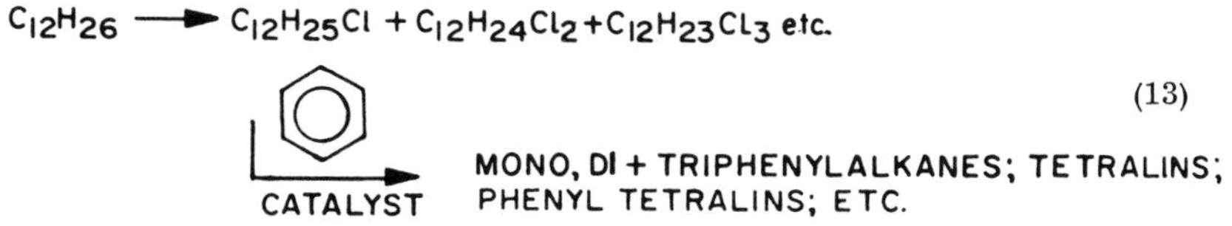

$$C_{12}H_{26} \longrightarrow C_{12}H_{25}Cl + C_{12}H_{24}Cl_2 + C_{12}H_{23}Cl_3 \text{ etc.}$$

(13)

$$\underset{\text{CATALYST}}{\bigcirc} \longrightarrow$$ MONO, DI + TRIPHENYLALKANES; TETRALINS; PHENYL TETRALINS; ETC.

SIDE CHAIN CHLORINATION

(14)

$$RCH_2\bigcirc + Cl_2 \longrightarrow R\underset{\underset{Cl}{|}}{CH}-\bigcirc \xrightarrow[\text{CATALYST}]{} \text{DIPHENYLALKANES + TETRALINS}$$

The alkylation reaction itself results in a number of important variations, which can be controlled by the proper choice of conditions. If the

aromatic compound is benzene, both mono- and dialkylbenzenes are produced, depending upon the mole ratio of benzene to alkylating agents [41].
Fragmentation of branched alkylating agents like propylene tetramer and
isomerization of linear groups can occur. Hydride-ion abstraction initiates
formation of compounds like branched alkanes and branched phenylalkanes,
aromatics and diolefins, cycloalkyl groups and diphenylalkanes. Highly unsaturated hydrocarbon-catalyst complexes must be removed carefully to
insure high-purity products [526].

Although the aromatic compound used for alkylation is generally benzene, others are also used. Toluene [42] and xylene react similarly to benzene. With naphthalene, both mono- and dialkylated products are produced.
Nonylnaphthalene from propylene trimer, and dibutylnaphthalenes are
typical products. A series of surfactants based upon diphenyl ether have
been developed [43]. Propylene tetramer is used to alkylate diphenyl ether
and the resulting aromatic alkyl compound is sulfonated as outlined in Eq.
(15).

$$C_{12}H_{24} + \text{(diphenyl ether)} \longrightarrow C_{12}H_{25}\text{(diphenyl ether)}$$

$$C_{12}H_{24}\text{(diphenyl ether)}-SO_3Na \xleftarrow[2.NaOH]{1.H_2SO_4} \tag{15}$$

Generally, alkylbenzenes are produced commercially by alkylation, but
other processes have been proposed. Synthesis of alkylbenzenes by dehy-
drocyclization of alkanes is an example [46, 47, 328, 351, 530]. The sim-
plicity of the process is intriguing, but satisfactory yields have not been
obtained.

$$RCH_2(CH_2)_6CH_3 \xrightarrow[\text{CATALYST}]{-H_2} \text{(aromatic ring with } CH_3, R) \tag{16}$$

The growth reaction of ethylene and tris (β-phenylethyl)aluminum produces linear alkylbenzene [44]. However, a wide homologous series is obtained. Similarly, telomerization of ethylene and toluene derivatives produces linear alkylbenzene [45]. In this case, branching is an additional
complication.

C. Sulfonate Production

In the 1960s most industrial research on the sulfonation for production of
alkylarylsulfonates has been devoted to the use of anhydrous SO_3. The basic

chemistry of SO_3 sulfonation is very similar to that with sulfuric acid or sulfur trioxide dissolved in sulfuric acid (oleum). However, since SO_3 is such a vigorous reactant, there is a great tendency for oxidative side reactions which lead to the formation of water and sulfur dioxide. These are, of course, undesirable since unstable, odorous, dark-colored by-products result. The water formed results in sulfuric acid and leads to unwanted inorganic salt in the sulfonate. Two basic procedures have evolved for controlling the destructive reactivity of sulfur trioxide.

If sulfur trioxide is dissolved in an inert solvent, it can be used for sulfonation much the same way that sulfuric acid is used. The problem is to find the inert solvent. Sulfur dioxide in the liquid state seems ideal for this purpose. It is inert and relatively easy to recover from the reaction mass. The need for refrigeration and pressure vessels adds to costs, however,

The other successful approach is to dilute the SO_3 with an inert gas such as air. By using high velocity between gas and liquid and by maintaining the liquid in a thin film to facilitate heat transfer and mixing, sulfonic acids of excellent quality are produced by SO_3 sulfonation of alkylbenzenes.

A novel process reacts aluminum alkyls and SO_3 [48]. Sodium octanesulfonate and sodium-2-phenylethanesulfonate were made in this manner.

Because surface-active agents are used in many critical applications where high purity or esthetic considerations are important, much effort has been devoted to obtaining light-colored, stable, odorless surfactants free of any extraneous impurities. Since the sulfonation reaction almost always results in some sulfuric acid being left in the sulfonic acid, many methods have been devised for removing sulfate from the finished sulfonate [506-517].

It has been found that various linear dodecylbenzenes sulfonate at different rates. The rate for the 6-isomer is 40% lower than that for 1-phenyldodecane. Also, hexylbenzene reacts faster than dodecylbenzene [518].

The typical commercial alkylbenzene used for sulfonation contains minor amounts of hydrocarbons that cannot be sulfonated. Also, during the sulfonation reactions, side reactions occur which produce oily by-products in small amounts. The composition of this so-called free oil or free fatty matter has been analyzed [49-51]. Unreacted alkylbenzenes and aliphatic hydrocarbons, diarylsulfones, and oxygenated compounds have been found. A method for removing the free oil has been devised [52].

Patents have been obtained for improving the color [53], odor, and stability [54] of alkylarylsulfonates. In order to stabilize the alkylarylsulfonic acid obtained by SO_3 sulfonation of dodecylbenzene during storage or transportation, enough sodium hydroxide is added to neutralize only the sulfuric acid in the sulfonic acid [55]. Other bases such as ammonia and sodium bicarbonate can also be used. Bleaching with hydrogen peroxide is described and an amine or urea are used to stabilize the bleached sulfonic acid [56]. Addition of a small amount of hydroxylamine with optionally some ethyl azodicarboxylate is claimed to stabilize the odor of alkylarylsulfonates [57].

Addition of a mixture of sodium phosphate and an antioxidant such as 2, 4, 6-trimethylphenol is claimed to improve the color, odor, and heat stability of alkylbenzenesulfonate surfactants [58].

If, in the chlorination route for producing linear alkylbenzenesulfonate, the n-paraffins contain substituted aromatic, the product may contain small amounts of by-product refractory chlorinated compounds. Passing the contaminated alkylate over calcium oxide at 400-410° C removes substantially all of the chloride [59].

Although most surfactants in this class are simple salts of metals and amines, sulfonamides have also been used. Dodecylbenzenesulfoamido-ethylsulfate was found to produce a nonphosphate detergent with poorer properties than a conventional LAS-phosphate formula [525].

D. Formulated Detergents

Alkylarylsulfonates are used for a great diversity of applications. The final surfactant or detergent is, therefore, produced in a wide variety of physical forms. Preparation of suitable liquids, powders, solutions, tablets, and composites have resulted in a considerable body of technology. Typical household laundry detergents are available as free-flowing hollow beads. These are produced by spray-drying of a slurry of surface-active agents plus builders and other additives [60, 61]. To avoid hydrolysis of polyphosphates by high temperatures in spray-drying, a fluidized bed drier has been proposed [62].

Other methods of making dry detergents are dry blending and the foam-drying process. The latter depends on heat of hydration of trimetaphosphate to foam the mass. After cooling and solidification, the porous detergent is ground to size [63-66].

Procedures for dry blending and for neutralization during spraying to make use of the heat of neutralization to evaporate water have also been described [67]. Conversion of borax and similar salts with alkali allows hydration of liquid water in detergent slurries and results in dry powders [68, 69].

In order to produce liquids it is not unusual to neutralize the sulfonic acid in the presence of other components of the finished liquid formulation. Light-duty liquids call for methods of solubilizing rather high concentrations of surface-active agents. It is in this application that the hydrotropic sulfonates such as sodium xylenesulfonate find widespread usage. Production of heavy-duty liquids for laundry detergents presents a somewhat different problem. It is necessary to find compatible systems of organic surfactants and inorganic builders, and again the hydrotropic sulfonates are useful.

Some other interesting physical forms of detergents have evolved in the last decade. Tablets [70, 71], encapsulated surfactants, and fabrics or paper products impregnated with surfactants are examples.

III. PROPERTIES OF ALKYLARYLSULFONATES

A. Chemical Properties

In general, the chemical properties of alkylarylsulfonates are of secondary importance compared to the physical effects of these substances. Chemical properties are of importance in the analysis of alkylarylsulfonates. There are a few covalent products that can be formed from sulfonates and, of course, there are many reactions in which these products participate as salts.

Several chemical characteristics of alkylarylsulfonates are used for their identification and estimation. Organic cations form complexes which are soluble in organic solvents and the alkylbenzenesulfonate content of most commercial products is assayed by cationic titration [72]. For analysis of spray-dried products, an automated procedure has been developed which depends upon ultraviolet absorption [73]. Characteristic infrared absorption bands have been used to determine the content and identify alkylarylsulfonates in slurries and liquid detergents [74, 75].

Mitigation of pollution of surface waters by alkylarylsulfonates has necessitated development of sophisticated analytical techniques for surfactants. The methylene blue cationic titration procedure has been used extensively. The role of interfering substances has been widely investigated for the methylene-blue procedure [76-78]. This method is still the one most widely used for quantitative determination of ABS and LAS in surface waters. However, for qualitative identification of alkylarylsulfonates, other methods are necessary. Procedures have been developed to isolate alkylarylsulfonates and subject them to hydrolysis or caustic fusion to produce the alkylaryl hydrocarbon or alkylated phenol. These volatile products can be identified by gas chromatography to determine the constitution of the original alkylarylsulfonate [79, 80]. In a similar procedure alkali fusion is used to recover the corresponding alkylphenol [81].

During the switchover from hard to soft alkylbenzenesulfonates, a procedure was developed for monitoring and policing the branched-chain alkylarylsulfonates [82, 83]. The alkylarylsulfonates were isolated by absorption on carbon. Infrared spectroscopy was then used to determine the amount of branched versus linear alkylaryl sulfonates. The IR bands characteristic of methyl and gem-dimethyl substitution were utilized.

The alkylaryl sulfonic acids can be characterized as nonoxidizing monobasic acids. Most salts with inorganic bases are water soluble. The alkyl part of the molecule must be quite high in molecular weight C_{16}-C_{18}) before the water solubility decreases. At alkyl chain lengths above C_{16}, oil solubility increases. Since commercial alkylarylsulfonates are almost invariably mixtures, they tend to be oils or amorphous solids. The alkylarylsulfonates can be absorbed by ion-exchange resins and eluted much the same way that inorganic salts are.

Sulfonic acids can form anhydrides with themselves and with other acids [84]. This tendency is the source of a problem with linear and branched alkylbenzene sulfonic acid produced by SO_3 sulfonation. Anhydrides of the sulfonic acid and SO_3 hydrolyze in water very slowly, especially when the solution is dilute or neutral. Thus, alkylarylsulfonate slurries are often subject to the phenomenon of pH drift. As the sulfonic acid anhydrides slowly hydrolyze, the aqueous solution becomes acidic and color degradation, odor formation, and corrosion of metal containers result. The problem can be solved very simply by adding a small amount of water [85] or other hydroxyl compounds such as alcohols [86] to the neat sulfonic acid. Under these conditions, hydrolysis is extremely rapid and the anhydrides are destroyed.

Alkylarylsulfonic acids and sulfonates can act as oxidizing agents [87]. In the presence of water at 600° F under autogenous pressure, p-toluene sulfonic acid is reported to oxidize aromatic compounds to the corresponding phenol. It is not unlikely that this reaction proceeds by hydrolysis of the sulfonate and oxidation of the hydrocarbon by the sulfate ion.

B. Physical Properties

The practical applications of alkylarylsulfonates depend almost entirely on the physical-chemical effects of these substances. Much effort has been devoted to understanding how the surface-active agents perform their various useful functions. Also, physical chemical effects are important in formulating alkylarylsulfonates into useful and convenient products. Since almost all applications of alkylarylsulfonates are in the form of aqueous solutions, this area has received the most study.

Alkylarylsulfonates exist in aqueous solution in various degrees of aggregation. Because of their dual hydrophilic-lipophilic nature, they concentrate and orient themselves at surfaces. In general, as the concentration of alkylarylsulfonate in water is increased, higher aggregates are produced. In very dilute solutions, alkylarylsulfonates may exist as single molecules. However, at low concentrations, complexes with integral ratios of o- and p-decylbenzenesulfonates have been identified. Stability of the discreet complexes was related to the cross-sectional area of the ortho and para isomers [88].

As concentration increases, micelles appear. These are laminar aggregates of alkylarylsulfonates which have important indirect effects on surfactant properties. The point at which micelles begin to form is determined by an inflection in the curve of surface activity vs concentration. Surface tension is a typical property measured to determine the critical micelle concentration (CMC).

In general, as the molecular weight of the alkylarylsulfonate increases, the CMC decreases [89, 90]. The structure of the alkyl group also has an effect. Because alkylarylsulfonate surfactants orient themselves, other solution effects are also evident.

Films containing oriented surfactant molecules are formed at interfaces [91, 92]. Branched alkylarylsulfonates such as ABS form membranes which are thin and in which the trapped aqueous solution has a low apparent viscosity. These films drain rapidly and tend to be unstable. Linear surfactants form more stable films because orientation is greater and the apparent viscosity of the trapped fluid is higher. Foams from linear surfactant films thus tend to be slow draining, thick and stable, or in the terms of the consumer, rich and luxurious. Since natural lipid-derived surfactants are linear and form slow-draining stable films, much effort has been devoted to synthesizing synthetic surfactants which have this property. Few alkylbenzenesulfonates form slow-draining films, but some of the more linear compounds do so [91]. In general, however, the practice is to use alcohol sulfates and similar surfactants to formulate products giving slow-draining films.

Other effects of structure on properties such as melting point, solubility, crystallization point, Krafft point, and spectroscopic properties have been studied [93-95].

Aqueous solutions of alkylarylsulfonates, as well as other surfactants, have the interesting property of solubilizing hydrocarbons and other organic substances. This is probably due to solution of the organic liquid in the layers of hydrocarbon formed within the oriented micelles. The effect is usually measured by use of a fat-soluble yellow dye. Solubilization increases with surfactant concentrations and molecular weight [96]. Electrolytes have an effect on solubilization and sodium carbonate has been found to increase the solubility of the dye. Although solubilization may have some importance in the laundering process, it is more likely that emulsification is the more important effect.

The viscosity of aqueous solutions of alkylarylsulfonates is important, primarily from a processing standpoint. In the detergent industry, it is common to formulate finished products containing alkylarylsulfonates from their slurry in water. A slurry normally contains from 40 to 50% of alkylarylsulfonate. It is important for the production, storage, and shipment of slurry that viscosity be controlled. Since alkylarylsulfonates tend to precipitate from hot aqueous solutions in the form of gels, the general method to control slurry viscosity is to have as much as possible of the sulfonate in crystalline form to prevent gel formation. Addition of inorganic water-soluble salts accomplishes this objective [97]. Also, if the temperature is kept low during neutralization, the concentration of alkylarylsulfonate in the liquid phase is low and consequently the product has a low viscosity. Control of viscosity in true solutions of surfactants is also important, but more so from the standpoint of the finished detergent product.

It is usually desirable to have as high a content of active substances as possible in a liquid detergent. Complex blends are generally employed.

First, the detergent formulator makes use of compatible blends to achieve the maximum solubility of a desired surface-active agent. Blends of ethylene oxide adducts of alcohols and alkylphenols, sulfated ethylene oxide adducts, and alkylarylsulfonates are common. Incorporation of foam boosting amine oxides and alkanolamides also aids in achieving maximum concentration of surfactants. The second method of keeping visocsity under control in liquid detergent is the use of hydrotropes. Solubilizing agents such as sodium toluene sulfonate, urea, and ethanol are commonly used to achieve solubilization. Since the consumer is accustomed to liquid detergents having a certain viscosity, occasionally thickening agents are added to very fluid aqueous solutions of alkylarylsulfonates. It is possible to choose an alkylarylsulfonate for its high aqueous solution viscosity to create the illusion of high content of surface-active agent in the formulation.

Attempts have been made to understand the basic viscosity behavior of alkylarylsulfonates [98, 99]. In general, however, because formulations are complicated, trial and error combined with intuition is used by the formulator to achieve satisfactory formulation viscosities.

Apart from the problem of formulating a liquid detergent with a convenient viscosity, it is necessary to solubilize sufficient surface-active agent to do the job intended. As mentioned above, hydrotropes are used to lower viscosity. They are also used to solubilize surface-active agents and alkylarylsulfonates are solubilized commonly by use of sodium and ammonium xylene- and toluenesulfonates. Partial phosphate esters [100] are claimed to solubilize alkylarylsulfonates. Polyvinyl acetate has been patented as a solubilizing agent for anionic surfactants [101]. Different alkylarylsulfonates require different formulations to achieve solubility. Thus, with ABS a different amount of urea is required than with LAS [102]. In some formulations of liquids, LAS with a high content of internal isomers is less soluble than LAS with more of the 2-phenyl isomer. Blends of ABS and LAS [103] and of nonionics and anionics [104] are examples of mixtures giving greater solubility than individual surface-active agents.

The physical characteristics of solid forms of alkylarylsulfonates have little to do with the actual functions of the molecules. However, physical characteristics do affect formulations of these products into usable household and industrial detergents. The goal of producing a noncaking, nondusting powdered detergent is a demanding one, and the physical properties of alkylarylsulfonates as well as formulation requirements work against this goal. Although pure alkylarylsulfonates are crystalline solids, balanced detergency requires use of mixtures. For example, better wetting is obtained with low-molecular-weight alkylarylsulfonates, while emulsification is promoted by the higher-molecular-weight products. However, addition of complimentary surfactants tends to lower the melting point of alkylarylsulfonates and their hydrates. Thus, by using broad homologous ranges and isomeric mixtures, and blends of different types of surfactants, problems of hygroscopocity and caking are commonly met.

A considerable body of technology has evolved to control the caking tendencies of dry detergent formulations. By controlling the size and shape of the individual particles, caking can be minimized. For example, proper spray drying forms selected sizes of hollow beads which cake less readily than a fine powder. Also, additives are used commonly to prevent caking.

It has been found that a number of sodium and potassium salts of low-molecular-weight organic acids function as anticaking agents for alkylaryl-sulfonate containing formulations. Benzene-, toluene-, and methylnaphthalenesulfonates are commonly cited [105-109]. Mixtures of water-soluble polymers which tend to coat the beads and toluenesulfonate have also been used [110]. Sodium sulfosuccinate [111, 112] and sodium and potassium phthaltes also function to reduce caking [113].

Another approach is to coat the sticky particles with a powder which prevents agglomeration. Various inorganic and anticaking agents such as aluminum silicates [114] have been used. Still another procedure is to coat the particle with the detergent builder [115].

Since addition of anticaking agents adds to the cost of the product without adding to the detergent performance, most formulations are produced without anticaking agents. This is accomplished by a careful choice of ingredients and manufacturing techniques and is an example of one of the many trade-offs in formulating useful detergent products.

C. Biological Properties

Among surfactants, the quaternary ammonium compounds are well known for their biological activity. Alkylarylsulfonates have also been found to have some bacteriostatic and fungicidal properties. Some interesting pharmacological effects have been discovered. Cleaning agents used in food preparation equipment have been found to have antimicrobial properties [116]. The complex alkaline silicates, phosphates, and other additives as well as sulfonates and other surfactants, when used in sufficiently high concentration showed antimicrobial activity. A screening of 24 anionic surfactants showed sodium dodecylbenzenesulfonate as well as several other anionic surfactants to have activity against staphylococcus aureus [117]. In acid solution [118], alkylarylsulfonate was found to be effective. When adsorbed on fabrics during laundering LAS is said to be an effective bacteriostat [119].

A number of surfactants, including alkylarylsulfonate, were found to act as fungicides. The mechanism appeared to be leakage of cell constituents caused by the surface-active agent [120].

The activity of anionic surfactants in tooth cleaning agents has been studied. For the aliphatic derivatives a chain length of C_{12} seemed to be optimum for foaming as well as bacteriostatic properties. Alkylarylsulfonates were enhanced in activity by the introduction of a halogen or nitro group into the ring [121]. Growth of organisms in raw starch [122] was found to be inhibited by addition of dodecylbenzenesulfonate.

Alkylarylsulfonates were found to have effects on plants and animals, and to be more of an antimitotic agent than sodium laurylsulfate. Cell division in bulbs and seeds was inhibited [123] by alkylarylsulfonates. Sodium tetrapropylbenzenesulfonate and a linear alkylbenzenesulfonate were found to inhibit contraction of frog muscle by acetylcholine [124]. Both sodium dodecylsulfate and sodium dodecylbenzenesulfonate were found to have strong antiphlogistic potencies similar to phenylbutazone and salicylate [125].

D. Surface-Active Properties

From the very classification of alkylarylsulfonates as surface-active agents, it can be concluded that this is the most important property of the substances. Surface activity of aqueous solutions of alkylarylsulfonates manifests itself in many ways. The individual effects, such as lowering of surface and interfacial tensions, wetting, foaming, emulsification, dispersion and adsorption, have been studied in detail. The practical effects of alkylarylsulfonates, for example as detergents in removing soils from solid surfaces, has also been the subject of close scrutiny.

Surface tension is often measured to determine critical micelle concentration. As more and more surfactant is added to the aqueous solution, surface tension decreases in a smooth curve until the CMC is reached. At this point, additional surfactant goes into micelle formation rather than into solution, and surface tension ceases to decrease. Except for solubilization, micelles probably do not contribute to the actual functioning of surface-active agents. Micelle formation must be taken into account because micelles act as surfactant-concentration buffers. It has been demonstrated that certain types of detergent action no longer increase with the addition of surfactant above the CMC [126].

The size and shape of alkylarylsulfonate molecules have a large effect on CMC. The lower-molecular-weight homologs such as dimethyl and diethyl benzenesulfonates do not form micelles [127] and in the case of very high molecular weights or highly crystalline materials with low water solubility, such as para-n-dodecylbenzenesulfonate [128], the solubility is below the CMC. The shape of micelles is affected by the chain length of the alkyl group. Alkylbenzenesulfonates with chain lengths of 10 to 12 form small spherical micelles which are highly hydrated. Alkyl groups of C_{13} to C_{16} form asymmetric micelles as determined by sedimentation and viscosity studies [129a]. In general, the more linear the alkylarylsulfonate molecule, the lower is the solubility, critical micelle concentration, and surface tension of an aqueous solution at a given concentration.

Micelle formation is affected by temperature [130a], shear rate [131a], and pH [132a]. Mixed micelles are also formed when different nonionic and anionic surfactants are in solution [133a]. Since there is less micelle formation at higher temperatures and shear rates, it would be expected that the higher concentration of active surfactant would cause the enhancement of practical effects at high temperatures and with good agitation.

Alkylarylsulfonates are used to enhance the wetting of solid surfaces by aqueous solutions, although nonionic surfactants are quite often more efficient for this purpose. Alkylarylsulfonates with short alkyl chains tend to be the best wetting agents [127]. Mono- and dialkylbenzene and naphthalenesulfonates with short alkyl chains are good wetting agents. Branched alkyl groups tend to be better than straight-chain alkyl groups. Surfactants with complicated shapes with the polar group in the center of the molecule or even with multiple polar groups, seem to be the most efficient wetting agents.

The surface to be wetted has a great deal to do with the choice of the wetting agent. In studying the penetrability of aqueous solutions, it was found that a homologous series of surfactants gave an inverse relation between paper and paraffin-impregnated paper [128]. Different wetting agents are chosen for wetting hydrophobic surfaces such as plastics [129] or minerals such as coal dust [130]. The surface characteristics of different minerals, and different crystal faces of the same mineral, vary greatly. Wetting agents must be chosen with this in mind [131]. Another example is the wetting of metal surfaces with oil. Here oil-soluble alkylarylsulfonates are chosen. Sulfonates with greater than C_{20} alkyl groups adsorb on the surface and prevent wetting by water. For metal degreasing, a low-molecular-weight sulfonate will displace the oil film and allow the surface to be wetted by water.

The most obvious characteristic of aqueous solutions of alkylarylsulfonates is their ability to form foam. Practical applications of foaming ability account for a minority of the surface-active agent usages. Froth flotation of minerals is probably the most important application. In the laundering process, foam can actually be detrimental and at best serves no useful function. However, foam is an important indicator of surfactant presence to the housewife. For this reason foaming of detergents has been studied a great deal. Two characteristics of detergent-solution foam are important, i.e., volume and stability. In comparison with soap and alcohol sulfate surfactants, alkylarylsulfonates tend to produce unstable foams.

It is common to use mixtures of surfactants to give a desirable foam profile [132, 133]. Some surfactants are especially useful for this purpose and are known as foam stabilizers. Dodecyldimethylamine oxide and lauric diethanolamide are examples of this type of surfactant. They are used to make the foam characteristics of shampoos, light-duty dishwashing detergents, and laundry detergents more desirable. A good foam is one which is

stable during the life of the washing cycle, is easily rinsed away, indicates
the correct volume of surface-active agent and is not unduly affected by soil
or temperature.

Alkylarylsulfonates in the dodecyl range are not as widely recognized
for their emulsifying ability as other types of surfactants. Although emulsification is important to the detergency process, investigations of this
phenomenon usually measure soil removal. Typical alkylarylsulfonate
emulsifiers are the oil-soluble products having alkyl chain lengths above
C_{16}. However, there are two important areas where detergent-range emulsifiers are used. These are in emulsion polymerization and emulsifiable
pesticide concentrates. Branched-chain dodecylbenzenesulfonates are preferred for emulsion polymerization of styrene and butadiene, although
short-chain di- and trialkylbenzenesulfonates can also be used [134-136].
The diphenylalkane by-products of detergent alkylate production have been
sulfonated and used as electrolyte-resistant emulsifiers for polymerization
[528]. Emulsification efficiency also depends on the oil phase [137, 138].
Addition of a solvent for a solid organic can aid in emulsification. Oils containing polar groups are easier to emulsify than straight hydrocarbons.

Alkylarylsulfonates are also active at interfaces between liquids and
solids. Alkylarylsulfonates adsorbed on surfaces such as lime-soap particles and clay alter the surfaces and render the particles more water dis-
persible [139, 140]. Theoretical studies of adsorption of alkylarylsulfonates
have been carried out [140, 141] and it has been found that many surfactants
behave ideally [142]. Adsorption of surfactants on the dropping mercury
electrode in the polarographic apparatus has been suggested as a method of
screening emulsifiers used for emulsion polymerization. Isotherms for
adsorption of alkylarylsulfonates on quartz sand have also been determined
[143].

Investigation of practical aspects of alkylarylsulfonate include adsorption on vegetables, fabrics, and activated sewage-sludge organisms. After
laundering, it is preferred that as little as possible of these surface-active
agents remain on the fabric. Rinsability is the term applied to this situation and it has been studied by use of radioactive tracers [144] and by extraction of the cloth [145]. Alkylarylsulfonates in the detergent range
adsorb to a lesser extent on fabrics than alcohol sulfates and some nonionic
surfactants. The amount of sodium dodecylbenzenesulfonate remaining on
dishes and vegetables washed in a solution of the surfactant was also determined by a radio-tracer technique. The amounts of surfactant adsorbed
were low but measurable [146].

The effect of sodium dodecylbenzenesulfonate on wetting of nylon-6
and polyethylene terephthalate has been studied [147]. Adsorption at the
solid-gas interface next to the drop of liquid was observed.

From a practical standpoint, detergency is the most important surfactant property of alkylarylsulfonates. Detergency involves removal of a
wide variety of soils from many different surfaces. Mechanical action,

temperature, solution pH, and hardness of the water also enter into the
process. Multiple interactions are encountered such as those between the
various types of surfactants and water hardness [148, 149], electrolytes
[150], soils [151, 152], mixtures of alkylbenzene sulfonates [153], builders
[154], and blends of surfactants [155]. The possible interaction of the many
variables affecting detergency should be kept in mind during the discussions
of specific effects of alkylarylsulfonate structure which follows.

The detergency of alkylbenzenesulfonates is determined by the size and
configuration of the alkyl chain, the point of attachment of the benzene ring
to the chain, the number of alkyl groups attached to the benzene ring, and
the orientation of the alkyl group ortho, meta, or para to the sulfonate group.
As might be gathered from the fact that the term dodecylbenzene is used
somewhat synonymously with alkylarylsulfonate or LAS, an alkyl group
averaging about 12 carbon atoms is usually found in the alkylarylsulfonates
of commerce.

Chain length does have important effects on detergency properties,
however. Examinations of detergency vs chain length of alkylbenzenesulfonates showed that detergency is negligible until the number of carbon atoms
in the alkyl group reaches 10. This generalization holds whether the alkyl
group is branched [156] or linear [157]. Detergency of a number of linear
alkylbenzenesulfonates prepared from linear olefins and linear chlorinated
paraffins was studied. On cotton, using a standard soil, it was concluded
that tridecylbenzene or mixtures of alkylarylsulfonates averaging tridecylbenzene were optimum [158, 159]. In another study [160], sodium p-n-
alkylbenzenesulfonates with chain lengths of C_8 and C_{18} were checked for
emulsifying ability, dispersing power with carbon black, and detergency.
Emulsion stability reached a peak at C_{14} and then decreased. However,
dispersing power and detergency continued increasing up to the highest
homolog checked. Since a temperature of 95° C was used for this study,
these results do not necessarily conflict with the previous reference where
tridecylbenzene was shown to be optimum.

Studies of three isomeric dodecylbenzenes [161, 162] illustrate the
effect of configuration on detergency properties. The highly branched
2,2',4,6,6'-pentamethyl-4-phenylheptanesulfonate, gave a lower surface
tension at 20°C than 1-phenyl- or 2-phenyldodecanesulfonates. 2-Phenyl-
dodecanesulfonate was the best detergent at 40° C, but at the boiling point
1-phenyldodecanesulfonate was the most efficient.

Linear alkylbenzenesulfonates were developed to solve the biodegradability problems of the branch-chain alkylbenzenesulfonates derived from
propylene tetramer. Attempts to substitute LAS into ABS formulations un-
covered differences in performance. Although detergency could be maximized by choosing the proper linear alkyl group, achieving the same foam
profile was much more difficult.

Figures 1 and 2 illustrate the problem that the detergent formulator
faces in substituting LAS for ABS so that the housewife would not notice a

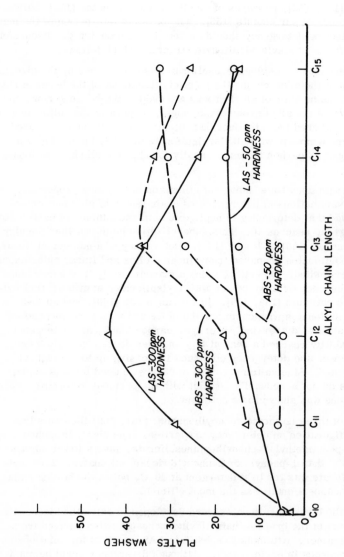

FIG. 1. Effect of alkyl chain length on foam stability [519], branched vs linear alkylarylsulfonates. Dishwashing foam stability test formula: 20% LAS or ABS, 50% STPP, 5% silicate, 15% Na_2SO_4, q.s. water. Concentration = 0.125%

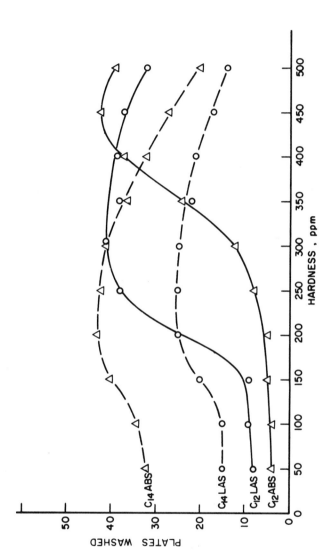

FIG. 2. Effect of hardness on foam stability [519], branched vs. linear alkylarylsulfonates. Dishwashing foam stability test formula: 20% LAS or ABS, 50% STPP, 5% silicate, 15% Na_2SO_4, q.s. water. Concentration = 0.125%

change in foam at various water hardnesses. The effect of water hardness on foam is greatly different for the branched and linear alkylarylsulfonates.

It was found that 2-phenylalkanes in general had foam profiles with concentration, temperature, and other formulation variables much different from the products derived from propylene tetramer. The internal isomers were much more similar and processes were developed to maximize production of internal phenylalkanes [163].

A series of alkylbenzenesulfonates with alkyl groups substituted in the number 2 position of the linear alkyl chain were synthesized and compared to other linear alkylbenzenesulfonates [164]. The slightly branched products tended to decrease in surface tension more rapidly than the corresponding n-alkylbenzenesulfonates. The linear products had better wetting power and were more soluble. The best wetting agent of the group was n-decylbenzenesulfonate. Foam stability was best with n-tetradecylbenzenesulfonate.

Although monoalkylbenzenesulfonate surfactants are much more common than dialkylbenzenesulfonates, the latter have been proposed as detergents [165-167]. Dihexyl-, diheptyl-, and dioctylbenzenesulfonates were shown to have interesting detergency properties.

As might be expected, orientation of the polar sulfonate group ortho, meta, or para to the lipophilic alkyl group on the benzene ring, does have some effect on detergency. Para derivatives were found to have better foaming action than ortho isomers, but ortho derivatives are better wetting agents [168, 169]. In a comparison of meta and para derivatives, the para isomer had better detergency, but the meta product was better at emulsification and suspending ability [170].

It is possible to alter the aromatic portion of the alkylarylsulfonate molecule without destroying the detergency properties. Alkylchlorobenzenesulfonates and alkyltetrahydronaphthalenesulfonates have been prepared and evaluated as detergents [171, 172]. Alkyldiphenylalkanesulfonates and diphenylalkanesulfonates have also been prepared [173, 174]. Although the monosulfonate of a compound like diphenyldodecane might be expected to act like a C_{14}-C_{18} alkylbenzenesulfonate, it is difficult to prepare pure monosulfonate. Disulfonation occurs and unreacted aromatic hydrocarbon remains until almost all of the product has been converted to disulfonate. Although the substituted arylsulfonates are interesting, their properties have not been sufficiently different to warrant commercial production. Alkylarylsulfonates, with functional groups in the alkyl chain, have been prepared. Presence of ether, hydroxyl, a cyclic group, or unsaturation lessened the lipophilic properties of the surfactant [175]. Longer alkyl chains can overcome this tendency. Similar results were obtained in a study of a series of compounds with ester and ketone groups in the alkyl chain [176]. Compared to tetrapropylbenzenesulfonate, sulfonates prepared from the benzene and toluene condensates with methyl undeceneoate had better tolerance to calcium ions.

It can be concluded that most any desirable surfactant or detergent property can be maximized by tailoring alkylbenzenesulfonate molecules. Changing the length and configuration of the alkyl group is the usual method of maximizing desired properties. The literature abounds with references to compounds which are better than "dodecylbenzene" in some characteristic or another. What must be remembered is that the formulator is trying to achieve an optimum balance among many, often conflicting, surfactant requirements, including the necessity to develop nonphosphate formulas [532].

E. Biodegradability

During the 1960s, biodegradability of detergents was a newsworthy topic. Much fundamental and practical work was done during this period which led to an understanding of detergent biodegradability and also to solution of the problem of environmental pollution by alkylarylsulfonates. Monitoring of branched and linear alkylbenzenesulfonate degradation continues in polluted rivers [521] and theoretical work has been reviewed recently [531].

Although ABS was very successful as a synthetic detergent ingredient, its large-volume usage soon pointed out a defect. The large volume of branched ABS and other hard surfactants entered surface waters in large amounts. Foaming was common in waters containing substantial quantities of sewage effluent because the rate and completeness of biological breakdown of ABS was too low. Biological degradation depends on the structure of the alkylarylsulfonate as well as upon the specific organisms and the environmental conditions. Microorganisms quite often require time to adapt to the assimilation of the new organic material [177, 178]. High initial concentration of surfactant also inhibits biodegradation, probably because of adsorption on the biological surfaces [179]. Adsorption of too much alkylarylsulfonate can seriously inhibit the anaerobic digestion of sewage sludge [180, 181]. Anaerobic biological processes are usually slower and there are fewer metabolic pathways, so that alkylarylsulfonates which are easily degraded under aerobic conditions may be slowly and incompletely degraded under anaerobic situations [182]. Septic-tank systems operating properly under aerobic conditions degrade alkylarylsulfonates quite well [183]. When the water table is high and the septic system is anaerobic, neither LAS nor other sewage is degraded completely.

Concern about detergent biodegradability includes not only rate but completeness. In a mixture of alkylarylsulfonates, some structures are degraded very rapidly, while others may degrade slowly or not at all. Researchers have also concerned themselves with metabolic products. The fate of the benzene ring has received much attention [184, 185]. Fragments have been identified, but the problem is a complicated one since the microorganisms are simultaneously synthesizing organic molecules while degrading others. There has been much debate over the proper analytical technique to determine the extent of biodegradability, but in the case of alkylarylsulfonates, the methylene-blue procedure, showing disappearance of the surfactant, is almost universally accepted.

Before proceeding to a discussion of structural effects on biodegradability of alkylarylsulfonates, and keeping in mind the problems in unequivocally assessing biodegradability, the following list of anionic surfactants in decreasing order of biodegradability is generally accepted:

a. Linear soaps and alcohol sulfates
b. Linear ether sulfates
c. Linear alkane and olefin sulfonates
d. Linear alkylbenzenesulfonates
e. Branched alcohol sulfates and soaps
f. Branched ether sulfates
g. Branched alkylbenzenesulfonates

The subject of surfactant biodegradation is covered in a recent book [186].

The effect of branching of the alkyl chain on biodegradability of alkylarylsulfonates has probably received more study than any other structural feature. One of the first studies [187] of the old hard ABS found that it was suitable as the sole carbon source for common water, soil, and activated sludge microorganisms. It was pointed out, however, that ABS was not completely decomposed and presence of a quaternary carbon atom in the hydrocarbon chain was indicated as the source of the nonbiodegradability. In later work [188] ω-trimethylalkylbenzenesulfonates were prepared and subjected to biodegradation. Alkylbenzenesulfonates with a quaternary atom at the point of attachment of the benzene ring were also studied. The compound with the quaternary carbon at the end of the chain strongly retarded biodegradation, while the other gave only a weak retarding effect. Cyclic compounds, such as those obtained in the kerylbenzene process by chlorination of kerosene, also are relatively nonbiodegradable [188-191]. More branching leads to less rapid and less complete biodegradation [192, 193]. The ratio of methyl group hydrogen to total alkyl group hydrogen, as determined by NMR spectroscopy has been quantitatively related to biodegradation [194], showing an inverse linear relationship. Useful bioresistant branched alkylarylsulfonates were also patented [195].

Even with completely linear alkyl chains, biodegradation is affected somewhat by the point of attachment of the phenyl group. Experiments in which nonbiodegraded alkylarylsulfonates were recovered and identified by microdesulfonation and gas-liquid chromatography showed that molecules with the longer unsubstituted alkyl chain degraded fastest. Biodegradation rate is very low when the number of carbon atoms between the phenyl and methyl group is less than six [196]. The same principle applies generally to shorter-chain alkylbenzenes. For example, hexylbenzene degrades quite slowly. The practical effect of these differences in biodegradability may be small since it has been shown that acclimation of the microorganisms decreases the difference in rates greatly [177, 178].

Di- and polyalkylbenzenesulfonates tend to be less biodegradable than the corresponding monoalkyl products. Although there is little difference between alkyltoluene and alkylbenzenesulfonates [197], compounds such as dihexylbenzenesulfonates do not degrade as rapidly [198]. Polyalkylbenzenesulfonates [199, 200] and diisobutylnaphthalenesulfonate [201] were reported to be nonbiodegradable.

Several studies have shown that the benzene ring in alkylarylsulfonates is subject to degradation [202-205]. It has also been shown that the short-chain alkylbenzenesulfonates, such as toluene- and xylenesulfonates, are readily biodegradable [198, 206]. Techniques used to illustrate this point are appearance of inorganic sulfate ion, disappearance of ultraviolet adsorption, and enzyme studies. It was shown that enzymes known to cleave aromatic rings were formed when biodegradable alkylarylsulfonates are added to a culture of microorganisms. From this induced formation of enzymes, the inference is that the benzene ring is cleaved [207]. Little is known about biodegradability of alkylnaphthalenesulfonates and other aromatic systems, but substitution of thiophene for benzene is reported to increase the rate of adaptation [198]. However, a study of the biodegradability of numerous sulfur analogues of dodecylbenzenesulfonate showed these compounds to offer no advantage [529].

F. Toxicological Properties

Because of widespread usage and their presence in surface waters, the toxicological properties of alkylarylsulfonates has been thoroughly investigated. Since these substances change the properties of water and adsorb on surfaces, including biological surfaces, it is not unexpected that they can affect the functioning of organisms. Under certain conditions, fish are particularly susceptible to toxic effects of alkylarylsulfonates.

Numerous investigations show that alkylarylsulfonates and other surfactants have lethal effects on fish at concentrations above a few milligrams per liter [208-217]. The effect of structure and size of linear alkylbenzenesulfonate on fish toxicity was the subject of an interesting study [214]. The LD_{50} of various pure linear alkylbenzenesulfonates on fish was determined. In general, LAS was found to be more toxic than ABS, which, in turn, was more toxic than C_{12} alkylsulfate. As molecular weight increases the LD_{50} decreases. The C_{16} homolog is anomalous in that it is intermediate between C_{12} and C_{13} in LD_{50}. Effects of positional isomers were also noted. With C_{12}, C_{14}, and C_{16} homologs, the isomers with the phenyl groups attached to the central carbon atom were more toxic than the 2-phenyl isomers. With C_8 and C_{10} homologs, the 2-phenyl isomers were most toxic. The toxicity order of four isomers of dodecylbenzene is: $2 > 3 > 6 > 4$. Toxic effects in the study was noted from concentrations of 0.2-180 mg per liter.

Feeding studies with rats indicate that the toxic dosage of ABS or LAS is quite high [218-221]. Adverse effects were found with 0.5% alkylarylsulfonate in the diet. Growth of the animals was retarded at 1 and 2% levels [222]. Although those fed 2% alkylbenzenesulfonates suffered adverse effects, no malformations were observed in the offspring [223]. Low orders of oral toxicity were observed in other animals also [224-228]. Toxic levels of alkylarylsulfonates have not been determined for man. However, these substances are regarded as relatively nontoxic and up to 0.5 mg per liter is regarded as safe for drinking water in the United States and the Soviet Union [229].

Alkylarylsulfonates interact with the skin to cause irritation [230] and increase permeation or adsorption by other compounds [231-234]. With the use of an elegant procedure for measuring the skin roughening effects of surfactants, it was found that a maximum is reached at 12 carbon atoms in the alkyl group for LAS as well as with a number of other surfactants [14]. Adsorption of LAS on the skin and other body surfaces is higher at lower pH, but no adverse effect on the skin was noted [522]. Inhalation toxicity is not normally a problem; but care should be exercised to minimize inhalation, for example, when using alkylarylsulfonates as dust suppressants in mines [235, 236].

Adverse effects of alkylarylsulfonates on plants, fungi, and microorganisms have been noted. Undegraded ABS causes undesirable changes in soil [237]. The surfactant not only caused water logging of the soil, but also decreased the water-holding capacity. Alkylarylsulfonates have been used ot enhance the effectiveness of herbicides [238]. Wetting ability and toxicity were found to depend on structure [239]. In general, branched-chain compounds were better wetters. Phytotoxicity of the homologus series usually passed through a maximum at C_{12} or lower.

The cationic surfactants are well known as algaecides and inhibit growth at concentrations of 0.1 mg per liter. Anionic surfactants, including alkylarylsulfonates, exert negative effects at 5-25 mg per liter and are lethal at 700-200 mg per liter [240]. The anionic surfactant, diisobutylnaphthalenesulfonate, was noted to act as a growth stimulant for algae at concentrations below 15 mg per liter [241]. As mentioned previously, high concentrations of LAS or ABS adsorbed on sewage sludge exhibit toxic effects and inhibit fermentation [242, 243]. At a use level of 0.2% ABS has been patented as mold preventive for cardboard [244].

IV. APPLICATIONS OF ALKYLARYLSULFONATES

A. Household Formulations

Alkylarylsulfonates are used as ingredients in almost every conceivable household cleanser application. The only major exception are the detergents for automatic dishwashers. For this purpose, alkylarylsulfonates foam too much and highly specialized nonionic surfactants which also act as defoamers are required. The largest area of usage of alkylarylsulfonates is in laundry detergents, but light-duty dishwashing compounds also consume major portions of LAS production. Alkylarylsulfonate surfactants derived from LAS are also used in scouring cleansers and numerous specialty cleaning products.

Laundry detergents are probably the most complex of all household formulations. Linear dodecyl-, tridecyl-, and tetradecylbenzenesulfonate mixtures are the LAS used in this application. They must operate efficiently in a variety of machines with different kinds of water, remove a mixture of soils from fabrics of divergent characteristics, while at the same time being safe for the user, the machine, and the fabric. Household detergent formulations, therefore, contain many ingredients to satisfy these specific requirements.

Although some ingredients perform multiple functions, they usually do one job better than others. Alkylarylsulfonates are generally blended with other surfactants and builders to give a balance of properties [245-247] and to help in removing, suspending, and emulsifying soils. Sodium tripolyphosphate is added to sequester calcium and magnesium ions, defloculate clays, and provide controlled alkalinity to the washing bath. With increasing concern about phosphate pollution, other sequestering agents such as nitrilotriacetic acid, polycarboxylates, and citric acid are being considered [248-252]. There has also been a return of the use of sodium carbonate and similar materials which precipitate hardness instead of sequestering it. Sodium silicates are added to provide corrosion protection for the washing machine parts. Antiredeposition agents such as carboxymethylcellulose and polyvinylalcohol are added to prevent the loosened soil particles from redepositing on the surfaces of the fabric [253-255]. Bleaches and enzymes are added to remove specific stains [256-260]. Fabric dyes which convert invisible ultraviolet light to visible light make the laundered fabrics appear cleaner and brighter. Soap and certain nonionic surfactants are often added to give alkylarylsulfonate formulations a more desirable foam profile [261, 246]. Borax is still a relatively common builder. The complex of an anionic and cationic surfactant with a total of 20 carbon atoms [262, 263] or, more commonly, alkanolamides are added as foam boosters. Detergent powders may be dyed to make them pleasing to the eye and fragrances are added to give the laundry a desirable fresh, clean smell.

Most laundry detergent is sold in the form of spray-dried powder but the convenience of liquids and tablets has attracted a sizeable following [264-271]. Formulation of liquid detergents is complicated by the necessity of getting all of the above ingredients into a highly concentrated compatible, stable, easily handled liquid [272, 273].

Light-duty detergent products usually contain little or none of the so-called detergent builders. The cleaning jobs they are designed to do are usually less stringent and contact with the hands precludes the use of alkaline materials. Blends of LAS with sulfated linear alcohol ethoxylates and alkanolamide or amine oxide foam boosters are often used in commercial products. Blends of surface-active agents are used to give desirable physical properties to the liquid as well as optimize cleaning and skin-mildness characteristics [274-280].

The LAS used in most light-duty liquid-detergent formulations has a molecular weight corresponding to undecyl- or dodecylbenzenesulfonate, whereas the LAS used for laundry detergents usually corresponds to dodecyl- or tridecylbenzenesulfonate. Depending upon the property measured, performance peaks at different molecular weights. Therefore the optimum product is usually a blend of several homologs. If, for example, C_{12} gives peak performance, a 50:50 mixture of C_{11} and C_{13} would give almost identical results. However, a blend of C_{10} and C_{14} is likely to show an appreciable decrease in desirable performance.

Formulations for cleaning hard surfaces often contain only small amounts of surface-active agents including LAS [281-284]. These materials are usually applied in highly concentrated form and rely on mechanical action, solvents, abrasives, or alkali to do a great deal of the cleaning job. Disinfectants are often added [285, 286]. Abrasive cleaners rely mainly on mechanical action but surface-active agents and bleaches certainly help.

In solvent cleaners, liquids like pine oil and butyl ethers of diethylene glycol greatly enhance the emulsifying ability of surfactants like LAS on greasy soils. Alkaline compounds like the phosphates are used in combination with LAS and other surfactants for demanding cleaning jobs such as cleaning floors.

In numerous other household products LAS has been incorporated as a surface-active agent. Examples are oven and carpet cleaners [287], bleach formulations [259], windshield cleaners [288], and sanitizers [289]. It is said to act also as a mothproofing agent [291]. In personal-care products other surface-active agents are more widely used, but LAS is frequently mentioned. Toilet bars are still largely made from soap because it is difficult to find synthetic surface-active agents which do not cause undesirable interactions with the skin. Certain alkylarylsulfonates tend to leave the skin with a sticky feeling. They are good lime-soap dispersants and are mentioned frequently in detergent-bar formulas [292-301]. Other types of skin cleansers also incorporate alkylarylsulfonates [302-304]. Also some permanent-wave formulations [305] and shaving-cream products [306] contain LAS.

B. Industrial Surfactants

Alkylarylsulfonates have been used in and proposed for an extremely large number of industrial applications. Most of the applications involve surface activity. Since the applications are so numerous and varied they are listed

with the appropriate references. Applications are classified according to
interactions at the various possible interfaces between solids, gases, and
liquids.

Application	Reference
Liquid-Liquid	
Emulsion polymerization of vinyl chloride, vinyl acetate, styrene, methyl methacrylate, vinylidine chloride, butadiene	307-322
Emulsion breaker for petroleum	323-326
Liquid ion exchange	327
Dehydrate and purified camphor	328, 329
Pesticide concentrate emulsifier	
Liquid-Gas	
Foam fractionation of zinc ion	330
Plastisol and polymer foams	331-333
Foaming agent for air drilling of oil wells	334
Foam for secondary petroleum recovery	335
Foaming of foundry-sand core slurry	336-338
Foams for lightweight ceramics, cement, and gypsum	339-342
Solid-Solid	
Antistatic electricity coatings for textiles	343-346
Anticaking agents	502-503
Antipilling finishes for textiles	347
Fiber lubricant	348
Solid-Liquid	
Polymer crystal modified	349
Crystal-growth modifiers for sulfur, gypsum, etc.	350-354
Increased rate of acidulation of phosphate rock	355-357
Lime dispersant	358
Fluid-loss additive and retarder for oil-well acidizing	359-360
Graphite and pigment dispersants	361-363
Dispersion of fluorocarbons	364

Application	Reference
Solid-Liquid (Continued)	
Wetting agent in paint strippers	365
Coagulation of limestone slurry	366
Wetting agent to speed removal of stamps	367
Improved oil-ammonium nitrate explosives by wetting action	368
Increased melting rate of snow by facilitating water drainage	369
Disperse tars in ethylene furnace effluent in quench water	370
Improved impregnation of surfaces with silicone	371
Wetting agent for concrete coatings	372
Wet coal dust	373
Disperse calcium stearate for waterproofing agent	374
Corrosion inhibitors for metals	375–383
Additive to oil-well drilling muds to increase drilling speed and decrease tendency of bit to stick	384–386
In photographic emulsions as precipitating agent and dispersing of color couplers	387–388
Coagulation of gelatin	389
Gelling of kerosene	390
Dye assistant for fabrics	391–397
Dairy cleaners	398–399
Leather-tanning additive	400–401
Phosphate-sludge coagulant	402
Rapid etching of zinc and magnesium	403
Brightener for aluminum	404–405
Additive for metal phosphating baths	406
Electroplating bath additive	407
Rust remover and brightener	408–409
Wetting agent at high pH	523

Application	Reference
Multi-Phase Interactions	
In secondary recovery of petroleum injection of an oil water emulsion followed by thickened water	410
Secondary recovery of petroleum	411–414
Froth flotation of mineral ores	415–426
Industrial Cleaning	
Starch	427
Metal cleaners	428–437
Textile cleaners	438–442
Drycleaning additives	443–446
Smokehouse cleaner	447
Car wash	448
Oil tanks	449–450
Miscellaneous Applications	
Felting and shrinkproofing of wool	451
Dispersion of water-closet contents	452
Reduction of viscosity of viscose-rayon baths	453
Rubber adjuvant	454
Manganese salt in light-resistant polyamide resin	455
Detergent polymer sizes	456
Treatment of waste waters	457
Shrinkproofing of textiles	458

C. Agricultural Applications

Alkylarylsulfonates perform a number of useful functions in agriculture. Their surface-active properties account for many of the applications. But in others, the mechanism of activity is obscure.

Various alkylarylsulfonates are used as anticaking agents in fertilizers. Sodium dodecylbenzenesulfonate, sodium butylnaphthalenesulfonate, and other surfactants mixed with polyvinyl acetate inhibit the caking of urea [459]. Ammonium nitrate coated with a mixture of silica-alumina and dimethyl naphthalene sulfonate is free-flowing [460]. Various alkylarylsulfonates are used to prevent caking of compounded fertilizers also [461-466].

Many pesticide formulations contain alkylarylsulfonates as emulsifiers and wetting agents [467-472]. In other applications, the alkylarylsulfonate acts synergistically with the active pesticide [473-474]. Mixtures of monosodium ethane-arsonate and hydrotropes such as sodium xylenesulfonate act synergistically on crab grass. Alkylarylsulfonates are included as adjuvants in insecticide formulations [475-476] and fungicides [477-478]. Purified sulfonates are used in sheep dip [479]. Dodecyl- and pentadecylbenzenesulfonates are claimed to act as nematocides [480] and against related soil organisms. An important application is the washing of fruits and vegetables to remove microorganisms [481].

V. HYDROTROPIC ALKYLARYLSULFONATES

Hydrotropes and coupling agents are used to increase the mutual solubility and miscibility of organic materials, with water and inorganic salts. The low-molecular-weight alkylarylsulfonates fit this classification. Although the hydrotropic sulfonates are not surface-active agents in the usual sense of the word, they are widely employed in the detergent industry. Since hydrotropes are used to improve the physical characteristics of the detergent products and add nothing to the performance of detergents, it is important to minimize the added cost by choosing the most efficient hydrotrope for a given system.

Considerable research has been done to understand the basic chemistry of sulfonation of short-chain alkylbenzenes. Alkylaryl hydrocarbons used for sulfonation to prepare hydrotropes include benzene, toluene, xylenes and ethyl benzene, cumene, alkylbenzenes with chains up to eight carbon atoms, and the methyl naphthalenes. Recent sulfonation studies have been concerned with toluene and xylene [482-484].

Sulfonation of toluene gives mainly the ortho and para isomers. Sulfonation experiments in homogeneous solutions [483] showed that the ratio of o- to p-toluene sulfonic acid changes from 0.276 to 1.08 by raising the concentration of sulfuric acid from 77.6 to 98.8% at 25° C. The yield of m-isomers was changed very little. m-Toluenesulfonic acid is produced by isomerization above 110° C. Hydrolysis of the less stable ortho and para isomers allows the concentration of the m-isomer to increase. Both m- and p-toluenesulfonic acids can be recovered in high purity by crystallization [482-485].

The same influences of sulfonation conditions apply to the xylenes but the number of isomers possible are much greater. In commercial practice, mixtures of xylene isomers and ethyl benzene are used. Since o-xylene, ethyl benzene and p-xylene are separated commercially from the equilibrium of C_8 aromatics, a variety of xylenes are available for sulfonation. m-Xylene is the preferred isomer [486].

There are two sulfonation by-products which have important negative effects on hydrotrope efficiency. Sulfone formation represents a loss of yield and for this reason is minimized. Although sulfones are separated during the purification, some sulfonated sulfones are also formed and these can cause problems because they remain in the hydrotrope product. Disulfonation is promoted by attempting too complete a conversion of the hydrocarbon to sulfonic acid; forcing conditions of high temperature and high concentrations of SO_3 in the sulfonating agent can also lead to disulfonates. Disulfonates are undesirable because they detract from hydrotropic properties [487-488].

Sulfonation processes for the commercial manufacture of toluene- and xylenesulfonate hydrotropes are dictated by the need for salt-free products of light color and low odor, containing little or no water-insoluble organic materials. The best-known sulfonation process employs 100% sulfuric acid or a lower concentration. The reaction is forced to completion by azeotropic distillation of water as it is formed by the sulfonation reaction. An excess of the hydrocarbon being sulfonated is used as the azeotropic agent [489-491].

By removing water as it forms, practically all of the sulfuric acid can be converted to sulfonic acid and a low salt content is achieved in the final product. The sulfonic acid solution is neutralized to prepare the desired sodium, potassium, or ammonium salts having a concentration of about 40% in water. The excess aromatic hydrocarbon extracts any sulfones or tars formed. The sulfonate solution may be steam stripped to remove residual odor and carbon treated to remove color bodies.

Other techniques are available for making sulfonates with low sulfate content. Sulfonation with anhydrous SO_3 dissolved in liquid sulfur dioxide has been described [492]. Addition of nitric acid and phosphoric acid to minimize sulfone formation has been patented [493]. Another process for using anhydrous SO_3 is cosulfonation [494]. Acid-washed cumene was mixed with acid-washed dodecylbenzene and sulfonated with SO_3 diluted aith air. Problems with this technique are that the short-chain alkylbenzenes used to make hydrotropes are volatile and tend to be swept out of the reaction, and it is almost impossible to get complete conversion so that the residual odor of the light hydrocarbon remains.

Another technique for removing residual salt involves precipitation of calcium or barium sulfate and filtration from the water-soluble calcium or barium sulfonates. The desired alkaline metal salts can then be made by ion exchange [495], or by addition of sodium carbonate to precipitate the insoluble alkaline earth carbonate [496].

An alternative to neutralization of the sulfonic acid with metal hydroxide is a high-temperature reaction with a metal chloride to form anhydrous HCl and a sulfonate salt. This process is probably not used to make the light-colored hydrotropic sulfonates [505].

Most alkylarylsulfonate hydrotropes are used in detergents. A few other uses include solubilizing dyes [497], solubilizing acrylonitrile in the electrolytic adiponitrile process [498], and as an aid in removal of free fatty acids from fats and oils [499]. Unrelated to hydrotropic properties is the usage of p-toluenesulfonic acid as an esterification catalyst.

Toluene and xylenesulfonate solutions can be analyzed by IR spectroscopy [500].

Use of hydrotropic sulfonates to formulate concentrated liquid detergent solutions is the classical detergent application for the hydrotropic sulfonates. Water-soluble organic compounds, such as urea and ethanol, are often used in combination with hydrotropic sulfonates to achieve optimum solubilization. Urea has the drawback of hydrolysis in alkaline solutions and ethanol is volatile and flammable. The choice of a hydrotrope for a detergent formulation depends on cost as well as effectiveness. One must consider the combination of surface-active agents, detergent builders, and other additives in choosing the optimum hydrotrope [501].

There are two other common applications of hydrotropic sulfonates in detergent formulas. Caking and tackiness in spray-dried detergent powders is lessened by the addition of sodium toluene- and xylenesulfonates [502-503]. Viscosity can be lessened and handling problems eased by inclusion of hydrotropes in the concentrates since concentrated slurries and solutions of surface-active agents are often prepared and shipped. Neutralization of sulfuric acid esters of alcohols and ethylene oxide adducts is often impossible without addition of hydrotropes. Viscosity may be so high that high temperatures must be used and local concentrations of acid result in autocatalytic hydrolysis back to the original alcohol [504].

Thus the low-molecular-weight alkylarylsulfonates, although they are not truly anionic surfactants, serve as important adjuncts to anionic surfactants.

ACKNOWLEDGEMENT

The assistance of the Continental Oil Co. Research Library in the literature search is greatly appreciated. The permission of Conoco Petrochemicals Management to write this chapter and the efforts of the secretarial staff is gratefully acknowledged.

REFERENCES

1. A. G. Monetov, Vses. Soveshch. Sintetich. Zhirozamenitelam. Poverkhnostnoaktivn. Veshchestvam Moyushchim Sredstvam, 3rd, Sb. Shebekino, 1965, 269-272; Chem. Abstr., 65, 19899.

2. G. I. Sorokin and M. M. Prokopets, Neft. Gaz. Prom., Nauchn.-Tekhn. Sb., 1, 43-47 (1962); Chem. Abstr., 59, 15101.

3. E. I. Kazakov, A. Ya. Larin, T. B. Voronina, Z. V. Lyubimova, and G. K. Goroshko, Tr. Inst. Goryuch. Iskop. Akad. Nauk. SSSR, 17, 169-173 (1962); Chem. Abstr., 58, 4342.
4. R. P. Govorova, Akad. Nauk. Ukr. RSR, Inst. Teploenerg., Zb. Prats, 25, 6267 (1962); Chem. Abstr., 58, 5904.
5. D. B. Orechkin, N. V. Popova, I. S. Rykova, and O. F. Shepet'ko, Khim. Tekhnol. Topliva Masel, 8, 27-30 (1963); Chem. Abstr., 58, 7773.
6. R. M. Agaeva, G. C. Ashumov, M. A. Ashimov, and S. E. Kanzaveli, Azerb. Neft. Kohz., 43 (12), 44-46 (1964); Chem. Abstr., 64, 5301.
7. A. I. Gershenovieh, Khim. Technol. Topliva Masel, 8, 14-20 (1957); Chem. Abstr., 52, 8527.
8. British Hydrocarbon Chemicals, Ltd., Belg. Pat. 623,466 (Apr. 10, 1963); Chem. Abstr., 61, 5912.
9. D. B. Orechkin, N. V. Popova, and O. F. Shepot'ko, Opyt Primeneniya Sintetich, Zhirozamenitclei V. Proizv. Myla Moyushchikh Sredstv, Moscow, Sb., 1962, 95-100; Chem. Abstr., 59, 15081.
10. British Hydrocarbon Chemicals Ltd., Belg. Pat. 623,466 (Apr. 10, 1963); Chem. Abstr., 61, 5912.
11. E. N. Rybko and O. K. Serdyukova, Dokl. L'vovsk, Politekhn. Inst., 5 (1-2), 84-87 (1963); Chem. Abstr., 61, 9684.
12. L. A. Potolovskii, A. I. Dolaudugin, and I. F. Blagovidov, Tr. Vses. Nauchn.-Issled. Inst. Prererabotke Nefti., 9, 110-169 (1963); Chem. Abstr., 60, 7031.
13. B. S. Nazarenko, D. Ya. Muchinskii, L. A. Potolvskii, and V. B. Serebryanyi, Neftekhim Akad. Nauk. Turkm. SSR, 1963, 233-239; Chem. Abstr., 60, 14418.
14. G. Imokawa, K. Sumura, and M. Katsumi, J. Am. Oil Chem. Soc., 52, 479-483 (1975).
15. V. A. Marushkina and A. Z. Bikkulov, Dokl. Neftekhim. Sekts. Bashkir. Respub. Pravl. Vses. Khim. Obshchest, 2, 216-218 (1966); Chem. Abstr., 67, 83239.
16. Thomas B. Hilton and Gilbert J. McEwan, U.S. Pat. 3,444,086 (1969), to Monsanto Co.; Chem. Abstr., 71, 14433.
17. Robert M. Engelbrecht, Raymond A. Franz, Richard N. Moore, James M. Schuck, and Robert George Schultz, U.S. Pat. 3,409,703 (Nov. 5, 1968), to Monsanto Co.; Chem. Abstr., 70, 21203.
18. Gilbert J. McEwan, U.S. Pat. 3,311,563 (March 28, 1967), to Monsanto Co.; Chem. Abstr., 67, 12861.
19. Mikulas Hrusovsky, Stella Foltanova and Karol Simko, Czech. Pat. 114,216 (April 15, 1965); Chem. Abstr., 64, 8507.
20. British Hydrocarbon Chemicals Ltd., Neth. Appl. 6,500,287 (July 19, 1965); Chem. Abstr., 64, 3861.
21. M. A. Ashimov, M. A. Israfilov, M. M. Rafiev, and Kh. M. Sultanova, Azerb. Neft. Kohz., 43 (4), 41-43 (1964); Chem. Abstr., 61, 9684.
22. G. G. Stepanova, Izvest. Akad. Nauk. Eston. SSSR, Ser. Fiz.-Mat. Tekh. Nauk, 10, 40-48 (1961); Chem. Abstr., 55, 24056.

23. S. Faingold, O. Kirret, M. Korv, H. Voore, and J. Joers, Chim. Phys. Appl. Prat. Ag. Surface, C. R. Cong. Int. Deterg., 5th, Sept. 9-13, 1968, 1, 201-208; Chem. Abstr., 74, 4904.

24. P. Hahn, Abh. Deut. Akad. Wiss. Berlin, Kl. Chem., Geol. Biol., 6, 128-138 (1966); Chem. Abstr., 68, 23278.

25. Chemische Werke Huels A.-G., Belg. Pat. 632,807 (Nov. 25, 1963); Chem. Abstr., 61, 2065.

26. Edwin Jones, Belg. Pat. 612,036 (Jan. 15, 1962); Chem. Abstr., 57, 9980.

27. Shuan K. Huang, Ger. Offen. 1,914,306 (Oct. 2, 1969), to Monsanto Co.; Chem. Abstr., 72, 14115.

28. Monsanto Co., Neth. Appl. 6,512,424 (March 28, 1966); Chem. Abstr., 65, 7477.

29. British Hydrocarbon Chemicals Ltd., Belg. Pat. 623,465 (Apr. 10, 1963); Chem. Abstr., 61, 5912.

30. Ioan Velea, Iosif Drimus, Ioan T. Ionescu, and Gheroghe Coianu, Rom. Pat. 48,911 (Nov. 10, 1967), to Romania Ministry of the Chemical Industry; Chem. Abstr., 69, 20639.

31. M. A. Ashimov, M. A. Mamedova, and S. A. Allakhverdieva, Izvest. Akad. Nauk Azerbaidzhan, SSR, 8, 31-38 (1956); Chem. Abstr., 51, 4299.

32. Bahi El-Din A. Gebril and Hamdy El-S. Abou-Zeid, Tenside, 6 (4), 194-197 (1969); Chem. Abstr., 71, 92946.

33. J. Balvay, Fr. Pat. 1,006,592 (Apr. 24, 1952), Standard Francaise des Petroles; Chem. Abstr., 51, 10932.

34. G. C. Feighner and B. L. Kapur, U.S. Pat. 3,316,294 (April 15, 1967), to Continental Oil Co.; Chem. Abstr., 67, 45105.

35. Paul Baumann, Chem. Phys. Appl. Surface Active Subst., Proc. Int. Congr., 4th, 1, 1-20 (1964); Chem. Abstr., 72, 14082.

36. Charles A. Cohen, U.S. Pat. 3,234,297 (Feb. 8, 1966), to Esso Research and Engineering Co.; Chem. Abstr., 64, 12976.

37. Lummus Co., Neth. Appl. 6,502,398 (Aug. 30, 1965); Chem. Abstr., 64, 8090.

38. Allied Chemical Corp., Brit. Pat. 1,138,118 (Dec. 27, 1968); Chem. Abstr., 70, 69484.

39. A. C. Olson, Ind. Eng. Chem., 52, 833-836 (1960); Chem. Abstr., 55, 7335.

40. Joseph Rubinfeld, U.S. Pat. 3,320,174 (May 16, 1967), to Colgate Palmolive Co.; Chem. Abstr., 62, 14938.

41. G. C. Feighner, O. C. Kerfoot, D. W. Marshall, and T. E. Howell, U.S. Pat. 3,391,210 (July 2, 1968), to Continental Oil Co.; corresp. pat., Fr. Pat. 5,467,786 (Jan. 27, 1967); Chem. Abstr., 68, 41338.

42. Sh. T. Akhmedov, S. G. Gadzhieva, M. N. Magerramov, G. C. Gadzhiev, and Sh. M. Omarov., Uch. Zap. Azerb. Gos. Univ. Ser. Khim. Nauk, 1, 49-53 (1966); Chem. Abstr., 67, 65755.

43. A. F. Steinhauer, U.S. Pat. 2,854,477 (Sept. 30, 1958), to Dow Chemical Co.; Chem. Abstr., 53, 15605.

44. Koster and Burger, Advan. Inorg. Radiochem., 7, 322 (1965); Br. Pat. 876,536 (appl. Apr. 28, 1959), to Esso Research and Engineering Co.; Chem. Abstr., 56, 6000.

45. G. G. Eberhardt and W. A. Butte, J. Org. Chem., 29, 2928-2932 (1964); Chem. Abstr., 60, 10616.

46. Arnaud De Garmont, Jean Maurin, and Joseph E. Weisang, Ger. Offen. 2,014,778 (Oct. 1, 1970), to Compagnie Francaise de Raffinage; Chem. Abstr., 74, 14422.

47. John Gilbert Hale, Peter T. White, and Frederick W. Porter, Br. Pat. 874,555 (Aug. 10, 1961), to British Petroleum Co. Ltd.; Chem. Abstr., 56, 12805.

48. Alfred J. Rutkowski and Albin F. Turback, U.S. Pat. 3,121,737 (Feb. 18, 1964), to Esso Research and Engineering Co.

49. Adriano Arpino, A. V. De Rosa, and Giovanni Jacini, Chem. Phys. Appl. Surface Active Subst. Proc. Int. Congr., 4th, 1, 507-513 (1964); Chem. Abstr., 70, 79401.

50. Kazuyoshi Itoh, Hiroo Sato, Atushi Shigetoshi, Masaya Ogawa, Toshio Yamamaka, and Haru Shibatani, Sekiya Gakkai Shi, 10 (2), 114-118 (1967); Chem. Abstr., 68, 115889.

51. Albert Metzger, Tenside, 3 (11), 381-386 (1966); Chem. Abstr., 66, 67042.

52. Kurt Kosswig, Manfred Sturm, Herbert Harting, and Hans Regner, Ger. Offen. 1,900,837 (Aug. 27, 1970), to Chemische Werke Huels A.-G.; Chem. Abstr., 73, 121754.

53. Atlantic Richfield Co., Br. Pat. 1,129,385 (Oct. 2, 1968); Chem. Abstr., 70, 30330.

54. Henkel and Cie., G.m.b.H., Fr. Demande, 2,012,791 (Mar. 20, 1970); Chem. Abstr., 73, 121749.

55. Alfred Davidson, Fr. Pat. 1,480,806 (May 12, 1967); Chem. Abstr., 67, 118456

56. William L. Groves, Jr. and Billy W. Terry, U.S. Pat. 3,068,279 (Dec. 11, 1962), to Continental Oil Co.; Chem. Abstr., 58, 10411.

57. Wolfgang K. Seifert, U.S. Pat. 3,351,655 (Nov. 7, 1967), to Chevron Research Co.; Chem. Abstr., 68, 4241.

58. Robert B. Doan, Robert C. Taylor, U.S. Pat. 3,436,351 (Apr. 1, 1969), to Atlantic Richfield Co.; Chem. Abstr., 70, 116464.

59. Robert W. Rosenthal, William L. Walsh, and John G. McNulty, U.S. Pat. 3,497,567 (Feb. 24, 1970), to Gulf Research and Development Co.; Chem. Abstr., 57, 16779.

60. John A. Monick, U.S. Pat. 3,055,835 (Sept. 25, 1962), to Colgate Palmolive Co.; Chem. Abstr., 57, 16779.

61. Marvin L. Mausner and Edwin T. Rainier, Fr. Pat. 1,501,615 (Nov. 10, 1967), to Witco Chemical Co., Inc.; Chem. Abstr., 71, 82915.

62. H. Stache, Chem. Process Eng., 47, (10), 68-69, 81 (1966); Chem. Abstr., 66, 4111.

63. Larry E. Meyer and Richard D. Walker, U.S. Pat. 3,345,297 (Oct. 3, 1967), to Procter & Gamble Co.; Chem. Abstr., 109952.

64. Harold E. Feierstein, Chung Yu Shen, and Robert R. Versen, U.S. Pat. 3,390,093 (June 25, 1968), to Monsanto Co.; Chem. Abstr., 69, 44743.

65. Ataru Amahane and Hiroshi Kasai, Japan Pat. 6,917,746 (Aug. 4, 1969), to Kao Soap Co. Ltd.; Chem. Abstr., 71, 126328.

66. Kenneth L. Shaner, Belg. Pat. 667,032 (Jan. 17, 1966), to Monsanto Co.; Chem. Abstr., 65, 7475.

67. Otto Pfrengle, Tenside, 2 (5), 146-150 (1965); Chem. Abstr., 63, 16619.

68. Walter J. Difiley, U.S. Pat. 2,791,562 (May 7, 1957); Chem. Abstr., 51, 10934.

69. Bernard B. Dugan, Ger. Pat. 1,144,866 (Mar. 7, 1963), to Chemical Services (Pty), Ltd.; Chem. Abstr., 58, 14323.

70. Robert L. Fuchs, U.S. Pat. 3,360,469 (Dec. 26, 1967), to FMC Corp.; Chem. Abstr., 68, 41343.

71. Societe anon. d'innovations chimiques Sinnova ou Saide, Fr. Pat. addn. 68,929, July 23, 1958; addn. to Fr. Pat. 1,111,670 and Fr. Pat. addn. 68,013; Chem. Abstr., 55, 5998.

72. E. W. Blank, Soap Chem. Specialties, 34 (1), 41-44, 107 (1958); Chem. Abstr., 52, 4215.

73. E. H. Brandli and R. M. Kelley, J. Amer. Oil Chem. Soc., 47 (6), 200-202 (1970); Chem. Abstr., 73, 26898.

74. V. A. Cirillo, Anal. Chem., 31, 959 (1959); Chem. Abstr., 53, 11092.

75. Kaname Abe, Hiromu Onzuka, and Shigeru Hashimoto, Kogyo Kagaku Zasshi, Japan, 69 (10), 201-219 (1966); Chem. Abstr., 66, 77262.

76. Pavel Pitter, Chem. Prumysl., 14, 320-321 (1964); Chem. Abstr., 61, 6770.

77. R. Wickbold, Mitt. Ver. Grosskesselbesitzer, 65, 106-107 (1960); Chem. Abstr., 56, 8476.

78. Zbigniew Grzbiela and Ewa Wysokinska, Zesz. Nauk. Politech. Slask. Chem., 34, 91-102 (1966); Chem. Abstr., 68, 15877.

79. R. D. Swisher, J. Am. Oil Chemists' Soc., 43 (3), 137-140 (1969); Chem. Abstr., 64, 15563.

80. E. A. Setzkorn and A. B. Carel, J. Am. Oil Chemists' Soc., 40, 57-59 (1963); Chem. Abstr., 58, 9337.

81. Shigeki Nagai, Kogyo Kagaku Zasshi, 69 (3), 452-457 (1966); Chem. Abstr., 69, 106087.

82. C. P. Ogden, J. L. Webster, and J. Halliday, Analyst, 86, 22-29 (1961); Chem. Abstr., 55, 21625.

83. Shintaro Fudano, Kazuo Konishi, J. Chromatogr., 51 (2), 211-218 (1970); Chem. Abstr., 73, 116179.

84. W. Flavell and N. C. Ross, J. Chem. Soc., 5474-5476 (1964); Chem. Abstr., 62, 6421.

85. J. C. Kirk and E. L. Miller, U.S. Pat. 2,928,867 (March 15, 1960), to Continental Oil Co.; Chem. Abstr., 54, 10355.

86. E. A. Fike, U.S. Pat. 2,940,936 (June 14, 1960), to Monsanto Chemical Co.; Chem. Abstr., 54, 16879.

87. William G. Toland, Jr., U.S. Pat. 3,060,232 (Oct. 24, 1962), to California Research Corp.; Chem. Abstr., 58, 5577.

88. F. F. Cheshko and Yu. A. Mirgorod, Zh. Obshch. Khim., 39 (2), 251-259 (1969); Chem. Abstr., 70, 118668.
89. O. S. Mochalova, R. M. Tomson, S. A. Nikitina, and S. I. Faingold, Khim. Tekhnol. Topl. Masel, 11 (12), 17-20 (1966); Chem. Abstr., 66, 67041.
90. Kenji Ohki and Fumikatsu Tokiwa, Nippon Kagaku Zasshi, 90 (11), 1114-1118 (1969); Chem. Abstr., 72, 57004.
91. Jacqueline W. Gershman, J. Phys. Chem., 61, 581-584 (1957); Chem. Abstr., 51, 13517.
92. J. M. Martinez Moreno, F. Catalina, and C. Gomez Herrera, Fette, Seifen, Anstrichmittel, 63, 915-917 (1961); Chem. Abstr., 56, 4116.
93. F. F. Cheshko, Yu. A. Mirgorod, and B. G. Distanov, Zh. Org. Khim., 4 (2), 307-312 (1968); Chem. Abstr., 68, 95579.
94. Fritz Pueschel and Obreten Todorov, Tenside, 5 (7-8), 193-198 (1968); Chem. Abstr., 69, 78644.
95. A. M. Kuliev, D. A. Akhmedzade, and K. I. Sadykhov, Tr. Azerbaidzhan. Gosudarst. Univ. S. M. Kirova, Ser. Khim., 1959, 59-70; Chem. Abstr., 55, 12337.
96. O. Serban and I. V. Nicolescu, Analele Univ., C. I. Parhon, Ser. Stiint Nat., 10 (30), 141-150 (1961); Chem. Abstr., 58, 7429.
97. Henry R. Nychka and Andrew Shultz, Fr. Pat. 1,532,735 (July 12, 1968), to Allied Chemical Corp.; Chem. Abstr., 71, 72233.
98. A. A. Shmidt, K. Kh. Paronyan, N. M. Kafiev, G. V. Mikhailina, A. A. Mikhailovskaya, and A. G. Anton, Maslo-Zhir. Prom., 34 (7), 26-28 (1968); Chem. Abstr., 69, 97880.
99. V. Kh. Paronyan and A. A. Shmidt, Maslo-Zhir. Prom., 34 (12), 24-27 (1968); Chem. Abstr., 70, 81109.
100. Jean Dupre and Richard M. Bassett, Fr. Pat. 1,397,604 (April 30, 1965), to Rohm & Haas Co.; Chem. Abstr., 63, 16621.
101. Iwao Maruta, Rumikatsu Tokiwa, Noboku Kusui, and Hajime Nakatani, Japan Pat. 17,181 (Sept. 11, 1967), to Kao Soap Co. Ltd.; Chem. Abstr., 68, 41342.
102. Cornelis Kortland and Pieter L. Kooijman, Dutch Pat. 88,982 (Aug. 15, 1958), to N. V. de Bataffsche Petroleum Maatschappij; Chem. Abstr., 53, 19978.
103. N. V. de Bataffsche Petroleum Maatschappij, Brit. Pat. 840,406 (July 6, 1960); Chem. Abstr., 55, 10390.
104. Imperial Chemical Industries of Australia and New Zealand Ltd., Australian Pat. 274,388 (Apr. 7, 1967); Chem. Abstr., 69, 53079.
105. Robert R. Fields, U.S. Pat. 3,491,030 (Jan. 20, 1970), to Union Carbide Corporation; Chem. Abstr., 72, 68520.
106. David M. Marquis, U.S. Pat. 3,356,709 (Dec. 5, 1967), to Chevron Research Co.; Chem. Abstr., 68, 41340.
107. Allen H. Lewis and Richard D. Stayner, U.S. Pat. 2,773,833 (Dec. 11, 1956), to California Research Corp.; Chem. Abstr., 51, 4132.
108. Martin D. Reinish, Fr. Pat. 1,569,133 (May 30, 1969), to Colgate Palmolive Co.; Chem. Abstr., 72, 68524.

109. Joachim Amende, Seifen-Oele-Fette-Wachse, <u>90</u> (17), 539-542 (1964); Chem. Abstr., <u>65</u>, 17213.

110. Helmut Stache, Tenside, <u>3</u> (10), 355-359 (1966); Chem. Abstr., <u>67</u>, 55344.

111. Chevron Research Co., Neth. Appl. 6,507,785 (Dec. 20, 1965); Chem. Abstr., <u>65</u>, 17213.

112. Marvin Mausner and Edwin Rainer, Soap Chem. Specialties, <u>44</u> (8), 34-37, 64, 66, 69; <u>44</u> (9), 56, 58, 100-101 (1968); Chem. Abstr., <u>69</u>, 107781.

113. David M. Marquis, U.S. Pat. 3,446,743 (May 17, 1969), to Chevron Research Co.; Chem. Abstr., <u>71</u>, 40631.

114. Louis G. Ricciardi, U.S. Pat. 2,770,600 (Nov. 13, 1956), to Colgate Palmolive Co.; Chem. Abstr., <u>51</u>, 4742.

115. Harold E. Feierstein and John T. Lewis, U.S. Pat. 3,329,616 (July 4, 1967), to Monsanto Co.; Chem. Abstr., <u>67</u>, 65800.

116. Sandor Monori and Erzsebet Varga, Budapesti Muszaki Egyet. Elelmkem. Tansz. Kozlemen, <u>1962</u>, 18-26; Chem. Abstr., <u>59</u>, 5535.

117. A. Delmotte and M. Emond, Therapie, <u>15</u>, 125-133 (1960); Chem. Abstr., <u>56</u>, 9226.

118. A. G. Yusifov, Tr., Vses. Nauch.-Issled. Inst. Vet. Sanit., <u>25</u>, 67-70 (1966); Chem. Abstr., <u>69</u>, 34915.

119. Monsanto Co., Br. Pat. 1,198,358 (July 15, 1970); Chem. Abstr., <u>73</u>, 100294.

120. F. R. Forsyth, Can. J. Botany, <u>42</u> (10), 1335-1347 (1964); Chem. Abstr., <u>61</u>, 16715.

121. Kohei Suzuki, Tokyo Yakka Daigaku Kenkyu Nempo, <u>7</u>, 61-80 (1957); Chem. Abstr., <u>53</u>, 6446.

122. Tetsuya Fujii, Kenkichi Oba, and Shiniehi Tomiyama, Agr. Biol. Chem., <u>25</u>, 735-736 (1961); Chem. Abstr., <u>55</u>, 27678.

123. Orfeo Turno Rotini, Nedo Guerrucci, and Luigi Maffei, Ricerca Sci., Rend., <u>1</u>, 30 (1961); Chem. Abstr., <u>57</u>, 2623.

124. Yasumi Ogura and Junko Tamura, Chiba Daigaku Fuhai Kenkyusho Kokoku, <u>20</u>, 95-99 (1968); Chem. Abstr., <u>70</u>, 45794.

125. Y. Mizushima, Acta Rheumatol. Scand., <u>9</u> (1), 33-40 (1963); Chem. Abstr., <u>59</u>, 15774.

126. C. P. Kurzendoerfer and Hermann Lange, Fette, Seifen, Anstrichmittel, <u>71</u> (7), 561-567 (1969); Chem. Abstr., <u>71</u>, 92949.

127. H. Koebel, D. Klamann, and E. Wagner, Vortraege Originalfassung Intern. Kongr. Grenzflaechenaktive Stoffe, Cologne, <u>3</u> (1), 27-30 (1960); Chem. Abstr., <u>57</u>, 7403.

128. Hideo Unuma, Hitoshi Uesugi, and Minoru Mitamura, Noyaku Seisan Gijutsu, <u>9</u>, 8-12 (1963); Chem. Abstr., <u>61</u>, 5911.

128a. Isao Hirose and Jiro Mikumo, Yushi Kagaku Kyokaishi, <u>3</u>, 107-108 (1954); Chem. Abstr., <u>53</u>, 8666.

129. Shigetaka Kuroiowa, Kogyo Kagaku Zasshi, <u>61</u>, 271-273 (1958); Chem. Abstr., <u>53</u>, 20893.

129a. Fumikatsu Tokiwa and Kensi Ohki, Kolloid-Z. Polym., 223 (1), 38-42 (1968); Chem. Abstr., 69, 70150.

130. S. S. Shanovskaya, Tr. Gos. Makevsk. Nauchn. Issled. Inst. Bezopasnosti Rabot Gorn Prom., 15, 206-240 (1963); Chem. Abstr., 62, 14392.

130a. O. S. Mochalova, R. M. Tomson, S. A. Nikitina, and S. I. Faingold, Khim. Tekhnol. Topl. Masel, 11 (12), 17-20 (1966); Chem. Abstr., 66, 67041.

131. M. T. Berkovich, Sbornik Rabot Silikozu, Adak. Nauk. SSSR, Ural. Filial, 1, 41-49 (1956); Chem. Abstr., 55, 24143.

131a. Armando Morpurgo, Atti Accad. Naz. Lincei, Rend., Cl. Sci. Fis. Mat. Nat., 41 (3, 4), 189-193 (1966); Chem. Abstr., 67, 34088.

132. R. D. Stayner, U.S. Pat. 2,863,836 (Dec. 9, 1958), to California Research Corp.; Chem. Abstr., 53, 7635.

132a. Miklos Gara and Viktor Schwarz, Kolor Ert., 9 (7-8), 197-208 (1967); Chem. Abstr., 67, 118442.

133. W. M. Sawyer, Jr. and F. M. Fowkes, J. Phys. Chem., 62, 159-166 (1958); Chem. Abstr., 52, 9715.

133a. Fumikatsu Tokiwa and Noboru Moriyama, J. Colloid Interface Sci., 30 (3), 338-344 (1969); Chem. Abstr., 71, 40613.

134. M. G. Zimina and N. P. Apukhtina, Kolloid. Zhur., 21, 181-188 (1959); Chem. Abstr., 54, 6178.

135. Teiichi Kaki, Japan Pat. 24,716 (Nov. 4, 1964), to Asahi Chemical Industry Co., Ltd.; Chem. Abstr., 62, 13358.

136. M. G. Zimina and N. P. Paukhtina, Kolloid. Zhur., 21 (1), 50-57 (1959); Chem. Abstr., 54, 6177.

137. J. T. Davies, Proc. Congr. Intern. Federation Soc. Cosmet. Chemists, 2nd, London, 1962, 201-208; Chem. Abstr., 61, 12207.

138. A. M. Mankowich, J. Chem. Eng. Data, 4, 254-256 (1959); Chem. Abstr., 55, 24056.

139. J. Alba Mendoza, Carlos Gomez Herrera, and Carmen Janer del Valle, Parfum. Cosmet. Savons, 11 (9), 383-385 (1968); Chem. Abstr., 70, 21181.

140. S. A. Alekperova, L. Z. Mamedova, G. A. Zaidova, and E. F. Khalilova, Uch. Zap. Azerb. Gos. Univ. Ser. Fiz.-Mat. Khim. Nauk, 4, 103-107 (1960); Chem. Abstr., 58, 3615.

140a. Walter J. Weber, Jr. and Thomas M. Keinath, Chem. Eng. Progr., Symp. Ser., 63 (74), 79-89 (1967); Chem. Abstr., 67, 55567.

141. E. S. Snavely, G. M. Schmid, and R. M. Hurd, Nature, 194, 439-441 (1962); Chem. Abstr., 57, 4061.

142. C. Arouri, Kon. Vlaam. Acad. Wetensch., Lett. Schone Kunsten Belg., Kl. Wetensch., Colloq. Grenslaagverschijnselen Vloeistoffilmen-Schuimen-Emulsies, Brussel, 1965, 144-159; Chem. Abstr., 66, 47580.

143. I. I. Kravchenko, V. A. Brovenko, and N. A. Rybinskaya, Primenenie Poverkhn.-Aktivn. Veshchestv Neft. Prom. Ufimsk. Neft. Nauchn.-Issled. Inst., Tr. Vtorogo Vses. Soveshch, 1963, 160-169; Chem. Abstr., 61, 9683.

144. B. E. Gordon, G. A. Gillies, W. T. Shebs, G. M. Hartwig, and G. R. Edwards, J. Am. Oil Chemists' Soc., 43 (4), 232-236 (1966); Chem. Abstr., 64, 20001.

145. Hans Schott, Text. Res. J., 37 (4), 336 (1967); Chem. Abstr., 67, 12843.

146. Kozo Shimoda, Koji Ito, Kenichi Yasuda, and Kyoji Kinoshita, Shokuhin Eiseigaku Zasshi, 3 (4), 387-391 (1962); Chem. Abstr., 59, 8983.

147. C. A. Smolders, Chem. Phys. Appl. Surface Active Subst., Proc. Int. Congr., 4th., 2, 343-349 (1964); Chem. Abstr., 72, 14107.

148. Procter and Gamble Co., Neth. Appl. 6,513,298 (April 18, 1966); Chem. Abstr., 65, 10813.

149. Yasushi Kimura, Shuhei Tanimori, and Terunosuke Shimo, Yukagaku, 13 (12), 656-661 (1964); Chem. Abstr., 63, 18484.

150. Stelian Anastasiu, Chem. Phys. Appl. Surface Active Subst. Proc. Int. Congr., 4th, 3, 165-175 (1964); Chem. Abstr., 72, 4547.

151. A. M. Mankowich, J. Am. Oil Chemists' Soc., 41 (1), 47-50 (1964); Chem. Abstr., 58, 11581.

152. Haruhiko Arai and Michio Aoki, Kogyo Kagaku Zasshi, 71 (1), 99-104 (1968); Chem. Abstr., 68, 97010.

153. G. A. Kral-Osikina, F. V. Nevolin, T. V. Zhavoronkova, and T. G. Laskina, Maslo-Zhir. Prom., 36 (7), 20-23 (1970); Chem. Abstr., 73, 89452.

154. Fumikatsu Tokiwa and Tetsuya Imamura, J. Amer. Oil Chem. Soc., 47 (11), 422-423 (1970); Chem. Abstr., 73, 132243.

155. S. M. Fadeeva and T. V. Yapryntseva, Kul't. Byt. Izdeliya, 1967, 77-82; Chem. Abstr., 70, 12839.

156. F. V. Nevolin, G. A. Kral-Osikina, and A. V. Yushkevich, Masloboina-Zhirovaya Prom., 23 (7), 33-34 (1957); Chem. Abstr., 51, 18658.

157. Helen A. Ludeman, John A. Balog, and J. C. Sherrill, J. Am. Oil Chemists' Soc., 35, 5-8 (1958); Chem. Abstr., 52, 3369.

158. A. Zanella and P. Peri, Riv. Ital. Sostance Grasse, 43 (9), 369-381 (1966); Chem. Abstr., 66, 12150.

159. P. Peri and A. Zanella, Riv. Ital. Sostanze Grasse, 42 (12), 573-583 (1965); Chem. Abstr., 64, 20003.

160. H. Koelbel, D. Klamann, and V. Hopp, Vortaege Originalfassung Intern. Kongr. Grenzflacchenaktive Stoffe, 3, Cologne, 1960, 4, 173-177; Chem. Abstr., 57, 9974.

161. L. I. Autus, F. V. Nevolia, G. I. Nikisiu, and A. D. Petrov, Parfums. Cosmet. Savons, 13 (1), 29-31 (1958); Chem. Abstr., 52, 11447.

162. L. I. Autus, F. V. Nevolia, G. K. Nikisiu, and A. D. Petrov, Rev. Chim., 8, 686-688 (1957); Chem. Abstr., 52, 10613.

163. Allied Chemical Corp., Neth. Appl. 6,603,138 (Sept. 12, 1966); Chem. Abstr., 66, 96502.

164. I. W. Griess, Fette Seifen, Anstrichmittel, 57, 24-32 (1955); Chem. Abstr., 51, 1879.

165. M. A. Ashimov and M. A. Mursalova, Dokl. Akad. Nauk. Azerb. SSR, 23 (8), 23-28 (1967); Chem. Abstr., 68, 70427.
166. Shinji Inagaki, Yasuji Izawa, and Yoshiro Ogata, Kogyo Kaguku Zasshi, 69 (11), 2161-2164 (1966); Chem. Abstr., 66, 77260.
167. F. V. Nevolin, G. I. Nikichin, A. D. Petrov, G. A. Kral-Isikina, and V. D. Vorobiev, Parfums, Cosmet. Savons, 2, 62-66 (1959); Chem. Abstr., 53, 10809.
168. Frederick W. Gray and Irving J. Krems, J. Org. Chem., 26, 200-212 (1961); Chem. Abstr., 55, 18646.
169. Frederick W. Gray, Irving J. Krems, and J. Fred Gerecht, J. Am. Oil Chemists' Soc., 42 (11), 998-1001 (1965); Chem. Abstr., 64, 3857.
170. Kongandra T. Achaya, Chim. Phys. Appl. Prat. Ag. Surface, C. R. Congr. Int. Deterg., 5th, Sept. 9-13, 1968, 1, 63-70; Chem. Abstr., 74, 14409.
171. M. A. Ashimov, M. A. Mamedova, and T. N. Mamed-Zade, (USSR), Sin. Prevrashch. Monomernykh Soedin, 1967, 76-86; Chem. Abstr., 70, 21188.
172. N. I. Shuikin, N. A. Pozdvnak, T. P. Dobyrynina, and A. Yu. Rabinovich, Izv. Akad. Nauk SSSR, Ser. Khim., 1966 (7), 1248-1251; Chem. Abstr., 65, 17213.
173. Yukio Kasai, Yasuji Izawa, and Yoshiro Ogata, Kogyo Kagaku Zasshi, 68 (11), 2073-2077 (1965); Chem. Abstr., 64, 12974.
174. Roger Perrus, Daniel Prace, and Marcel Prillieux, Fr. Pat. 1,345,126 (Dec. 6, 1963), to Esso Standard Societe Anon. Francaise; Chem. Abstr., 60, 16107.
175. P. A. Demchenko and A. V. Dumanskii, Kolloid. Zhur., 22, 272-276 (1960); Chem. Abstr., 55, 8997.
176. K. T. Achaya and P. K. Saraswathy Amma, Ind. J. Chem., 5 (3), 109-114 (1967); Chem. Abstr., 67, 45100.
177. W. A. Sweeney, Soap Chem. Specialties, 40 (3), 45-47, 190 (1964); Chem. Abstr., 60, 13456.
178. Ciattoni and Scardigno, Riv. Ital. Sostanze Grasse, 45 (1), 15-26 (1968); Chem. Abstr., 69, 11631.
179. V. P. Kaplin, N. G. Fesenko, A. I. Fomina, and L. S. Zhuravleva, Nauch. Tr., Akad. Kommunal. Khoz., 37, 97-110 (1965); Chem. Abstr., 66, 79389.
180. A. M. Bruce, J. D. Swanwick, and R. A. Ownsworth, Inst. Sewage Purif. J. Proc., 1966 (5), 427-442; Chem. Abstr., 66, 118609.
181. John W. Hernandez and Don E. Bloodgood, J. Water Pollution Control Fed., 32, 1261-1268 (1960); Chem. Abstr., 55, 7721.
182. P. Pitter, Fortschr. Wasserchem. Ihrer Grenzgeb., 3, 36-41 (1965); Chem. Abstr., 68, 60767.
183. P. H. McGauhey and Stephen A. Klein, Purdue Univ. Eng. Bull. Ext. Ser., 117 (Part 1), 1-8 (1964); Chem. Abstr., 66, 5588.
184. R. H. Bogan, Sewage Ind. Wastes, 30, 208-214 (1958); Chem. Abstr., 53, 621.

185. R. Krueger, Feete, Seifen Anstrichmittel, 66, 217-221 (1964); Chem. Abstr., 61, 5909.

186. Robert D. Swisher, Surfactant Science Series, Vol. 3, Marcel Dekker, Inc., New York, 1970.

187. Ross E. McKinney and James M. Symons, Sewage Ind. Wastes, 31, 549-556 (1959); Chem. Abstr., 53, 18348.

188. Herbert Koelbel, Peter Kurzendoerfer, and Claus Werner, Tenside, 4 (2), 33-40 (1967); Chem. Abstr., 66, 116936.

189. R. C. Tarring, Air Water Pollution, 9 (9), 545-552 (1965); Chem. Abstr., 64, 3858.

190. Pavel Pitter, Chem. Prumysl, 13 (6), 284-287 (1963); Chem. Abstr., 59, 11764.

191. K. Offhaus, Juenchner Beitr. Abwasser Fisch. Flussbiol., 9, 64-77 (1967); Chem. Abstr., 69, 5054.

192. Ray C. Allred and R. L. Huddleston, Southwest Water Works J., 49 (2), 26-28, 30 (1967); Chem. Abstr., 67, 57126.

193. B. Berger, Ind. Chim., 51 (569), 421-431 (1964); Chem. Abstr., 62, 13392.

194. Yasuo Fujiwara, Tetsuya Takezono, Saburo Kyono, Sadao Sakayanagi, Kazuhide Yamasato, and Hiroshi Iizuka, Yukagaku, 17 (7), 396-399 (1968); Chem. Abstr., 69, 53074.

195. Raymond C. Allred and Robert L. Huddleston, U.S. Pat. 3,138,543 (June 23, 1964), to Continental Oil Co.; Chem. Abstr., 61, 7667.

196. R. L. Huddleston and R. C. Allred, Develop. Ind. Microbiol., 4, 24-38 (1962); Chem. Abstr., 61, 11738.

197. P. Pitter, Sb. Vysoke Skoly Chem.-Technol. Praze, Technol. Vody, 8 (2), 13-39 (1965); Chem. Abstr., 64, 19180.

198. C. Borstlap and C. Kortland, J. Am. Oil Chemists' Soc., 44 (5), 295-297 (1967); Chem. Abstr., 67, 12844.

199. B. L. Lipman, Nauch. Tr. Akad. Kommun Khoz., 37, 14-25 (1965); Chem. Abstr., 66, 77259.

200. V. D. Artemenko, Nauch. Tr. Akad. Kommun Khoz., 37, 125-130 (1965); Chem. Abstr., 66, 79388.

201. Werner Winter, Wasserwirtsch.-Wassertech., 14 (12), 369-372 (1964); Chem. Abstr., 65, 11966.

202. Robert D. Swisher, Purdue Univ. Eng. Bull., Ext. Ser., No. 129 (Pt. 1), 375-392 (1967); Chem. Abstr., 70, 108980.

203. Robert D. Swisher, Ind. Chim. Belge, 32 (Spec. No.), 719-722 (1967); Chem. Abstr., 70, 60612.

204. R. D. Swisher, Develop. Ind. Microbiol., 9, 270-279 (1967); Chem. Abstr., 69, 20634.

205. Melvin A. Benarde, Bernard W. Koft, Raymond Horvath, and Louis Shaulis, Appl. Microbiol., 13 (1), 103-105 (1965); Chem. Abstr., 62, 4370.

206. R. L. Huddleston and E. A. Setzkorn, Soap Chem. Specialties, 41 (3), 63-64, 120-121 (1965); Chem. Abstr., 62, 14935.

207. R. B. Cain and D. R. Farr, Biochem. J., <u>106</u> (4), 859-877 (1968); Chem. Abstr., <u>68</u>, 93673.
208. John Cairns, Jules J. Loos, Proc. Pa. Acad. Sci., <u>40</u> (2), 47-52 (1967); Chem. Abstr., <u>69</u>, 84542.
209. V. M. Brown, V. V. Mitrovic, and G. T. C. Stark, Water Res., <u>2</u> (4), 255-263 (1968); Chem. Abstr., <u>69</u>, 42052.
210. H. Mann, Muenchner Beitr. Abwasser.-Fisch.-Flussbiol., <u>9</u>, 131-138 (1967); Chem. Abstr., <u>68</u>, 107734.
211. Thomas O. Thatcher and Joseph F. Santner, Purdue Univ. Eng. Bull. Ext. Ser., No. 121, 996-1002 (1966); Chem. Abstr., <u>67</u>, 57103.
212. G. A. Holland, J. E. Lasater, E. D. Neumann, and W. E. Eldridge, Wash. Dept. Fisheries, Res. Bull. No. 5, 278 (1960); Chem. Abstr., <u>62</u>, 2019.
213. R. Marchettin, Riv. Ital. Sostanze Grasse, <u>41</u> (10), 533-542 (1964); Chem. Abstr., <u>62</u>, 10895
214. Erich Hirsch, Vom Wasser, <u>30</u>, 249-259 (1963); Chem. Abstr., <u>61</u> 16498.
215. J. Wurtz-Arlet, Vortraege Originalfassung Intern. Kongr. Grenz-flaechenaktive Stoffe, 3, Cologne, <u>3</u>, 329-331 (1960); Chem. Abstr., <u>57</u>, 10374.
216. O. J. Schmid and H. Mann, Nature, <u>192</u>, 675 (1961); Chem. Abstr., <u>56</u>, 6490.
217. K. Opitz and A. Loeser, Experientia, <u>20</u> (5), 277-278 (1964); Chem. Abstr., <u>61</u>, 1015.
218. William R. Michael, Toxicol. Appl. Pharmacol., <u>12</u>, 473-485 (1968); Chem. Abstr., <u>69</u>, 42460.
219. Bernard L. Oser and Kenneth Morgareidge, Toxicol. Appl. Pharmacol., <u>7</u> (6), 819-825 (1965); Chem. Abstr., <u>64</u>, 652.
220. T. W. Tusing, O. E. Paynter, and D. L. Opdyke, Toxicol. Appl. Pharmacol., <u>2</u>, 464-473 (1960); Chem. Abstr., <u>54</u>, 21510.
221. J. H. Kay, F. E. Kohn, and J. C. Calandra, Toxicol. Appl. Pharmacol., <u>7</u> (6), 812-818 (1965); Chem. Abstr., <u>64</u>, 2651.
222. Orville E. Paynter and Robert J. Weir, Jr., Toxicol. Appl. Pharmacol., <u>2</u>, 641-648 (1960); Chem. Abstr., <u>55</u>, 3849.
223. Yoshihito Omori, Tsukasa Kuwamura, Kunio Kawashima, and Shinsuke Nakaura, Shokuhin Eiseigaku Zasshi, <u>9</u> (6), 473-480 (1968); Chem. Abstr., <u>70</u>, 76083.
224. H. Havermann and K. H. Menke, Landwirtsch. Forsch. Sonderh., <u>9</u>, 140-146 (1957); Chem. Abstr., <u>51</u>, 14980.
225. Fumimasa Yanagisawa and Tatunori Yamagishi, Tokyo Ritsu Eisei Kenkyusho Nenpu, <u>18</u>, 105-111 (1966); Chem. Abstr., <u>69</u>, 9437.
226. Akira Takase and Tomoko Bito, Koshu Eiseiin Kenkyu Hokoku, <u>12</u>, 237-243 (1963); Chem. Abstr., <u>62</u>, 2148.
227. Yoshio Ikeda, Yoshito Omori, Tokuichi Itomine, Hiroshi Muto, and Hamako Yoshimoto, Shokuhin Eiseigaku Zasshi, <u>3</u>, 399-401 (1962); Chem. Abstr., <u>59</u>, 9232.

228. T. L. Chrusciel, Z. Kleinrock, M. Chrusciel, R. Brus, W. Janiec, and H. Trzeciak, Med. Pracy., 17 (2), 104-108 (1966); Chem. Abstr., 66, 64165.

229. V. F. Garshenin, Nauch. Tr. Akad. Kommun. Khoz., 37, 111-124 (1965); Chem. Abstr., 66, 88572.

230. Marton Valer, Munkavedelem, 14 (1-3), 38-50 (1968); Chem. Abstr., 69, 97876.

231. F. Ray Bettley, Br. J. Dermatol., 77, 98-100 (1965); Chem. Abstr., 63, 7545.

232. Leonard J. Vinson and B. R. Choman, J. Soc. Cosmetic Chemists, 11, 127-137 (1960); Chem. Abstr., 54, 15721.

233. James Scala, Don E. McOsker, and Herbert H. Reller, J. Invest. Dermatol., 50 (5), 371-379 (1968); Chem. Abstr., 69, 113020.

234. Jan E. Wahlberg, Acta Dermato-Venereol., 45 (5), 335-343 (1965); Chem. Abstr., 64, 11706.

235. T. Chrusciel, M. Chrusciel, Z. Kleinrok, Z. Herman, A. Kruszyna, and L. Samochowiee, Med. Pracy., 10, 235-246 (1959); Chem. Abstr., 54, 13425.

236. A. L. Reshetyuk, L. S. Shevchenko, Gig. Tr. Prof. Zabol, 12 (8), 43-36 (1968); Chem. Abstr., 70, 31441.

237. P. R. den Dulk, Neth. J. Agr. Sci., 8, 139-148 (1960: Chem. Abstr., 60, 16448.

238. L. L. Jansen, Weeds, 13 (2), 117-123 (1965); Chem. Abstr., 63, 3555.

239. C. G. L. Furmidge, J. Sci. Food Agr., 10, 419-425 (1959); Chem. Abstr., 54, 805.

240. D. Matulova, Vodni Hospodarstvi, 14 (10), 377-378 (1964); Chem. Abstr., 62, 10210.

241. D. Matulova, Sb. Vysoke Skoly Chem. Technol. Praze, Technol. Vody, 8 (2), 251-301 (1965); Chem. Abstr., 62, 10210.

242. Ludwig Harmann, Gesundh. Ing. Z., 87 (10), 301-305 (1966); Chem. Abstr., 66, 88481.

243. C. H. Wayman, J. B. Robertson, and H. G. Page, U.S. Geol. Surv. Profess. Paper, 475-B, 205-208 (1963); Chem. Abstr., 59, 13134.

244. Procter and Gamble Co., Belg. Pat. 654,738 (Feb. 16, 1965); Chem. Abstr., 64, 19977.

245. Procter and Gamble Co., Belg. Pat. 665,954 (Dec. 27, 1965); Chem. Abstr., 65, 923.

246. Marles-Kuhmann-Wyandotte, Neth. Appl. 6,602,590 (Oct. 10, 1966); Chem. Abstr., 66, 47605.

247. Societe Anon. d'Innovations Chimiques "Sinnova" ou "Sadic", Fr. Pat. 1,508,266 (Jan. 5, 1968); Chem. Abstr., 70, 39117.

248. Burton H. Gedge, III, Fr. Pat. 1,398,753 (May 14, 1965), to Procter and Gamble Co.; Chem. Abstr., 64, 3860.

249. Jerald A. Cavataio, Raymond C. Odioso, Ger. Offen. 1,804,872 (June 26, 1969), to Colgate-Palmolive Co.; Chem. Abstr., 71, 40632.

250. Richard P. Carter, Jr., Fr. Pat. 1,529,466 (June 14, 1968), to Monsanto Co.; Chem. Abstr., 71, 14431.

251. Unilever N. V., Neth. Appl. 67,09,714 (Jan. 15, 19680; Chem. Abstr., 69, 20640.

252. Procter and Gamble Co., Neth. Appl. 6,501,871 (Aug. 16, 1966); Chem. Abstr., 66, 4122.

253. Karl Herrle, Ludwig Zuern, Owsald Schmidt, Heinrich Mertens, and Sigismund Heimann, Ger. Pat. 1,134,786 (Aug. 16, 1961), to Badische Anilin-& Soda-Fabrik A.-G.; Chem. Abstr., 57, 16779.

254. Fumikatsu Tokiwa, Kenji Ohki, Takedo Wada, Tetsuya Imamura, and Shokichi Igarashi, Ger. Offen. 1,957,456 (July 9, 1970), to Kao Soap Co., Ltd.; Chem. Abstr., 73, 67911.

255. Jack T. Inamorato, Ger. Pat. 1,145,735 (Mar. 21, 1963), to Colgate-Palmolive Co.; Chem. Abstr., 58, 14323.

256. Arnord S. Roald, Nicolaas T. Deoude, and Charles B. McCarth, Fr. Pat. 1,561,078 (March 28, 1969), to Procter and Gamble Co.; Chem. Abstr., 72, 14117.

257. Arnvid S. Roald, and Nicolaas T. Deoude, Fr. Pat. 1,520,262 (April 5, 1968), to Procter and Gamble European Technical Center; Chem. Abstr., 70, 14444.

258. Unilever N. V., Neth. Pat. 113,890 (Dec. 15, 1967); Chem. Abstr., 68, 97027.

259. Marvin M. Crutchfield and Riyad R. Irani, U.S. Pat. 3,297,578 (Jan. 10, 1967), to Monsanto Co.; Chem. Abstr., 66, 47606.

260. Procter and Gamble Co., Belg. Pat. 621,901 (Dec. 14, 1962); Chem. Abstr., 58, 14324.

261. Daniel M. Van Kampen, Frederik J. Kerkhove, and Willem Van der Star, Ger. Pat. 1,277,496 (Sept. 12, 1968); Chem. Abstr., 70, 12851.

262. Alexander Ritchie, Br. Pat. 1,050,791 (Dec. 7, 1966), to Procter and Gamble Ltd.; Chem. Abstr., 66, 47598.

263. Norman R. Smith, Ger. Pat. 1,170,574 (May 21, 1964), to Procter and Gamble Co.; Chem. Abstr., 61, 5913.

264. Mojmir Ranny, Josef Kvapilik, and Milos Nigrin, Czech. Pat. 102,442 (Jan. 15, 1962); Chem. Abstr., 59, 5372.

265. Harold E. Wixon, U.S. Pat. 3,075,922 (Jan. 29, 1963), to Colgate-Palmolive Co.; Chem. Abstr., 58, 40411.

266. David C. Steer and Norman R. Smith, U.S. Pat. 3,085,982 (April 16, 1963), to Procter and Gamble Co.; Chem. Abstr., 58, 14323.

267. Robert Paul Davis and Frank Joseph Mueller, U.S. Pat. 3,503,889 (March 31, 1970), to Procter and Gamble C0.; Chem. Abstr., 72, 134462.

268. David F. Bath, U.S. Pat. 3,407,144 (Oct. 22, 1968), to Procter and Gamble Co.; Chem. Abstr., 70, 30329.

269. George C. Hampson and William Rickatson, U.S. Pat. 3,185,649 (May 25, 1965), to Lever Brothers Co.; Chem. Abstr., 63, 4519.

270. Daniel M. van Kampen, Frederick J. Kerkhoven, and Willem van der Star, U.S. Pat. 3,370, 015 (Feb. 20, 1968), to Lever Brothers Co.;

271. Jerome S. Schrager and Harold E. Wixon, U.S. Pat. 3,247,123 (April 19, 1966), to Colgate-Palmolive Co.; Chem. Abstr., 65, 4116.

272. Theodore L. Treitler, U.S. Pat. 3,393,154 (July 16, 1968), to
     Colgate-Palmolive Co.; Chem. Abstr., 69, 60208.
273. Procter and Gamble Co., Neth. Appl. 6,511,599 (March 7, 1967);
     Chem. Abstr., 67, 55353.
274. Benjamin R. Briggs, U.S. Pat. 3,321,408 (May 23, 1967), to Purex
     Corp., Ltd.; Chem. Abstr., 34096.
275. Marvin L. Mausner and Arnold H. Dater, Fr. Pat. 1,501,661
     (Nov. 10, 1967), to Witco Chemical Co., Inc.; Chem. Abstr., 69,
     88203.
276. Hill M. Priestley and James H. Wilson, U.S. Pat. 3,317,430 (May 2,
     1967), to Lever Brothers Co.; Chem. Abstr., 67, 34099.
277. James H. Wilson, U.S. Pat. 3,150,098 (Sept. 22, 1964), to Lever
     Brothers Co.; Chem. Abstr., 62, 737.
278. John M. Blakeway and Ian D. Burgess, Br. Pat. 921,036 (March 13,
     1963), to Colgate-Palmolive Ltd.; Chem. Abstr., 58, 14323.
279. Shell Internationale Research Maatschappij N.V., Belg. Pat. 610,211
     (May 14, 1962); Chem. Abstr., 57, 6779.
280. Howard F. Drew and Roger E. Zimmere, U.S. Pat. 3,001,945
     (Sept. 21, 1961), to Procter and Gamble Co.; Chem. Abstr., 56, 560.
281. Ronald L. Hemingway, Ger. Pat. 1,130,546 (May 30, 1962), to Proc-
     ter and Gamble Co.; Chem. Abstr., 57, 9979.
282. Thomas G. Jones and David W. Stephens, U.S. Pat. 3,281,367
     (Oct. 25, 1966), to Lever Brothers Co.; Chem. Abstr., 66, 4124.
283. Unilever N.V., Neth. Appl. 6,610,250 (Jan. 23, 1967); Chem.
     Abstr., 67, 12865.
284. William Chirash and Robert H. Trimmer, Fr. Pat. 1,523,007
     (April 2, 1968), to Colgate-Palmolive Co.; Chem. Abstr., 71, 31661.
285. Morris V. Shelanski and Murray W. Winicov, U.S. Pat. 2,977,278
     (Mar. 28, 1961), to West Laboratories, Inc.; Chem. Abstr., 55,
     12753.
286. P. M. Borick, P. M. Brown, F. H. Dondershine, and R. A. Hollis,
     Chem. Specialties Mfrs. Assoc. Proc. Ann. Meeting, 54, 110–114
     (1967); Chem. Abstr., 69, 54273.
287. Christopher M. Brooke and Richard T. Littleton, Ger. Offen.
     1,930,636 (July 9, 1970), to Albright and Wilson Ltd.; Chem. Abstr.,
     71, 57448.
288. Gerald W. Wyant, Br. Pat. 1,078,554 (Aug. 9, 1967); Chem. Abstr.,
     67, 83144.
289. Pennsalt Chemicals Corp., Br. Pat. 917,432 (Feb. 6, 1963); Chem.
     Abstr., 58, 10411.
291. Daniel Stewart, Br. Pat. 973,670 (Oct. 28, 1964), to Scottish Oils
     Ltd.; Chem. Abstr., 62, 737.
292. Werner Prosch, Ger. Pat. 1,141,040 (Dec. 13, 1962), to Chemische
     Werke Witten G.m.b.H.; Chem. Abstr., 58, 5905.
293. Russell E. Compa, Charles F. Fischer, Robert Tweedy Hunter, Jr.,
     and Raymond C. Odioso, S. African Pat. 6,806,822 (April 21, 1970),
     to Colgate-Palmolive Co.; Chem. Abstr., 73, 89467.

294. Sintex Organico-Industrial, S.A., Span. Pat. 341,838 (July 1, 1968); Chem. Abstr., 70, 12850.
295. Colgate-Palmolive Co., Fr. Pat. 1,576,756 (Aug. 1, 1969); Chem. Abstr., 72, 91674.
296. Francis S. K. MacMillan, U.S. Pat. 3,226,329 (Dec. 28, 1965), to Procter and Gamble Co.; Chem. Abstr., 64, 8503.
297. W. A. Kelly, Belg. Pat. 616,497 (Oct. 16, 1962), to Unilever N.V.; Chem. Abstr., 58, 10411.
298. Gordon W. Bell, Jr., U.S. Pat. 3,383,320 (May 14, 1968), to Avisun Corp.; Chem. Abstr., 69, 28790.
299. Louis McDonald, U.S. Pat. 3,278,444 (Oct. 11, 1966), to Kelite Chemicals Corp.; Chem. Abstr., 66, 47591.
301. John C. Wittwer, Fr. Pat. 1,351,411 (Feb. 7, 1964); Chem. Abstr., 61, 2066.
302. Jean V. Morelle, Fr. Pat. Addn. 79,425 (Nov. 30, 1962); Chem. Abstr., 58, 14324.
303. Chemische Fabrik Grunau Akt.-Ges., Ger. Pat. 968,127 (Jan. 16, 1958); Chem. Abstr., 54, 7988.
304. Edward Eigen and Sidney Weiss, S. African Pat. 67,00,945 (Aug. 17, 1968), to Colgate-Palmolive Co.; Chem. Abstr., 70, 79412.
305. Curt Wholfarth (VEB Farbenfabrik Wolfen), Ger. (East) Pat. 31,898 (April 26, 1965); Chem. Abstr., 63, 11246.
306. Tikitaka Shiba, Japan Pat. 6150 (Aug. 10, 1957); Chem. Abstr., 52, 9532.
307. Masao Gotoda, Impregnated Fibrous Mater. Rep. Study Group, 1967, 107-114; Chem. Abstr., 70, 98067.
308. Raymond C. De Wald, U.S. Pat. 3,370,028 (Feb. 20, 1968), to Firestone Tire & Rubber Co.; Chem. Abstr., 68, 69567.
309. M. A. Ashimov, Yu. D. Bukh, S. I. Sadykh-Zade, and S. S. Shchegol, Vses. Soveshch Sin. Zhirozamen., Poverkhnostnoaktiv. Veshchestvam Moyushch. Sredstvam, 3rd, Shebekino, 1965, 293-296; Chem. Abstr., 66, 76710.
310. Maurice Morton and B. Das, Am. Chem. Soc. Div. Polymer Chem., Preprints, 7 (2), 544-547 (1966); Chem. Abstr., 66, 76694.
311. Maurice Morton, Irja Piirma, and B. Das, Rubber Plastics Age, 46, 404-409 (1965); Chem. Abstr., 65, 15630.
312. Edwin S. Smith and James E. Sell, U.S. Pat. 3,226,530 (Dec. 18, 1965), to Goodyear Tire & Rubber Co.; Chem. Abstr., 64, 14375.
313. Robert O. Symcox, Brit. Pat. 998,345 (July 14, 1965), to Monsanto Chemicals Ltd.; Chem. Abstr., 63, 8562.
314. C. E. McCoy, Jr., A. S. Teot, and W. H. Cass, Paint Technol., 26 (9), 36-40 (1962); Chem. Abstr., 57, 12752.
315. Reginald D. Singer and Eric T. Yarwood, Brit. Pat. 976,086 (Nov. 25, 1964), to Dunlop Rubber Co. Ltd.; Chem. Abstr., 62, 4194.
316. Paul Wicht, Swiss Pat. 373,184 (Dec. 31, 1963), to Lonza Ltd.; Chem. Abstr., 60, 16067.

317. The Dow Chemical Co., Br. Pat. 912,068 (Dec. 5, 1962); Chem. Abstr., 58, 7042.
318. Thomas Love and James R. Wallace, Br. Pat. 882,535 (Nov. 15, 1961); Chem. Abstr., 57, 2428.
319. Alexi Trofimow, Philip K. Isaacs, and Donald Goodman, Ger. Pat. 1,112,832 (Aug. 17, 1961), to W. R. Grace & Co.; Chem. Abstr., 56, 10406.
320. Tadashi Ikemura and Kooji Ikeda, Nippon Daigaku Kogaku Kenkyusho Iho, 15, 238-241 (1957); Chem. Abstr., 53, 10829.
321. Tadaki Bito and Hiromi Yamakita, Yukagaku, 8, 22-27 (1959); Chem. Abstr., 54, 14769.
322. Francesco Siclari, Ital. Pat. 532,304 (Aug. 22, 1955), to SNIA VIS-COSA Societa Nazionale Industria Applicazioni Viscosa Societa per azioni; Chem. Abstr., 52, 17805.
323. I. N. Bybachok, USSR Pat. 170,602 (April 23, 1965); Chem. Abstr., 63, 9726.
324. M. P. Bortsova, G. D. Pavlov, R. A. Filina, R. A. Martirosov, N. P. Shpichko, and M. I. Rebeza, Tr. Grozmensk. Neft. Nauch.-Issled. Inst., 15, 34-41 (1963); Chem. Abstr., 60, 11809.
325. A. A. Petrov, N. P. Borisova, Tr. Gos. Inst. Proekt. Issled. Rab. Neftedobyvayushchei Prom., 10, 88-95 (1967); Chem. Abstr., 70, 12843.
326. E. A. Myshkin, Neft. Khoz., 44 (5), 55-58 (1966); Chem. Abstr., 65, 7473.
327. David A. Ellis, U.S. Pat. 3,219,422 (Nov. 23, 1965), to Dow Chemical Co.; Chem. Abstr., 64, 9283.
328. Masatoshi Nakayama, Japan Pat. 8722 (July 2, 1960), to Asahi Chemical Industry Co., Ltd.; Chem. Abstr., 55, 7766.
329. Masatoshi Nakayama, Japan Pat. 8721 (July 7, 1960), to Asahi Chemical Industry Co., Ltd.; Chem. Abstr., 55, 7767.
330. Ksamu Koizumi, Kogyo Kagaku Zasshi, 70 (10), 1641-1646 (1967); Chem. Abstr., 69, 5676.
331. O. G. Tarakanov and E. G. Dubyaga, Vses. Soveshch. Sintetich. Zhirozamenitelyam, Poverkihnostuoacktivn. Veshchestvam Moyushchim Sredstvam, 3rd, Sb., Shebekino, 1965, 387-388; Chem. Abstr., 66, 4100.
332. Rudolf Kern and Wilhelm N. Grohs, U.S. Pat. 2,891,016 (June 16, 1959), to Rhein-Chemie G.m.b.H.; Chem. Abstr., 53, 17555.
333. Continental Oil Co., Fr. Pat. 1,313,443 (Dec. 28, 1962); Chem. Abstr., 58, 14242.
334. Achyut K. Phansalkar and Jack L. Brown, U.S. Pat. 3,269,468 (Aug. 30, 1966), to Continental Oil Co.; Chem. Abstr., 66, 4682.
335. A. I. Gorbatova, Yu. V. Solyakov, and I. A. Shvestov, Nauch-Tekh. Sb. Dobyche Nefti, Vses. Nefte-Bazov. Nauch.-Issled. Inst., 1968, 32, 75-82; Chem. Abstr., 69, 78909.
336. Central Scientific Research Institute of Technology and Machine Building., Fr. Pat. 1,441, 044 (June 3, 1966); Chem. Abstr., 66, 88365.

337. C. Cosneanu, E. Cohn, and L. Soforni, Metalurgia (Bucharest), 19 (10), 545-548 (1967); Chem. Abstr., 69, 21470.

338. Carl Rauh, Alois Hewel, and D. Bahner, Giesserei, 54 (14), 374-378 (1967); Chem. Abstr., 68, 5906.

339. Joseph R. Parsons, U.S. Pat. 3,150,989 (Sept. 29, 1964), to Chicago Fire Brick Co.; Chem. Abstr., 61, 14329.

340. Stanislaw Bastian and Malgorzata Gruner, Pol. Pat. 46,430 (Nov. 12, 1962); Chem. Abstr., 59, 3638.

341. Masatane Kokubu and Masaki Kobayashi, J. Fac. Eng. Univ. Tokyo, Ser. B, 29 (4), 247-260 (1968); Chem. Abstr., 70, 40440.

342. Chuyo Hisatsune and Akina Shimizu, Sekko To Sekkai, 37 (10-17) (1958); Chem. Abstr., 53, 677.

343. Toyo Rayon Co., Ltd., Fr. Pat. 1,526,402 (May 24, 1968); Chem. Abstr. 70, 116170.

344. Alfred Reichle, Martin Wandel, Ernst Gutschik, and Dietrich Glabisch, Ger. Pat. 1,213,943 (April 7, 1966), to Farbenfabriken Bayer A.-G.; Chem. Abstr., 64, 20004.

345. Werner Blank, Alfred Reichle, and Martin Wandel, Ger. Pat. 1,209, 542 (Jan. 27, 1966), to Farbenfabriken Bayer A.-G.; Chem. Abstr., 64, 12874.

346. Chemische Werke Huels A.-G., Belg. Pat. 658,964 (May 17, 1965); Chem. Abstr., 64, 8415.

347. Julian J. Hirshfeld and Edward Szlosberg, Belg. Pat. 632,531 (Nov. 18, 1963), to Monsanto Co.; Chem. Abstr., 61, 4544.

348. Werner Roth, Annemarie Jung, and Erika Berkoben, Ger. (East) Pat. 41,701 (Oct. 5, 1965); Chem. Abstr., 64, 8373.

349. O. G. Tarakanov and A. I. Demina, Vysokomolekul Soedin., 7 (2), 224-225 (1965); Chem. Abstr., 62, 13321.

350. Heinz Nemitz, Gerhard Grosse, Reinhard Hoerl, Hans U. Lindner, Edward Puesche, and Peter Zwicker, Ger. (East) Pat. 57,112 (Aug. 5, 1967); Chem. Abstr., 68, 97184.

351. Paul H. Ralson, U.S. Pat. 3,353,927 (Nov. 21, 1967), to Calgon Corp.; Chem. Abstr., 68, 23231.

352. Peter Urban and William K. T. Gleim, U.S. Pat. 2,998,304, to Universal Oil Products Co.; Chem. Abstr., 56, 2152.

353. Wm. C. Bauer, Allen P. McCue, and Kennerth C. Rule, U.S. Pat. 2,954,282 (Sept. 27, 1960), to Food Machinery and Chemical Corp.; Chem. Abstr., 55, 4902.

354. Yukio Okamoto, Itaru Hiraki, Michio Sekiya, Goro Yamaguchi, and Shoichiro Nagai, Sekko To Sekkai, 1, 1425-1429 (1957); Chem. Abstr., 51, 17114.

355. Giovanni Petrosini and Geltrude Gugnoni, Univ. Studi Perugia, Ann. Fac. Agrar., 12, 17-32 (1956); Chem. Abstr., 52, 18987.

356. James E. Seymour, Congr. mondial detergence et prod. tensioactifs, 1st, Paris, 3, 1127-1138 (1954); Chem. Abstr., 52, 634.

357. Jay C. Harris, U.S. Pat. 2,844,455 (July 22, 1958), to Monsanto Chemical Co.; Chem. Abstr., 52, 18990.

358. Keith L. Johnson, U.S. Pat. 3,294,703 (Dec. 27, 1966), to Swift & Co.; Chem. Abstr., 66, 39074.

359. John A. Knox, U.S. Pat. 3,374,835 (March 26, 1968), to Halliburton Co.; Chem. Abstr., 69, 116199.

360. Deral D. Knight, U.S. Pat. 3,353,603 (Nov. 21, 1967), to Byron Jackson Inc.; Chem. Abstr., 68, 41959.

361. John Chor, Jr. and Lee N. Hodson, U.S. Pat. 3,341,454 (Sept. 12, 1967), to Hodson Corp.; Chem. Abstr., 67, 118914.

362. James Hossack, Br. Pat. 1,072,702 (June 21, 1967), to Geigy (U.K.) Ltd.; Chem. Abstr., 67, 91783.

363. Harold M. Schmidt, Volmey Tullsen, Leon Katz, and Lawrence D. Lytle, U.S. Pat. 3,022,299 (Feb. 20, 1966), to General Aniline & Film Corp.; Chem. Abstr., 57, 11345.

364. Albert W. Bauer and Howard E. Phillips, U.S. Pat. 3,502,586 (March 24, 1970), to E. I. du Pont de Nemours & Co.; Chem. Abstr., 73, 5245.

365. Diversey (U.K.) Ltd., Br. Pat. 860,373 (Feb. 1, 1961); Chem. Abstr., 55, 12707.

366. Zbigniew Mikolajewski, Zenon Magulewski, and Tadeusz Puzak, Pol. Pat. 50,615 (Feb. 10, 1966), to Cementownie "Podgrodzie"; Chem. Abstr., 66, 68655.

367. Aoki Keisho, U.S. Pat. 3,441,509 (April 29, 1969), to Hagaron Co., Ltd.; Chem. Abstr., 71, 51504.

368. S. A. Mel'nikova, Vzryvnoe Delo, 1968, 65 (22), 22-24.; Chem. Abstr., 70, 116761.

369. N. G. Wahlberg, Swed. Pat. 206,893 (Aug. 23, 1966), to Mo och Domsjo Aktiebolag; Chem. Abstr., 68, 22932.

370. David P. Keckler and Ronald W. Pokluda, U.S. Pat. 3,338,984 (Aug. 20, 1967), to Diamond Alkali Co.; Chem. Abstr., 67, 110345.

371. Hugo Arens, Ger. Pat. 963,552 (May 9, 1957), to Goldschmidt Akt.-Ges.; Chem. Abstr., 53, 12707.

372. Henri Clemencot, Fr. Pat. 1,474,102 (March 24, 1967); Chem. Abstr., 67, 46903.

373. S. S. Shanovskaya, Materialy XV (Pyatnadsatogo) Plenuma Resp. Komis. Bor'be Silikozom (Kiev: Akad. Nauk Ukr. SSR) Sb., 1963, 35-39; Chem. Abstr., 62, 12939.

374. Amil Mares, Mila Pleva, Antonin Prikryl, and Vladimir Jancik, Czech. Pat. 121,244 (Dec. 15, 1966); Chem. Abstr., 67, 119985.

375. James R. Stanford, U.S. Pat. 3,412,024 (Nov. 19, 1968), to Nalco Chemical Co.; Chem. Abstr., 70, 30654.

376. Louis A. Joo and Robert C. Kimble, U.S. Pat. 3,267,038 (Aug. 16, 1966), to Union Oil Co. of California; Chem. Abstr., 65, 15011.

377. Arthur F. Wirtel and Charles M. Blair, Jr., U.S. Pat. 2,888,399 (May 26, 1959), to Petrolite Corp.; Chem. Abstr., 53, 22872.

378. Willi Presting, Ilse Steiner, Klaus Rockstroh, Erich Bischof, Manfred Poelzing, Heinz Mueller, Heinz Kohlhase, and Rosemarie Fege, Ger. (East) Pat. 45,452 (Jan. 20, 1966); Chem. Abstr., 65, 11878.

379. Robert C. Kimble, John B. Braunwarth, and Louis A. Joo, U.S. Pat. 3,247,124 (April 19, 1966), to Union Oil Co. of California; Chem. Abstr., 65, 4115.

380. Atlantic Refining Co., Br. Pat. 973,595 (Oct. 28, 1964); Chem. Abstr., 62, 4220.

381. Hayward R. Baker and Curtis R. Singleterry, U.S. Pat. 3,138,558 (June 23, 1964), to U.S. Dept. of the Navy; Chem. Abstr., 61, 7239.

382. I. Atanasiu, L. Blum-Lazar, and M. Constantinescu, Rev. Chim. (Bucharest), 13, 452-457; But. Inst. Politeh. Bucuresti, 23 (3), 63-78 (1961); Chem. Abstr., 58, 4229.

383. D'Arcy A. Shock and John D. Sudbury, U.S. Pat. 2,840,477 (June 24, 1958), to Continental Oil Co.; Chem. Abstr., 52, 15900.

384. E. F. Epshtein, L. V. Korchagin, N. A. Dudlya, and N. M. Gavrilenko, Izv. Vyssh. Ucheb. Zaved., Geol Razved., 12 (2), 143-146 (1969); Chem. Abstr., 72, 91670.

385. James H. Norton, U.S. Pat. 3,425,940 (Feb. 4, 1969), to Esso Research and Engineering Co.; Chem. Abstr., 70, 79738.

386. James L. Lummus and Arthur Park, U.S. Pat. 3,328,295 (June 27, 1967), to Pan American Petroleum Corp.; Chem. Abstr., 67, 75124.

387. I. A. Rozelman, V. I. Lipchanskaya, S. M. Gabdullina, and P. N. Shemshurenko, USSR Pat. 165,643 (Oct. 12, 1964); Chem. Abstr., 62, 4817.

388. Hans Glockner and Ernst Meier, Ger. Pat. 1,143,707 (Feb. 14, 1963), to Perutz Photowerke G.m.b.H.; Chem. Abstr., 58, 9788.

389. Yu. B. Vilenskii, N. A. Petrova, and V. N. Dolbin, Zh. Nauch. Prikl. Fotogr. Kinematogr., 11 (6), 412-416 (1966); Chem. Abstr., 66, 42306.

390. Paul Gibson and Raymond W. Starmann, U.S. Pat. 2,982,724 (May 2, 1961), to Swift & Co.; Chem. Abstr., 56, 10464.

391. John R. Caldwell, U.S. Pat. 3,432,472 (Mar. 11, 1969), to Eastman Kodak Co.; Chem. Abstr., 70, 97912.

392. Pompelio A. Ucci, Belg. Pat. 640,144 (May 19, 1964), to Monsanto Co.; Chem. Abstr., 63, 726.

393. Monsanto Co., Neth. Appl. 6,401,227 (Aug. 17, 1964); Chem. Abstr., 62, 1784.

394. Monsanto Chemical Co., Br. Pat. 918,518 (Feb. 13, 1963); Chem. Abstr., 58, 10340.

395. Teddy G. Traylor and Ardy Armen, U.S. Pat. 3,043,811 (July 10, 1962), to Dow Chemical Co.; Chem. Abstr., 57, 14018.

396. Yoshikane Inoue, Japan. Pat. 12,239 (1961); Chem. Abstr., 56, 6210.

397. Osamu Ishiwari and Kojin Matsui, Japan Pat. 6644 (June 3, 1961), to Toyo Spinning Co., Ltd.; Chem. Abstr., 57, 6167.

398. Fany Fulga, Dorothea Rosu, and Ana Cristescu, Ind. Aliment. (Bucharest), 21 (4), 187-192 (1970); Chem. Abstr., 73, 78816.

399. W. C. T. Major, Queensland J. Agr. Sci., 19, 107-112 (1962); Chem. Abstr., 58, 3616.

400. Willi Hessler, Ger. Pat. 1,122,646 (Jan. 25, 1962), to Werner &
     Mertz, G. m. b. H.; Chem. Abstr., 56, 13066.
401. Friedrich Schmitt (Alice Schmitt, heir), U.S. Pat. 2,973,240 (Feb. 28,
     1961), to Bohme Fetterhemie G. m. b. H.; Chem. Abstr., 55, 10936.
402. Heinz Theobald, Gerhard Biberacher, and Horst Moegling, Ger. Pat.
     1,098,527 (Feb. 2, 1961), to Badische Anilin- & Soda Fabrik A. -G.;
     Chem. Abstr., 56, 13799.
403. Abraham I. Sherer and John Ruzicka, U.S. Pat. 3,074,836 (Jan. 22,
     1963), to Ball Brothers Co., Inc.; Chem. Abstr., 58, 11068.
404. Pyrene Co. Ltd., Br. Pat. 1,024,620 (March 30, 1966); Chem.
     Abstr., 64, 17208.
405. A. O. Minklei, Belg. Pat. 643, 925 (June 15, 1964), to Société Con-
     tinentale Parker.; Chem. Abstr., 63, 9620.
406. Chester W. Smith, U.S. Pat. 2,857,298 (Oct. 21, 1958); Chem.
     Abstr., 53, 3023.
407. Jiri Czerny, Czech. Pat. 102,953 (March 15, 1962); Chem. Abstr.,
     59, 2490.
408. Wataru Funabashi, Nagoya Kogyo Gijutsu Shikensho Hokoku, 12,
     15-18 (1963); Chem. Abstr., 58, 13533.
409. Leslie Kelenyi (1/2 to Magda Zolna), Can. Pat. 631,065 (Nov. 14,
     1961); Chem. Abstr., 57, 520.
410. Claude P. Coppel, U.S. Pat. 3,261,399 (July 19, 1966), to Marathon
     Oil Co.; Chem. Abstr., 65, 12037.
411. George P. Ahearn and Albin F. Turbak, U.S. Pat. 3,283,812 (Nov. 8,
     1966), to Esso Production Research Co.; Chem. Abstr., 66, 12742.
412. Howard H. Ferrell and David E. Baldwin, Jr., U.S. Pat. 3,366,174
     (Jan. 30, 1968), to Continental Oil Co.; Chem. Abstr., 68, 70976.
413. Howard H. Ferrell, George J. Heuer, Jr., David E. Baldwin, Jr.,
     George C. Feighner, and William L. Groves, Jr., Br. Pat. 1,065,292
     (April 12, 1967), to Continental Oil Co.; Chem. Abstr., 68, 106645.
414. David E. Baldwin, Jr., George C. Feighner, Howard H. Ferrell,
     William L. Groves, Jr., and George J. Heuer, Jr., Ger. Pat.
     1,232,534 (Jan. 19, 1967), to Continental Oil Co.; Chem. Abstr.,
     66, 97231.
415. V. S. Alekseev, N. A. Aleinikov, B. E. Chistyakov, G. G. Morozov,
     and L. E. Antonento, USSR Pat. 202,804 (Sept. 28, 1967); to Mining-
     Metallurgical Inst., Kola Branch; Chem. Abstr., 69, 21283.
416. V. S. Alekseev, G. G. Morozov, L. E. Antonensko, and B. E. Chist-
     yakov, Tsvet. Metal, 41 (3), 4406 (1968); Chem. Abstr., 69, 4441.
417. M. C. Fuerstenau and D. A. Elgillani, Trans. AIME, 235 (4), 405-
     413 (1966); Chem. Abstr., 66, 47878.
418. J. Petroviky, Rudy (Prague), 13 (9), 25-27 (1965); Chem. Abstr.,
     64, 3108.
419. John P. Neal, U.S. Pat. 3,214,018 (Oct. 26, 1965), to Feldspar
     Corp.; Chem. Abstr., 64, 334.
420. Luigi Usoni, Anna Maria Marabini, and Maria Grazia Beltrani, Ric.
     Sci. Rend., Sez. A, 8 (2), 271-272 (1965); Chem. Abstr., 63, 14432.

421. T. G. Gutsalyuk, M. P. Korablina, and M. A. Sokolov, Tr. Inst. Met. Obagashch. Akad:, Nauk Kaz. SSR, 9, 3-7 (1964); Chem. Abstr., 62, 7425.

422. James E. Seymour, U.S. Pat. 3,164,549 (Jan. 5, 1965), to Armour & Co.; Chem. Abstr., 62, 4945.

423. I. N. Plaksin, A. M. Okolovich, and N. A. Kryukova, Tsvetn. Metal., 37 (1), 7-12 (1964); Chem. Abstr., 61, 338.

424. V. I. Klassen and N. S. Vlasova, Akad. Nauk SSSR Inst. Gorn. Dela, 1956, 48-60; Chem. Abstr., 55, 10842.

425. G. N. Nazarova, A. N. Okolovich, and I. N. Plaksin, Akad. Nauk SSSR Inst. Gorn. Dela, 1960, 159-167; Chem. Abstr., 55, 3350.

426. Shigeru Mukai and Centaro Kano, Nippon Kogyo Kaishi, 74, 297-302 (1958); Chem. Abstr., 53, 1592.

427. Tetsuya Fujii, Dempunto Gijutsu Kenkyu Kaiho, 23, 11-27 (1961); Chem. Abstr., 60, 8216.

428. Jarsolav Gilbert, Frantisek Hruska, Vladimir Lelek, and Sandze Unkurov, Czech. Pat. 103,715 (May 15, 1962, Apr. 15, 1961); Chem. Abstr., 59, 14204.

429. P. A. Galaktionov, L. F. Dribin, M. V. Sardanovskii, E. A. Gurkov, N. G. Grechukhia, and E. V. Gannushkina, USSR Pat. 114,044 (July 30, 1958); Chem. Abstr., 53, 18843.

430. John W. Stilwell, U.S. Pat. 3,114,675 (Dec. 17, 1963); Chem. Abstr., 60, 10347.

431. Abraham Mankowich, U.S. Pat. 3,484,379 (Dec. 16, 1969), to U.S. Dept. of the Army; Chem. Abstr., 71, 68517.

432. Hubert Claude, Fr. Pat. 1,535,662 (Aug. 9, 1968); Chem. Abstr., 71, 51503.

433. D. M. Abramov, B. V. Evdokimov, A. Yu. Alekperova, and S. Yu. Akhundova, Azerb. Khim. Zh., 1968 (4), 89-90; Chem. Abstr., 70, 49654.

434. O. S. Mochalova and A. D. Karaseva, USSR Pat. 225,362 (Aug. 29, 1968); Chem. Abstr., 70, 12863.

435. Gerhard Moser, Hans Wassermann, and Gerhard Walter, Ger. Pat. 1,192,025 (Apr. 29, 1965), to VEB Fettchemie; Chem. Abstr., 63, 4520.

436. Irwin R. Ehren and David G. Ellis, U.S. Pat. 3,166,444 (Jan. 19, 1965), to Lubrizol Corp.; Chem. Abstr., 62, 10190.

437. Heinz Keller, Richard Tuch, and Willi Werner, Ger. Pat. 1,069,984 (Nov. 26, 1959), to Metallgesellschaft A.-G.; Chem. Abstr., 55, 11710.

438. Sewaryn Witkowski, Henryk Stefanski, Wladyslaw Rybak, and Zdzishaw Diszynski, Pol. Pat. 46,828 (Apr. 20, 1963), to Pabianckie Zaklady Przemyslu Bawelnianego; Chem. Abstr., 61, 5913.

439. N. V. Industrieele en Handelmaatschappij "Senzora" voorheen A. J. Shoemaker & Zonen, Neth. Appl. 6,517,034 (June 19, 1967); Chem. Abstr., 83248.

440. Boehme Fettchemie G.m.b.H., Fr. Pat. 1,407,383 (July 30, 1965); Chem. Abstr., 65, 17216.

441. Anon., Melliand Textilben., 42, 1032 (1961).

442. E. A. Korol'kova and V. G. Chasova, Nauchn. Issedovatel. Tr. Nauchn. Issedovatel. Inst. Mekhovoi. Prom., 6, 72-82 (1955); Chem. Abstr., 54, 10359.

443. H. Wedell, Melliand Textilber., 40, 798-801 (1959); Chem. Abstr., 53, 19400.

444. Richard Hess, Ger. Pat. 1,234,907 (Feb. 23, 1967), to Chemische Fabrik Stockhausen & Cie.; Chem. Abstr., 66, 86899.

445. Elzbieta Matusewica and Hanna Szczepanowska, Pol. Pat. 49,439 (May 31, 1965), to Centrala Wytworczo-Uslugowa "Libella"; Chem. Abstr., 64, 6914.

446. Edwin B. Michaels and Clayton A. Wetmore, U.S. Pat. 3,349,038 (Oct. 24, 1967), to Stamford Chemical Industries, Inc.; Chem. Abstr., 68, 4251.

447. Homer E. Crotty, Charles R. Coffey, and Thomas C. Tesdahl, S. African Pat. 69/03,496 (Dec. 19, 1969), to W. R. Grace and Co.; Chem. Abstr., 73, 5265.

448. Knut Oppenlaender and Friederich Fucks, Ger. Pat. 1,298,077 (June 26, 1969), to Badische Anilin- & Soda Fabrik A.-G.; Chem. Abstr., 71, 62488.

449. N. A. Gerchenova, L. M. Gurvich, O. I. Zelenskaya, A. F. Korets-kii, P. A. Machul'skii, M. P. Nesterova, E. P. Selivanova, A. B. Taubaman, and E. V. Fulrava, Br. Pat. 1,086,284 (Oct. 4, 1967), to State Planning Design and Scientific Research Institute of Sea Trans-portation; Chem. Abstr., 68, 4252.

450. State Design-Construction and Scientif Research Institute of Moscow Transportation, Neth. Appl. 6,506,203 (Nov. 15, 1966); Chem. Abstr., 66, 86890.

451. H. D. Feldtman and J. R. McPhee, Textile Res. J., 36 (1), 48-55 (1966); Chem. Abstr., 64, 11369.

452. Rene Derouineau, Fr. Pat. 1,433,168 (March 25, 1966); Chem. Abstr., 65, 18320.

453. John D. Brandner, U.S. Pat. 2,966,422 (Dec. 27, 1960), to Atlas Power Co.; Chem. Abstr., 55, 6879.

454. Rudolf Kern, Ger. Pat. 923,219 (Feb. 7, 1955), to Rhein Chemie G.m.b.H.; Chem. Abstr., 52, 780.

455. Keizo Ueda and Satoshi Ando, Japan Pat. 7706 (March 29, 1967), to Kanegafuchi Spinning Co., Ltd.; Chem. Abstr., 68, 40919.

456. Tetsuya Matsukawa, Kogyo Kagaku Zasshi, 63 (6), 1030-1035 (1960); Chem. Abstr., 56, 15693.

457. L. Argiero and A. Paggi, Health Phys., 14 (6), 581-592 (1968); Chem. Abstr., 69, 12776.

458. Boleslaw Tarchalski, Jan Adamkiewicz, Stefan Paliga, Eugeniusz Furmanczyk, and Wladyslaw Jasionowicz, Pol. Pat. 43,638 (Feb. 27, 1961), to Centralne Laboratorium Przemyslu Bawelnianego; Chem. Abstr., 58, 3617.

459. Iwao Maruta, Haruhiko Arai, Akhiko Iida, and Shoji Horin, Kogyo
     Kagaku Zasshi, 70 (12), 2248-2280 (1967); Chem. Abstr., 68, 97007.
460. Joseph F. Wilson, U.S. Pat. 3,190,774 (June 22, 1965), to Phillips
     Petroleum Co.; Chem. Abstr., 63, 5441.
461. Edgar W. Sawyer, Jr. and Homer A. Smith, U.S. Pat. 3,169,053
     (Feb. 9, 1965), to Minerals & Chemicals Philipp Corp.; Chem.
     Abstr., 62, 11107.
462. Jaroslav Nyvlt, Jaroslav Gottfried, and Jaroslav Krickova, Chem.
     Prumsyl, 14 (5), 242-244 (1964); Chem. Abstr., 61, 4916.
463. Joseph F. Wilson, Van C. Vives, and John C. Hillyer, U.S. Pat.
     3,116,185 (Dec. 31, 1963), to Phillips Petroleum Co.; Chem.
     Abstr., 60, 10259.
464. Van C. Vives, U.S. Pat. 3,034,858 (May 15, 1962), to Phillips Pet-
     roleum Co.; Chem. Abstr., 57, 7660.
465. A. Jouhki, Suomen Kemistilehti, 31A, 232-241 (1958); Chem. Abstr.,
     53, 13467.
466. Paolo Peri, Antonio Montagna, and Giorgio Morandi, Ital. Pat.
     739,090 (Jan. 16, 1967), to Societa Edison; Chem. Abstr., 69, 11639.
467. Hideo Amatsu and Toshiro Karazawa, Japan Pat. 10,789 (Dec. 24,
     1956), to Nissan Chemical Industries, Ltd.); Chem. Abstr., 52,
     14958.
468. Siegfried Altscher and Thomas F. Groll, Jr., U.S. Pat. 3,172,750
     (March 9, 1965), to NOPCO Chemical Co.; Chem. Abstr., 62, 12392.
469. Paul L. Lindner, U.S. Pat. 2,976,208 (March 21, 1961), to Witco
     Chemical Co., Inc.; Chem. Abstr., 55, 15822.
470. Thomas W. Sauls, U.S. Pat. 2,897,114 (July 28, 1959), to Tennessee
     Corp.; Chem. Abstr., 53, 22714.
471. Vincent J. Keenan, U.S. Pat. 2,862,848 (Dec. 2, 1958), to Atlantic
     Refining Co.; Chem. Abstr., 53, 5578.
472. A. Davidsohn, Israel Pat. 9224 (Aug. 7, 1956), to "Dahlia" Kibbutz
     Hashomer Hazair; Chem. Abstr., 52, 646.
473. R. D. Ilnicki, W. H. Tharrington, J. F. Ellis, and E. J. Visinski,
     Proc. Northeast. Weed Control Conf., 19, 295-299 (1965); Chem.
     Abstr., 62, 11084.
474. Victor H. Unger and Robert E. Wolfrom, U.S. Pat. 3,307,931
     (March 7, 1967), to Rohm & Haas Co.; Chem. Abstr., 66, 94153.
475. Albert A. Schreiber, U.S. Pat. 2,842,476 (July 8, 1958), to McLaugh-
     lin Gormley King Co.; Chem. Abstr., 52, 20868.
476. Rokuro Sato and Kiroshi Kubo, Botyu-Kagaku, 24, 89-93 (1959);
     Chem. Abstr., 60, 7384.
477. Jean N. Duperray, Fr. Pat. 1,317,304 (Feb. 8, 1963), to Chimio-
     technic, Union Chimique du Nord et du Rhone; Chem. Abstr., 59,
     4497.
478. Jean N. Duperray, Fr. Pat. 1,326,412 (May 10, 1963), to Chimio-
     technic, Union Chimique du Nord et du Rhone; Chem. Abstr., 59,
     6932.
479. James C. Reid, Austral. Pat. 239,148 (June 26, 1962), to Hart and
     Co. Pty Ltd.; Chem. Abstr., 63, 542.

480. Geoffrey H. Beames, Everett E. Gilver, Curtis Richardson, and Benjamin Veldhuis, U.S. Pat. 3,034,952 (May 15, 1962), to Allied Chemical Corp.; Chem. Abstr., 57, 6366.

481. A. Horubala, Zeszyty Nauk. Szkoly Glownej. Gospodarst. Wiejskiego Warszawie, Techno. Rolno-Spozyucza, 2, 79-92 (1962); Chem. Abstr., 65, 4114.

482. A. A. Spryskov, Izv. Vysshikh Uchebn. Zavedenii, Khim. Khim. Tekhnol., 4, 981-984 (1961); Chem. Abstr., 57, 16464.

483. H. Cerfontain, F. L. J. Sixma, and L. Volbracht, Rec. Trav. Chim., 82 (7), 659-670 (1963); Chem. Abstr., 59, 12685.

484. H. Cerfontain and A. Telder, Rec. Trav. Chim., 84 (11), 1613-1616 (1965); Chem. Abstr., 64, 12483.

485. Chemetron Corp., Br. Pat. 967,874 (Aug. 26, 1964); Chem. Abstr., 61, 13244.

486. H. Stache, Fette, Seifen, Anstrichmittel, 71 (5), 381-386 (1969); Chem. Abstr., 71, 40616.

487. Witco Chemical Co., Inc., Fr. Pat. 1,349,806 (Jan. 17, 1964); Chem. Abstr., 61, 844.

488. A. A. Spryskov and T. I. Potapova, Izv. Vysshikh Uchebn. Zavedenii, Khim. Khim. Tekhnol., 5, 280-283 (1962); Chem. Abstr., 57, 16464.

489. Witco Chemical Co., Inc., Neth. Appl. 290,738 (June 25, 1965); Chem. Abstr., 64, 3860.

490. Carlo Giraudi and Albert Sharphouse, Br. Pat. 963,787 (July 15, 1964), to Witco Chemical Co., Inc.; Chem. Abstr., 61, 12211.

491. Carlo Giraudi and Albert Sharphouse, Ger. Pat. 1,248,646 (Aug. 31, 1967), to Witco Chemical Co., Inc.; Chem. Abstr., 67, 109957.

492. Samuel L. Norwood, Walter H. C. Rueggeberg, and Thomas W. Sauls, U.S. Pat. 2,798,089 (July 2, 1957), to Tennessee Corp.; Chem. Abstr., 52, 1231.

493. American Cyanamid Co., Br. Pat. 949,852 (Feb. 19, 1964); Chem. Abstr., 60, 15781.

494. Edwin J. Eccles, Jr., Richard T. Haynes, Leo J. Weaver, and Lloyd E. Weeks, U.S. Pat. 3,174,935 (March 23, 1965), to Monsanto Co.; Chem. Abstr., 62, 14938.

495. George W. Ayers, and William A. Krewer, U.S. Pat. 3,496,224 (Feb. 17, 1970), to Union Oil Co. of California; Chem. Abstr., 72, 113088.

496. J. C. White and H. L. Holsopple, U.S. Atom. Energy Comm. ORNL-TM-183 (1962); Chem. Abstr., 59, 5053.

497. I. H. Rath, J. Rau, and D. Wagner, Melliand Textilber., 43, 718-723 (1962); Chem. Abstr., 57, 9286.

498. Imperial Chemical Industries Ltd., Neth. Appl. 6,508,492 (Jan. 3, 1966); Chem. Abstr., 64, 15750.

499. Adolf Koebner and Thomas Thornton, Ger. Offen. 1,808,443 (July 10, 1969), to Albright and Wilson Ltd.; Chem. Abstr., 71, 82900.

500. S. D. Kullbom and H. F. Smith, Anal. Chem., 35, 912-914 (1963); Chem. Abstr., 59, 4550.

501. Marvin Mausner and Paul Sosis, Soap Chem. Specialties, 38, 47-50, 105-106 (1962); Chem. Abstr., 56, 11735.

502. M. A. Phillips, Mfg. Chemist, 34 (12), 575-577 (1963); Chem. Abstr., 60, 8236.

503. Marvin Mausner and Edwin Rainer, Soap Chem. Specialties, 44 (8), 34-37, 64, 66, 69; 44 (9), 56, 58, 100-101 (1968); Chem. Abstr., 69, 107781.

504. Miklos Szeplaky and Janos Simonek, Novenyolaj Haztartasi Vegyipari Kutatointezet Kozlemenyei, 1961, 25-28; Chem. Abstr., 61, 7237.

505. Otte Schere and Peter P. Rammert, Ger. Pat. 1,126,381 (Mar. 29, 1962), to Farbwerke Hoechst A.-G.; Chem. Abstr., 57, 9744.

506. Continental Oil Co., Br. Pat. 773,423 (Apr. 24, 1957); Chem. Abstr., 51, 12482.

507. Monsanto Chemical Co., Br. Pat. 761,095 (Nov. 7, 1956); Chem. Abstr., 51, 10932.

508. Kenneth R. Gerhart and Edward J. Karwacki, U.S. Pat. 2,820,056 (Jan. 14, 1958), to Continental Oil Co.; Chem. Abstr., 52, 5860.

509. Peter J. Pengilly, Br. Pat. 820,340 (Sept. 16, 1959), to Thomas Hendley & Co. Ltd.; Chem. Abstr., 54, 2792.

510. Allen H. Lewis, U.S. Pat. 2,897,156 (July 28, 1959), to California Research Corp.; Chem. Abstr., 54, 4009.

511. Gunter Spengler and Josef Gansheimer, Ger. Pat. 1,050,329 (Feb. 12, 1959), to Gunter Spengler; Chem. Abstr., 55, 4433.

512. Gilles Chomel, Fr. Pat. 1,269,832 (Nov. 30, 1961), to Societe Anon. des Produits Chimiques-Shell-Saint-Gobain; Chem. Abstr., 56, 15365.

513. Innovations Chimiques "Sinnova" or "Sadic", Fr. Pat. 1,313,905; Chem. Abstr., 59, 1558.

514. Shell Internationale Research Maatschappij N.V., Belg. Pat. 618,971 (Dec. 14, 1962); Chem. Abstr., 58, 14323.

515. Societa Italiana Resine S.p.A., Belg. Pat. 664,316 (Sept. 16, 1965); Chem. Abstr., 65, 3798.

516. Cornelis Kortland and Pieter L. Kooijman, Br. Pat. 813,971 (May 27, 1959), to N. V. de Bataafsche Petroleum Maatschappij; Chem. Abstr., 54, 4008.

517. Edwin J. Eccles and Robert J. O'Neill, U.S. Pat. 2,926,142 (Feb. 23, 1960), to Monsanto Chemical Co.; Chem. Abstr., 54, 21801.

518. Yasushi Kimura, Shuhei Tanimori and Terunosuke Shimo, Yukagaku, 13 (2), 63-66 (1964); Chem. Abstr., 63, 18485.

519. J. J. Langford, Continental Oil Co., unpublished data.

520. D. L. Crain, J. Catalysis, 12, 110-113 (1969); Chem. Abstr., 70, 77199.

521. I. Ihara et al., Yukagaku, 19, 1043-1048 (1970).

522. M. Iimori, Yakagaku, 20, 91-94 (1971).

523. A. Benson and G. Karg, U.S. Pat. 3,617,207 (Nov. 2, 1971).

524. D. Gasztych et al., Tluzcze, Srodki Piorace, Kosmet. 15 (5), 3-15
       (1971).
525. R. P. Languth, T. C. Campbell, and H. R. Alun, Soap Cosmetics
       Chem. Specialties, 49 (2), 50-62 (1973).
526. O. C. Kerfoot and G. W. Duchmann, U.S. Pat. 3,703,559 (Nov. 21,
       1972).
527. A. E. Straus, U.S. Pat. 3,816,354 (June 11, 1974).
528. R. Stechler, J. M. Folliot, and M. J. Warren, U.S. Pat. 3,816,354
       (June 11, 1974).
529. M. R. Rogers and A. M. Kaplan, J. Am. Oil Chemists' Soc., 51 (12),
       519-574 (1974).
530. C. Marty, J. Maurin, and J. E. Weisang, U.S. Pat. 3,843,564
       (Oct. 22, 1974).
531. R. Wickbold, Tenside Detergents, 12 (1), 35-38 (1975).
532. F. G. Flynn and P. H. Jones, U.S. Pat. 3,846,325 (Nov. 5, 1974).